*The Institute of British Geographers*
*Special Publications Series*

32 Ecological Relations in Historical Times

ROYAL GEOGRAPHICAL SOCIETY
(WITH THE INSTITUTE OF BRITISH GEOGRAPHERS)
*Special Publications Series*



23 Rural Change in Tropical Africa
*David Siddle and Kenneth Swindell*

24 Teaching Geography in Higher Education
*John R. Gold, Alan Jenkins, Roger Lee, Janice Monk, Judith Riley, Ifan Shepherd and David Unwin*

25 Wetlands: A Threatened Landscape*
*Edited by Michael Williams*

26 The Making of the Urban Landscape*
*J. W. R. Whitehand*

27 The Impact of Sea-level Rise on European Coastal Lowlands
*Edited by M. J. Tooley and S. Jelgersma*

28 The Challenge for Geography*
*Edited by R. J. Johnston*

29 Geography and National Identity*
*Edited by David Hooson*

30 Geography and Empire*
*Edited by Anne Godlewska and Neil Smith*

31 Diffusing Geography
*Edited by A. D. Cliff, P. R. Gould, A. G. Hoare and N. J. Thrift*

32 Ecological Relations in Historical Times
*Edited by Robin A. Butlin and Neil Roberts*

*In preparation*

Space and Social Theory
*Edited by Georges Benko and Ulf Strohmayer*

Salt Marshes and Coastal Wetlands
*Edited by D. R. Stoddart*

Demographic Patterns in the Past
*Edited by R. Smith*

The Geography of the Retailing Industry
*John A. Dawson and Leigh Sparks*

Asian-Pacific Urbanization
*Dean Forbes*

The Rise of the South China Coast
*Carolyn Cartier*

* Available in paperback editions.

*For a complete list see pp. 343–344*

# Ecological Relations in Historical Times

## Human Impact and Adaptation

*Edited by* Robin A. Butlin and Neil Roberts

on behalf of the Biogeography and Historical Geography Research Groups

BLACKWELL
*Oxford UK & Cambridge USA*

First published 1995

Blackwell Publishers Limited
108 Cowley Road
Oxford OX4 1JF
UK

Blackwell Publishers Inc.
238 Main Street
Cambridge, Massachusetts 02142
USA



*British Library Cataloguing in Publication Data*

A CIP catalogue record for this book is available from the British Library.

*Library of Congress Cataloging-in-Publication Data*

Ecological relations in historical times : human impact and adaptation / edited by
Robin A. Butlin and Neil Roberts.
p. cm. – (The Institute of British Geographers special publications series : 32)
Includes bibliographical references and index.
ISBN 0–631–19506–8 (hbk : acid-free paper)
1. Man–Influence on nature. 2. Biotic communities–History. 3. Landscape changes–History. 4. Agriculture ecology. 5. Paleoecology. I. Butlin, R. A. (Robin Alan), 1938– II. Roberts, Neil, 1953– . III. Series: Special publications series (Institute of British Geographers) : 32.
GF75.E25 1995
304.2–dc20 94–38784
CIP

Typeset in 10 on 12 pt Plantin
by Apex Products, Singapore
Printed in Great Britain by
T.J. Press (Padstow) Ltd, Padstow, Cornwall.

This book is printed on acid-free paper

# Contents

# List of Figures

# List of Plates

# List of Tables

# List of Contributors

P. W. Abrahams, Institute of Earth Sciences, University of Wales, Aberystwyth

Allan C. Ashworth, Department of Geography, University of North Dakota, Fargo, USA

J. J. Blackford, Department of Geography, University of Durham

Mark A. Blumler, Department of Geography, State University of New York, Binghamton, USA

Barbara A. Brayshay, Department of Archaeology and Prehistory, University of Sheffield

Mark Brayshay, Department of Geographical Sciences, University of Plymouth

A. G. Brown, Departments of Geography, University of Exeter

Paul C. Buckland, Department of Archaeology and Prehistory, University of Sheffield

Robin A. Butlin, Department of Geography, Loughborough University

Christian Carpelan, Department of Archaeology, University of Helsinki, Finland

S. Crane, Department of Archaeology, University of Leicester

Janice P. Darch, Environmental Change Unit, University of East Anglia

R. A. Dodgshon, Institute of Earth Sciences, University of Wales, Aberystwyth

Andrew J. Dugmore, Department of Geography, University of Edinburgh

Kevin J. Edwards, Department of Archaeology and Prehistory, University of Sheffield

Jane A. Entwistle, School of Geography, Kingston University, London

Peter A. Furley, Department of Geography, University of Edinburgh

David D. Gilbertson, Institute of Earth Sciences, University of Wales, Aberystwyth

Richard Gulliver, Faculty of Education, University of Strathclyde, Glasgow

Sheila Hicks, Department of Geology, University of Oulu, Finland

M. Kent, Department of Geographical Sciences, University of Plymouth

H. F. Lamb, Institute of Earth Sciences, University College of Wales, Aberystwyth

David M. Munro, Department of Geography, University of Edinburgh

D. O'Sullivan, Department of Archaeology, University of Leicester

Roland R. Randall, Girton College, Cambridge

Alasdair Roberts, Northern College, Aberdeen

Neil Roberts, Department of Geography, Loughborough University

Jon P. Sadler, School of Geography, University of Birmingham

Jean-Luc Schwenninger, Department of Archaeology and Prehistory, University of Sheffield and Department of Geography, Royal Holloway College, London

Don W. Schwert, Department of Geography, University of North Dakota, Fargo, USA

Peter Skidmore, Doncaster Museum and Art Gallery

Guðrún Sveinbjarnardóttir, National Museum of Iceland, Rejkjavik, Iceland

K. Walsh, Department of Archaeology, University of Leicester

P. A. Wathern, Department of Biological Sciences, University of Wales, Aberystwyth

David Watts, School of Geography and Earth Resources, University of Hull

R. Weaver, Department of Geographical Sciences, University of Plymouth

Graeme Whittington, Department of Geography and Geology, University of St Andrews

Andrew Williams, Department of Geographical Sciences, University of Plymouth

R. Young, Department of Archaeology, University of Leicester

# Preface

The interest of geographers, archaeologists, anthropologists, ecologists and historians in the relations between human beings and their physical environments has a long history, reaching peaks of interest on at least six occasions in modern times. The emergence of distinctive academic disciplines during and after the later nineteenth century meant that subsequent research programmes and paradigms concerning human–environmental interactions tended to be focussed through individual disciplinary perspectives, and this unidisciplinary focus is still common. There has also been a tendency for specialists within academic disciplines – for example, human and physical geographers – to ignore some of the more interesting problems at the human/environmental interface.

In 1976 a special theme number of the *Transactions* (New Series, Vol. 1, number 3) of the Institute of British Geographers was published, under the guidance of L. F. Curtis and I. G. Simmons, with the title *Man's Impact on Past Environments*. That publication included eight essays on various impacts of humans on environments in prehistoric and historic time, with the case studies drawn from Britain (the one exception was a study of pollen analysis in north-east Finland by Sheila Hicks, who also is a contributor to this book). It demonstrated the advantages to be gained from approaches to palaeoenvironmental research problems which drew on different methodological and disciplinary skills, although with an emphasiz on the prehistoric period and on palaeo-ecological reconstruction through pollen analysis.

The present volume of essays seeks in part to build on and develop this tradition, by widening the range of disciplinary perspectives involved, but also by focusing specifically on the historical time period during which the range of potential data sources is at its broadest. The volume derives, in large part, from a two-day session of the Institute of British Geographers annual conference held at Royal Holloway College, London, in January 1993. The session was convened by the editors on behalf of the Biogeography and Historical Geography Research Groups. Sixteen papers were presented at the meeting, and of these, 13 are included here, along with two others which it was not possible to include in the original programme. All have been revised to take account of referees' comments and sequenced to produce a balance of regional and topical interests. We are greatly encouraged by the number of papers which are the results of interdisciplinary team research, and hope that the contributions to the book will in turn stimulate further debates and research.

We wish to acknowledge assistance with the production of a number of the maps and diagrams for this book by Erica Milwain and Peter Robinson of the cartography unit of the Geography Department of Loughborough University. The index was complied by J. Roberts. We are most grateful to the scholars who kindly assisted us in this venture, notably by chairing sessions at the original meeting or by agreeing to referee the papers. They include F. W. Barnes, S. Bendall, J. Böchur, K. R. Boucher, R. Bradshaw, F. Chambers, J. Chapman, C. David, J. Dearing, G. Dickenson, A. Dugmore, K. Edwards, M. T. Greenwood, J. Grove, M. Jones, D. Lowenthal, W. R. Mead, A. V. Morgan, S. O'Hara, P. J. Osborne, O. Rackham, R. Randall, B. K. Roberts, J. Sheail, T. C. Smout, A. C. Stevenson, R. Suffling, D. Watts, C. Watkins, R. Whittaker, G. Whittington and M. Williams.

Robin A. Butlin and Neil Roberts
Loughborough University

1

# Ecological Relations in Historical Times: an Introduction

Neil Roberts and Robin A. Butlin

The destruction of natural habitats such as wetlands and forests, and their replacement by semi-natural or agricultural ecosystems, are issues of pressing importance for the conditions of human existence, for science, and for natural resource management. Increasing human impact on ecological systems also threatens the traditional small-scale adaptations by many pre-modern cultures to their local environments. Human-induced transformation of the natural world is proceeding at different rates and in different ways over the face of the Earth, but its overall pace has conspicuously accelerated during the last two millennia. For historical time it is possible to document environmental impact and landscape change by means of a wide range of human and natural 'archives'. The timespan and precision of these different data sources strongly influence the ways in which individual environmental histories can be reconstructed.

Over timescales ranging from decades to centuries, documentary archives have traditionally provided the primary means of accessing the past, and upon this complex array of sources have been built the scholarly and scientific enterprises of historical geography (Harley 1982, Butlin 1993), historical ecology (Sheail 1980) and – most recently – environmental history (Worster 1988, Brimblecombe and Pfister 1990, Williams 1994). Historic maps, diaries, tax records, estate papers, travellers' and explorers' accounts, official state surveys, ships' logs and similar archival sources have provided an important focus for research on geomorphological and climatic change over these 'intermediate' timescales (Hooke and Kain 1982, Bradley and Jones 1992). For longer time periods, such as the Holocene, proxy data sources have to be employed (Roberts 1989). Palaeoecology and archaeology, for example, are concerned, respectively, with the reconstruction of biotic and cultural change, and they share a common reliance on dating methods such as radiocarbon ($^{14}C$) assay. While the main applications of these methods remain on a millennial timescale, there has been a growing realization that they can also produce valuable insights into shorter-term environmental processes, such as cultural eutrophication (Oldfield 1983) and the effects of Little Ice Age climatic change (Grove and Battagel 1981,

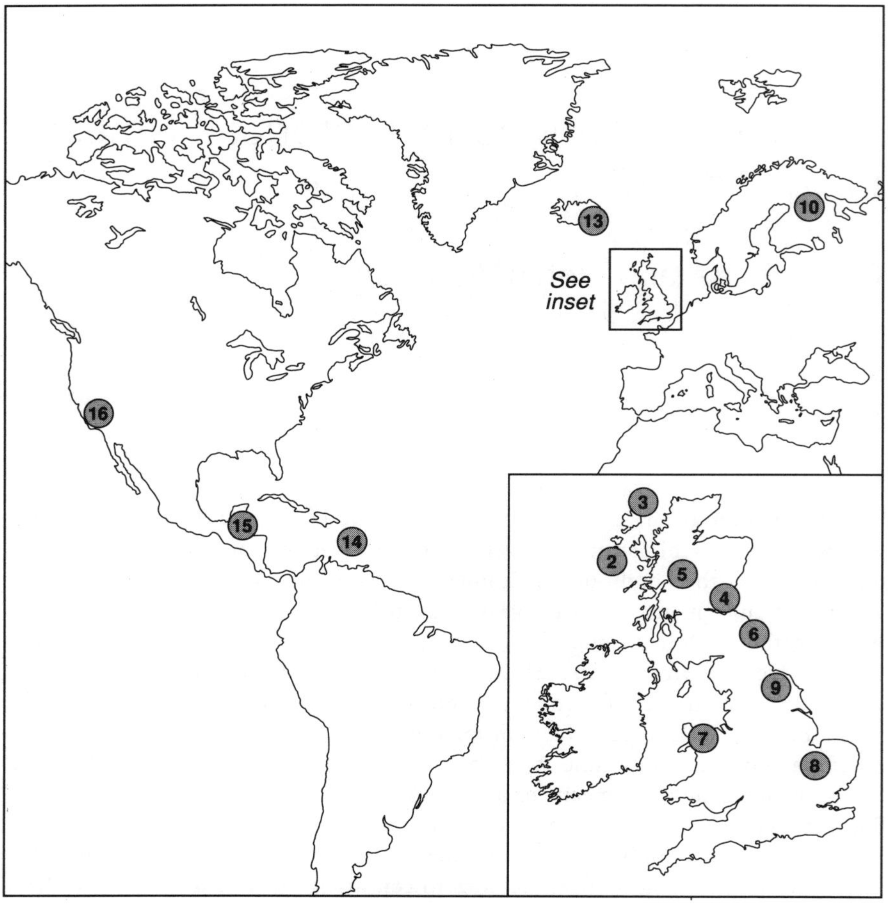


Figure 1.1 The locations of the studies in this volume: numbers denote individual chapters.

Campbell and McAndrews 1993). As a result, these disciplines have a potentially significant role to play in elucidating historical as well as prehistoric changes in the environment, especially in regions, such as North America, where the depth of the historical record is less than it is in Europe. A third data source comes from present-day observations, such as those on the biogeographical distribution of individual taxa, or on the character of human–environmental relations as observed through directly and remotely sensed relict features of former habitation and cultivation practices and through study of residual largely non-westernized societies. The present day represents an essential point of departure for any investigation into the environmental past and also provides the uniformitarianist basis (the assumption that the same processes operating in the present were also active in the past) for palaeoecological analysis. Present-day sources do, however, need to be interrogated as critically as those from any

Table 1.1 Data sources used in the studies in this book

| *Author(s)* | *Proxy* | *Archival* | *Contemporary* |
|---|---|---|---|
| ch. 2, Gilbertson et al. | xx | x | xx |
| ch. 3, Entwistle et al. | xx | x | |
| ch. 4, Whittington and Edwards | xx | x | |
| ch. 5, Roberts | | xx | x |
| ch. 6, Walsh et al. | xx | xx | |
| ch. 7, Brayshay and Williams | | xx | x |
| ch. 8, Butlin | | xx | x |
| ch. 9, Gulliver | | x | xx |
| ch. 10, Carpelan and Hicks | xx | x | x |
| ch. 11, Sadler and Skidmore | xx | x | xx |
| ch. 12, Buckland, Ashworth and Schwert | x | xx | x |
| ch. 13, Buckland et al. | xx | x | |
| ch. 14, Watts | | xx | x |
| ch. 15, Furley et al. | x | xx | x |
| ch. 16, Blumler | x | x | xx |

xx = Main data source; x = supplementary source.
Proxy = palaeoecology and bioarchaeology.
Archival = historical documentation.
Contemporary = biogeographic and genetic distribution, species composition.

other period, for human–environmental relationships which exist today cannot be assumed always to have been characteristic of former times.

The studies which form this volume concern the human impact upon, and interaction with, changing natural environments during – at most – the last two thousand years. They focus geographically on north-west Europe (especially Britain) and North America (including the Caribbean), and several chapters deal specifically with relations between the two sides of the North Atlantic Ocean (figure 1.1). This regional and temporal frame means that all the authors are able to draw upon on historical documentation for their reconstruction of former environmental relations. In addition, as can be seen from table 1.1, they also all employ either proxy or contemporary data, or both of these. This use of multiple data sources offers a perspective on the past which is potentially richer and more holistic than reliance on any single one. Pollen analysis provides one mirror in which environmental history is reflected, archival records provide another, and the omissions and distortions in one source may be counterbalanced by the data and trends provided and indicated by others. The discovery in the 1950s that the seemingly 'natural' Norfolk Broads of eastern England were in fact flooded medieval peat workings is but one illustration of the success that can be achieved by combining multiple data sources, in this case from historical geography, archaeology and ecology (Lambert et al. 1960).

It is vital that our attempts at reconstructions of the major human and environmental features of past times should employ as diverse a range of sources and perspectives as possible (Jäger 1969, Cronon 1992). This in turn points to the advantages of multidisciplinary team approaches to particular research sites and problems, an approach much in evidence in the essays that follow. That is not to say that the path of multidisciplinary studies of environmental relations is, however, an easy one to tread. The traditions of individual scholarship within social and economic history and historical geography are not readily given up, although an increasing number of exemplary combined studies by historians, geographers and archaeologists point to the rewards of such enterprises.

## MULTIPLE DATA SOURCES: CONFLICT AND COMPLEMENTARITY

A basic requirement of multidisciplinary work is the need for accurate correlation between different data sources, in space as well as through time. Such correlation is hindered by problems of commensurability; that is, whether or not like is being compared with like. These problems include matching continuous with discontinuous time series, the dating of proxy data, uneven taxonomic resolution, biased representation (e.g. in palynology), and the frequent lack of locational correspondence between different data sets. Some of the most informative insights can occur when different data sets fail to agree, but there is always a strong temptation to ascribe such divergences simply to incommensurability of the data.

Proxy records, such as lake sediment archives, commonly form a continuous time series but with a temporal resolution which is more likely to be decadal than annual. With a sediment accumulation rate of 1 m per 1000 years, a 1 cm thick sample of lake mud would aggregate data over a 10 year period, although in reality, sediment mixing at the mud–water interface is likely to reduce this resolution somewhat. Smoothed data have many advantages, notably in cutting out high-frequency 'noise', but they may not be easy to compare directly with archival data, which may represent conditions in a particular year or season. Temporal precision is improved in annually laminated lake sediments and in tree-ring sequences, and consequently these may be more directly commensurable with historic records. Dendrochronological data can provide excellent data on climate history (e.g. Briffa et al. 1990; Till and Guiot 1990) but are less informative regarding human impact, except for fire frequencies. Lakes containing laminated sediments are, by contrast, an extremely effective means of tracing and quantifying human-induced landscape change (O'Sullivan 1983, McAndrews 1988, Card 1994). Laminated sediments have been found in some unpromising settings in recent years, such as the freshwater coastal 'lagoon' of Loe Pool in Cornwall, England (Simola et al. 1981), and the use of freeze-core technology holds out the promise that more will be discovered in the future.

Data from historical archives have a high temporal resolution (sometimes daily), but are often discontinuous or semi-continuous in character, and rarely form a time series spanning more than a few centuries. Data from old maps or surveys form snapshots at particular moments in time, but their frequency declines as one goes back in time. The Domesday survey of England in 1086 is an early and very informative example of an important and well-explored historical data source, but in many regions, such as North America, archival data are absent prior to the era of European contact from the fifteenth century. A major uncertainty with historic documentation is the interpretation of gaps or silences in the archival record. Alasdair Roberts (chapter 5) finds no convincing reference to the biting midge in the Scottish Highlands prior to the middle of the nineteenth century, and on that basis argues that it has become much more common since that time. But does absence of evidence mean evidence of absence? In a telling contribution to the meeting at which most of these chapters were originally presented, Roger Suffling recalled an early Canadian travel diary with the remains of biting insects still preserved in the book's spine, notwithstanding the total lack of reference to them in the written text! Additional complex difficulties of interpretation attach to historical documentary records, including terminology, variation in units of, for example, land measurement, the accuracy and provenances of historical maps, and the determination of the degree to which they reflect the located experiences of a broad spectrum of society, rather than simply of those of a privileged elite. The human ecologies of peasant farming in the past have differed from those of large-scale early capitally intensive farming, and there have sometimes been clashes between the two sets of interests, most frequently recorded in official accounts with distinct biases.

Archival data are, by their very nature, usually dated, but the same cannot be said of most proxy data, especially during the last two millennia when dating problems are – perversely – often more acute than earlier in the Holocene. While radiocarbon calibration does not lead to systematic divergences from true (or calender) ages during this time period, differences between the two timescales still exist, notably during $^{14}C$ 'wiggles' when dating resolution may be reduced to ± 100 years at best. Even more problematic is the question of radiocarbon reliability during the historical time period, related to uptake of 'old' carbon in lakes receiving hardwater or experiencing catchment erosion. Even timbers can give false ages, as Paul Buckland et al. show in chapter 13, in this case because wood appears to have drifted ashore onto the North Atlantic island of Papey from Siberia. There is a clear need for alternative forms of dating to complement radiocarbon during the last 2000 years. Some of these, such as lead-210 and caesium-137, are well established, in this case for the last two centuries; while others such as luminescence have undergone new methodological development. However, the application of additional dating techniques sometimes appears to create more, not less, chronological uncertainty. Jane Entwistle et al. (chapter 3) use the changing atmospheric flux in long-distance pollutants such as lead to help overcome some of the problems of providing a time frame based on other radiometric dating techniques. The advantages of dating by means of tree rings and laminated sediments have already been alluded to above, but another increasingly important means of dating (or, more strictly, correlation) is via the use of volcanic tephras as marker horizons, such as those from Hekla in Iceland. In eastern North America, another valuable marker horizon in sediment core records is provided by the sharp rise in *Ambrosia* (ragweed) pollen which accompanied the initial wave of European forest clearance, and which can usually be tied to a known historical date at each location.

Correlation between records has a spatial as well as a temporal dimension, because different data types are unlikely to derive from identical locations or areas. It is obviously important that the different component parts of a broader multidisciplinary project take place in close proximity to each other. This was the case for the pioneering Fenland Research Committee in eastern England, established in 1932, which combined the expertise of archaeology, historical geography and palaeoecology (Phillips 1951, Godwin 1978). A more recent example of the same kind of regional landscape evolution study is the Ystad project in southern Sweden (Berglund 1991). Two landscape units with well-defined natural boundaries, and which therefore merit special attention are island and lake-catchment ecosystems. This volume includes two case studies based on small offshore islands, Lindisfarne (Holy Island) in Northumbria, England (Walsh et al. chapter 6), and Papey, off Iceland (Buckland et al, chapter 13), and another two based on lake drainage basins, Kilconquhar Loch and Loch Stiapavat, both in Scotland (Whittington and Edwards, chapter 4; Entwistle et al, chapter 3).

## THE CREATION AND MODIFICATION OF CULTURAL LANDSCAPES

Palaeoecological, bioarchaeological and historical records together document the transformation of northern temperate woodlands and grasslands to agricultural and semi-natural ecosystems during the second half of the Holocene (Roberts 1989). This in turn had permanent consequences for soil erosion and landscape stability (Dearing 1994). For many of the areas of best agricultural potential within north-west Europe, this transformation originally took place in prehistoric rather than historical times. For instance, the evidence provided by one of the earlier major archival sources in England, the Domesday Book of AD 1086, suggests that only 15 per cent of the country then remained under woodland, including wood-pasture, not all of which would have been primary (Rackham 1986: 76).

In those areas cleared of their original forest cover in later prehistory, the major changes during the historical period involved shifts between different types of human land use and tenure. These changes have been documented largely via historical archives (Reed 1984). The private and general parliamentary acts and the resultant allocations or awards employed for the enclosure of common land in England and Wales, for example, which reached a peak in the later eighteenth and early nineteenth centuries, have been extensively quarried (Tate and Turner 1978, Turner 1980, 1984), as also have the awards and tithe files of the commissioners appointed for the commutation of tithes in England and Wales in consequence of the Tithe Commutation Act of 1836 (Kain 1984, 1986, Kain and Prince 1985). The agricultural censuses of Great Britain are major sources for the reconstruction of agricultural characteristics and change since their inception in 1866 (Coppock 1984). Many older forms of agricultural land use are still evident in the modern landscape via such relict features as field systems (Hoskins 1955, Baker and Butlin 1973, Dodgshon 1980, Beckett 1989).

Relatively subtle land-use changes, such as between pastoral and arable, cannot be detected easily or unambiguously by palaeoecological methods such as pollen analysis (David 1991; see also the relatively 'insensitive' pollen records from Lewis and Papey in chapters 3 and 13). Traditionally, the main approach adopted by palynologists has been to use indicator species such as ribwort plantain (*Plantago lanceolata*), in this case for pastureland (Behre 1981), but quantitative estimation of human impact on the landscape is now being attempted via methods such as correspondence analysis (Birks et al. 1988). In their study of sediment cores from Kilconquhar Loch in eastern Scotland (chapter 4), Graeme Whittington and Kevin Edwards develop this more rigorous and comprehensive approach to reconstructing changes in rural land use from pollen data. Pollen from this lake includes notable amounts of cultural type indicators such as cereals and hemp, and by employing numerical analyses of pollen spectra it has been possible to establish weed–crop associations and evidence of crop rotation, and hence gain an improved understanding of changing land-use practices within the catchment area. Where it is possible to compare pollen data directly with archival records of past land use, some measure of the sensitivity of the palaeoecological record may be achievable (e.g. David and Roberts 1990).

For much of the circum-North Atlantic region, however, the dominant process of the last millennium has been the European expansion into regions which, although within the ecumene, had previously been perceived as marginal to a nascent Eurocentric 'global' economic system (Wallerstein 1974, 1980, 1989, Nitz 1993), a system which reached its full effective development during the nineteenth century. Above all, this early modern expansion involved lands *beyond* Europe, notably in the 'New' World of the Americas; but expansion into 'marginal' lands also occurred *within* Europe, reflecting a marriage of economic ambitions of growing nation–states, population expansion and technological innovation. In a case study from Snowdonia in North Wales, for example, Mark Brayshay and Andrew Williams (chapter 7) show how the cultural landscape changed over the last four centuries as the local farming economy increasingly became integrated within regional and national economic networks. Their study was based on documentary records from two parishes close to Llanberis Pass, the physical environment – and hence land values – of which change over short distances. In upland areas such as Snowdonia or Scotland's Highlands and Islands, land clearance and subsequent agricultural intensification typically led to soil erosion and nutrient depletion. Unless counterbalanced by manuring and other conservation methods, this progressively lowered the carrying capacity of what was already relatively infertile land, and further reduced the self-sufficiency and economic independence of marginal rural communities (Dodgshon 1988). In chapter 3, Jane Entwistle et al. present data from lake sediment cores taken adjacent to abandoned farms on the islands of Skye and Lewis in the Scottish Hebrides, the land-use history of which since the eighteenth century is recorded in archival sources. Geochemical and magnetic analyses indicate an increase and subsequent (twentieth-century) decline in soil inwash caused by catchment disturbance.

Both in Europe and beyond it, post-medieval expansionary processes led to culture contact between societies with indigenous cultural ecologies (e.g. the *Saami* hunter–fisher–gatherers of Finnish Lapland's Boreal forests discussed by Christian Carpelan and Sheila Hicks in chapter 10), and populations from the heartlands of Europe who were economically and technically more powerful. This process was almost always to the detriment of the former and typically led to a breakdown in traditional relations between cultural groups and their natural environments. One of the marginal environmental types subject to extensive transformation in historical times has been wetlands, discussed in this volume in studies of the Fens of Eastern England by Robin Butlin (chapter 8) and of Belize in the Caribbean by Peter Furley and others (chapter 15). Drainage of wetlands has progressively reduced the habitat and numbers of hydrophilous plants and animals. The costs of this have been social as well as environmental, because biotic resources such as wildfowl and fish were almost always important to the indigenous inhabitants. In the case of the fen and fen-edge region of eastern Cambridgeshire in the nineteenth century, the poorest landholders and agricultural labourers found their living conditions deteriorating as the loss of their traditional resources through drainage and enclosure led to intensification of economic marginalisation, poverty and landlessness. In the case of Belize, the most ecologically sustainable human use of the interior wetlands appears to have occurred during the Mayan period, although – perhaps significantly – this system of drained-field agriculture had already fallen into disuse before European contact.

## EUROPE'S BIOLOGICAL IMPERIALISM

The European exploration and 'colonization' of the North Atlantic region from *c.*AD 1450, brought together two sets of biota which had been separated for millions of years (Crosby 1986, Bray 1993). Some taxa, notably domesticated plants and animals, were deliberately introduced from the Old to the New World (Unwin 1991) or from the New to the Old, but many more – including human disease pathogens – were introduced unintentionally. In a number of cases the history of introduction and spread is well documented – for example, the diffusion of the potato through central Europe (Deneke 1976, McNeill 1991) – but in many more cases it is not. This volume includes two studies of Coleoptera – that most taxonomically diverse of all groups (chapters 11 and 12) – and another of mediterranean-type grasses (chapter 16). Both of these types of organism display the well-attested tendency for a less than balanced exchange of species between New and Old Worlds, with Europe as the main exporter and North America the main importer. The reasons for this imbalance are not entirely clear, although Mark Blumler argues that the invasion of annual Old World grasses into California was not simply due to its association with pastoralism and other cultural activities, but may have occurred in advance of European settlement (i.e. pre-1769). California's valley grassland has undergone one of the most dramatic transformations of any plant community on a continental land mass, with alien (primarily Mediterranean) species accounting for 80–90 per cent of the vegetative cover even where 'undisturbed'.

North, Central and South America were certainly no pristine 'Garden of Eden' at the time of first contact, with human-induced landscape change already well under way before the European encounter (Denevan 1992). In regions such as central Mexico, pre-Hispanic land degradation is now well attested (O'Hara et al. 1993). The demise of much of the native Amerindian population in the years immediately following AD 1492 led to an end to indigenous woodland or grassland management systems. In the Caribbean, the ecosystem shock between 1492 and *c.*1524 led to a re-expansion of forests, as a result of the eradication of the indigenous peoples by contact with Hispanic colonizers. A similar expansion of secondary woodland has occurred along the eastern seaboard of North America. Other changes in the northeastern United States during the last five centuries include a decline in tree taxa such as beech which are fire-sensitive, and those prone to insect infestation or pathogen attack, such as fir and chestnut (Russell et al. 1993). These changes in forest composition and structure often created suitable habitats for immigrating European beetle species (Buckland et al. chapter 12). On the other hand, European plants and animals did not always displace native ones so easily. In the island Caribbean, the invasion of weedy plants was associated with the development of sugar cane estates after 1645, but, as David Watts shows in chapter 14, only 14 per cent of the common weeds on Barbados today are of Old World origin; most have derived from elsewhere in the Caribbean region. The European impact on New World ecosystems was thus regionally variable and associated with a complex environmental response.

## CONTEMPORARY BIOGEOGRAPHY AND ENVIRONMENTAL CONSERVATION

There has been a significant impetus in the last two decades towards making historical scholarship sensitive to 'green' issues (Williams 1990, Sheail 1993). No less important is the need for contemporary ecological management to include an historical dimension, especially if the goal is environmental sustainability. 'Sustainability' has little meaning without an understanding of long-term ecosystem trajectories and a knowledge of baseline conditions, if they ever existed. Recent research programmes on the history of freshwater acidification provide one example of the successful application of palaeolimnological techniques to establish baseline environmental conditions (Charles et al. 1989, Battarbee 1990). Significantly, this research led not only to a better understanding of the causes of freshwater acidification within the scientific community, but also had a major influence on public policy towards 'acid rain' as an environmental issue.

Worster (1988) has gone further and suggested that one of the goals of environmental history should be to establish how ecosystems functioned before human societies entered and rearranged them. In reality – and with the exception of isolated environments such as previously uninhabited oceanic islands – this is almost certainly a false goal, given the long-standing and often subtle anthropisation of ecosystems during the Holocene (Roberts 1989). For example, charcoal and fire scar records may provide data on fire histories prior to European contact, but it is less easy to establish the relative importance of climate change and indigenous burning practices as causes of forest fires within particular regions of North America or Australia (MacDonald et al. 1991, Suffling 1992, Kershaw 1986).

A number of studies in this volume combine historical and contemporary biogeographic or ecological information in order to tackle the problem of separating human from natural causation. Jon Sadler and Peter Skidmore (chapter 11), for example, use the fossil record to test the role of European, particularly Norse, migrations in determining the modern distribution of insect faunas on North Atlantic islands such as the Faeroes and Iceland. Richard Gulliver (chapter 9) employs species richness and relict landscape features to identify ancient woodland in North Yorkshire although, as a number of studies have shown (e.g. Hannon and Bradshaw 1989), 'ancient' may not be the same as 'primary'. The recent origin of some apparently natural ecosystems is reinforced by David Gilbertson et al. (chapter 2) for the distinctive machair vegetation found on the western side of the Outer Hebrides of Scotland. Stratigraphic records indicate that machair vegetation is only maintained by a combination of human and natural re-working of largely aeolian soils and sediments.

## SOME CONCLUSIONS

An important and distinctive feature of the studies which follow is that natural and cultural changes have been reconstructed and analysed over the same, not over different timescales. Nature does not create landscapes, stop its work, and then hand over to human agency to complete the transformation: the two sets of processes overlap and interact (cf. the approaches adopted by Wooldridge and Linton 1955, and Darby 1951, 1952). In this context it is also interesting to note that recent ideas in cultural geography, which emphasize the significance of the cultural construction of landscapes by human imagination and agency, are now being questioned in relation to their apparent underestimation of the role of physical environments in people–environmental relations. Thus Demeritt (1994: 164) has questioned the schools of thought that 'make landscapes malleable cultural projections, whose shape and meaning are determined ultimately by the linguistic and social contexts associated with them', and encouraged an fresh emphasis on 'nature as a real, autonomous actor in human history' (Demeritt 1994: 172). This book attempts to examine the balance of this interaction, and is therefore concerned not only with human impact on, and adaptation to, natural environments, but also with the set of dialectical or confrontational ecological relations which have linked nature and society through time.

REFERENCES

Baker, A. R. H. and Butlin, R. A. (eds) (1973) *Studies of Field Systems in the British Isles* (Cambridge University Press, Cambridge)

Battarbee, R. W. (1990) 'The causes of lake acidification with special reference to the role of acid deposition', *Philosophical Transactions of the Royal Society, London B* 327: 339–47

Beckett, J. V. (1989) *A History of Laxton: England's Last Open-field Village (Blackwell, Oxford)*

Behre, E.-K. (1981) 'The interpretation of anthropogenic indicators in pollen diagrams', *Pollen et Spores* 23: 225–45

Berglund, B. E. (ed.) (1991) 'The cultural landscape during 6000 years in southern Sweden', *Ecological Bulletins* 41: 1–495

Birks, H. J. B., Line, J. M. and Persson, T. (1988) 'Quantitative estimation of human impact on cultural landscape development', in Birks, H. H., Birks, H. J. B., Kaland, P. E. and Moe, D. (eds) *The Cultural Landscape – Past, Present and Future* (Cambridge University Press, Cambridge) pp. 229–40

Bradley, R. S. and Jones, P. D. (eds) (1992) *Climate since AD 1500* (HarperCollins, London)

Bray, W. (1993) 'Crop plants and cannibals: early European impressions of the New World', in Bray, W. (ed.) *The meeting of Two Worlds: Europe and the Americas 1492–1650* (The British Academy and Oxford University Press, London and Oxford) pp. 289–326

Briffa, K. R., Bartholin, T. S., Eckstein, D., Jones, P. D., Karlen, W., Schweingruber, F. H. and Zetterberg, P. (1990) 'A 1,400-year tree-ring record of summer temperatures in Fennoscandia', *Nature* 346: 434–39

Brimblecombe, P. and Pfister, C. (eds,) (1990) *The Silent Countdown: Essays in European Environmental History* (Berlin)

Butlin, R. A. (1993) *Historical Geography: Through the Gates of Space and Time (Edward Arnold, London)*

Campbell, I. D. and McAndrews, J. H. (1993) 'Forest disequilibrium caused by rapid Little Ice Age cooling', *Nature* 366: 336–8

Card, V. (1994) 'The diatoms and pollen of Big Watab Lake: can varved sediment be unlaminated (or vice versa)?' *13th AMQUA abstracts*, University of Minnesota, p. 68

Charles, D. F., Battarbee, R. W., Renberg, I., van Dam, H. and Smol, J. P. (1989) 'Palaeoecological analysis of lake acidification trends in North America and Europe using diatoms and chrysophytes', in Norton, S. A., Lindberg, S. E. and Page, A. L. (eds.) *Acid Precipitation*, vol. 4, *Soils, Aquatic Processes and Lake Acidification* (Springer-Verlag, New York) pp. 207–76

Coppock, J. T. (1984) 'Mapping the agricultural returns: a neglected tool of historical geography', in Reed, M. (ed.) *Discovering Past Landscapes* (Croom Helm, Beckenham) pp. 8–55

Cronon, W. (1992) 'A place for stories: nature, history and narrative', *Journal of American History* 78: 1342–75

Crosby, A. W. (1986) *Ecological imperialism. The biological expansion of Europe 900–1900* (Cambridge University Press, Cambridge)

Darby, H. C. (1951) 'The changing English landscape', *Geographical Journal* 117: 377–98

Darby, H. C. (1952) *The Domesday Geography of Eastern England* (Cambridge University Press, Cambridge)

David, C. A. (1991) 'Contemporary and historical pollen recruitment to a small lowland English lake', unpubl. Ph.D. thesis, Loughborough University

David, C. A. and Roberts, N. (1990) 'Vegetation change and pollen recruitment in a lowland lake catchment: Groby Pool, Leics. (England)', *Hydrobiologia* 214: 305–10

Dearing, J. (1994) 'Reconstructing the history of soil erosion', in Roberts, N. (ed.) *The Changing Global Environment* (Blackwell, Oxford) pp. 242–61

Demeritt, D. (1994) 'The nature of metaphors in cultural geography and environmental history', *Progress in Human Geography* 18(2): 163–85

Deneke, D. (1976) 'Innovation and diffusion of the potato in central Europe in the seventeenth and eighteenth centuries', in Buchanan, R. H., Butlin, R. A. and McCourt, D. (eds) *Fields, Farms and Settlement in Europe* (Ulster Folk and Transport Museum/HMSO, Holywood, Co. Down) pp. 60–96

Denevan, W. M. (1992) 'The pristine myth: the landscape of North America on the eve of contact', *Annals, Association of American Geographers* 82: 369–85

Dodgshon, R. A. (1980) *The Origin of British Field Systems: an Interpretation* (Academic Press, London)

Dodgshon, R. A. (1988) 'The ecological basis of Highland peasant farming, 1500–1800 AD', in Birks, H. H., Birks, H. J. B., Kaland, P. E. and Moe, D. (eds) *The Cultural Landscape – Past, Present and Future* (Cambridge University Press, Cambridge) pp. 139–51

Godwin, H. (1978) *Fenland: its ancient past and uncertain future* (Cambridge University Press, Cambridge)

Grove, J. M. and Battagel, A. (1981) 'Tax records as an index of Little Ice Age environmental and economic deterioration, from Sunnfjord Fogderi, western Norway 1667–1815', in Delano Smith, C. and Parry, M. (eds) *Consequences of climatic change* (Department of Geography, University of Nottingham, Nottingham) pp. 70–87

Hannon, G. E. and Bradshaw, R. H. W. (1989) 'Recent vegetation dynamics on two Connemara lake islands, western Ireland', *Journal of Biogeography* 16: 75–81

Harley, J. B. (1982) 'Historical geography and its evidence: reflections on modelling sources', in Baker, A. R. H. and Billinge, M. (eds) *Period and Place: Research Methods in Historical Geography* (Cambridge University Press, Cambridge) pp. 261–73

Hooke, J. M. and Kain, R. J. P. (1982) *Historical Change in the Physical Environment* (Butterworth, London)

Hoskins, W. G. (1955) *The Making of the English Landscape* (Hodder and Stoughton, London)

Jäger, H. (1969) *Historische Geographie* (Georg Westermann Verlag, Braunschweig)

Kain, R. J. P. (1984) 'The tithe files of mid-nineteenth-century England and Wales', in Reed. M. (ed.) *Discovering Past Landscapes* (Croom Helm, Beckenham) pp. 56–84

Kain, R. J. P. (1986) *An Atlas and Index of the Tithe Files of Mid-nineteenth-century England and Wales* (Cambridge, Cambridge University Press)

Kain, R. J. P. and Prince, H. C. (1985) *The Tithe Surveys of England and Wales* (Cambridge, Cambridge University Press)

Kershaw, A. P. (1986) 'Climatic change and Aboriginal burning in north-east Australia during the last two glacial/interglacial cycles', *Nature* 322: 47–9

Lambert, A. M., Jennings, J. N., Smith, C. T., Green, C. and Hutchinson, J. N. (1960) *The Making of the Broads* (Royal Geographical Society, London)

McAndrews, J. H. (1988) 'Human disturbance of North American forests and grasslands:

the fossil pollen record', in Huntley, B. and Webb, T. (eds) *Vegetation History* (Kluwer, Dordrecht) pp. 672–97

MacDonald, G. M., Larson, C. P. S., Szeics, J. M. and Moser, K. A. (1991) 'The reconstruction of boreal forest fire history from lake sediments: a comparison of charcoal, pollen, sedimentological and geochemical indices', *Quaternary Science Reviews* 10: 53–71.

McNeill, W. H. (1991) 'American food crops in the Old World', in Viola, H. J. and Margolis, C. (eds) *Seeds of Change* (Smithsonian Institution Press, Washington and London) pp. 43–59

Nitz, H.-J. (ed.) (1993) *The Early Modern World-system in Geographical Perspective* (Franz Steiner Verlag, Stuttgart)

O'Hara, S. L., Street-Perrott, F. A. and Burt, T. P. (1993) 'Accelerated soil erosion around a Mexican highland lake caused by by prehispanic agriculture', *Nature* 362: 48–51

Oldfield, F. (1983) 'Man's impact on the environment: some recent perspectives', *Geography* 68: 245–56

O'Sullivan, P. E. (1983) 'Annually-laminated lake sediments and the study of Quaternary environmental changes – a review', *Quaternary Science Reviews* 1: 245–313

Phillips, C. W. (1951) 'The Fenland Research Committee, its past achievements and future prospects', in Grimes, W. F. (ed.) *Aspects of Archaeology in Britain and Beyond* (Edwards, London) pp. 258–73

Rackham, O. (1986) *The History of the Countryside* (Dent, London)

Reed. M. (ed.) (1984) *Discovering Past Landscapes* (Croom Helm, Beckenham)

Roberts, N. (1989) *The Holocene: an Environmental History* (Blackwell, Oxford)

Russell, E. W. B., Davis, R. B., Anderson, R. S., Rhoades, T. E. and Anderson, D. S. (1993) 'Recent centuries of vegetational change in the glaciated north-eastern United States', *Journal of Ecology* 81: 647–66

Sheail, J. (1980) *Historical Ecology: the Documentary Evidence* (Institute of Terrestrial Ecology, Cambridge)

Sheail, J. (1993) 'Green history – the evolving agenda', *Rural History* 4: 209–23

Simola, H. L. K., Coard, M. A. and O'Sullivan, P. E. (1981) 'Annual laminations in the sediments of Loe Pool, Cornwall', *Nature* 290: 238–41

Suffling, R. (1992) 'Climate change and boreal forest fires in Scandinavia and central Canada', in Boer, M. M. and Koster, E. A. (eds) *Greenhouse Impact* (Catena supplement) 22: 111–132

Tate, W. E. and Turner, M. E. (1978) *A Domesday of English Enclosure Acts and Awards* (The Library, University of Reading, Reading)

Till, C. and Guiot, J. (1990) 'Reconstruction of precipitation in Morocco since 1100 AD based on *Cedrus atlantica* tree-ring widths', *Quaternary Research* 33: 337–51

Turner, M. E. (1980) *English Parliamentary Enclosure* (Dawson, Folkestone)

Turner, M. E. (1984) *Enclosures in Britain 1750–1830* (Macmillan, London)

Unwin, T. (1991) *Wine and the Vine: an Historical Geography of Viticulture and the Wine Trade* (Routledge, London)

Wallerstein, I. (1974) *The Modern World-system I: Capitalist Agriculture and the Origins of the European World-economy in the Sixteenth Century* (Academic Press, London)

Wallerstein, I. (1980) *The Modern World-system II: Mercantilism and the Consolidation of the European World-economy* (Academic Press, London)

Wallerstein, I. (1989) *The Modern World-system III: the Second Era of Great Expansion of the Capitalist World-economy, 1730–1840s* (Academic Press, London)

Williams, M. (ed.) (1990) *Wetlands: a Threatened Landscape* (Blackwell, Oxford)

Williams, M. (1994) 'The relations of environmental history and historical geography', *Journal of Historical Geography* 20: 3–21

Wooldridge, S. W. and Linton, D. L. (1955) *Structure, Surface and Drainage in South-east England* (George Philip, London)

Worster, D. (ed.) (1988) *The Ends of the Earth, Perspectives on Modern Environmental History* (Cambridge University Press, Cambridge)

# PART I

# Scotland

2

# The Machair Vegetation of South Uist and Barra in the Outer Hebrides of Scotland: its Interacting Ecological, Geomorphic and Historical Dimensions

David D. Gilbertson, M. Kent, Jean-Luc Schwenninger, P. A. Wathern, R. Weaver and Barbara A. Brayshay

ABSTRACT

Floristic studies of the machair vegetation of parts of South Uist and Barra have demonstrated the presence of eight main plant communities, essentially agreeing with previous research. The prime environmental controls are shown to be soil moisture and soil organic matter, with soil pH of secondary importance. The studies re-emphasize the uniformity of the machair grassland plant community under a variety of land-use conditions such as rabbit grazing and burrowing, sheep grazing, cattle grazing, or patch cultivation by spade or plough for potatoes, or mixed crops of oats and rye. This uniformity appears to be in part the result of selection in favour of perennial species capable of growing up through accreting sand. In part, it can also be explained by the frequent stirring of the viable seed bank, along with sediment and soil by wind, animals and land-use practices. This re-working is evidenced by the presence of a type of sedimentary stratification, termed here 'machair stratification' which appears to be invariably associated with machair grassland. The machair stratification is nearly always underlain by a palaeosol. Field description and micromorphological studies show the machair stratification to be the result of the re-deposition of (1) calcareous shell debris in various stages of weathering, together with organic material from soils and exposed surfaces elsewhere in the machair dune complex, and (2) a thin layer of *in situ* and re-worked organic debris from plants and animals on the former land surface. The deflation and re-working of sediment and inferred re-working of viable seed is believed to extend, as a result of the impacts of major storms, to the depth of a resistant soil or archaeological surface. Geomorphological, archaeological and historical surveys in the study area suggest that 'machair stratification' is geologically 'young' – the feature occurring as a surface blanket over underlying deposits. This type of stratification is younger than *c.*AD 1900 on the Eoligarry Isthmus on Barra, and younger than Iron Age and Bronze Age archaeological sites on South Uist. Re-working may be a useful key to the biogeographical as well as the geomorphic landscapes of the machair of the study area.

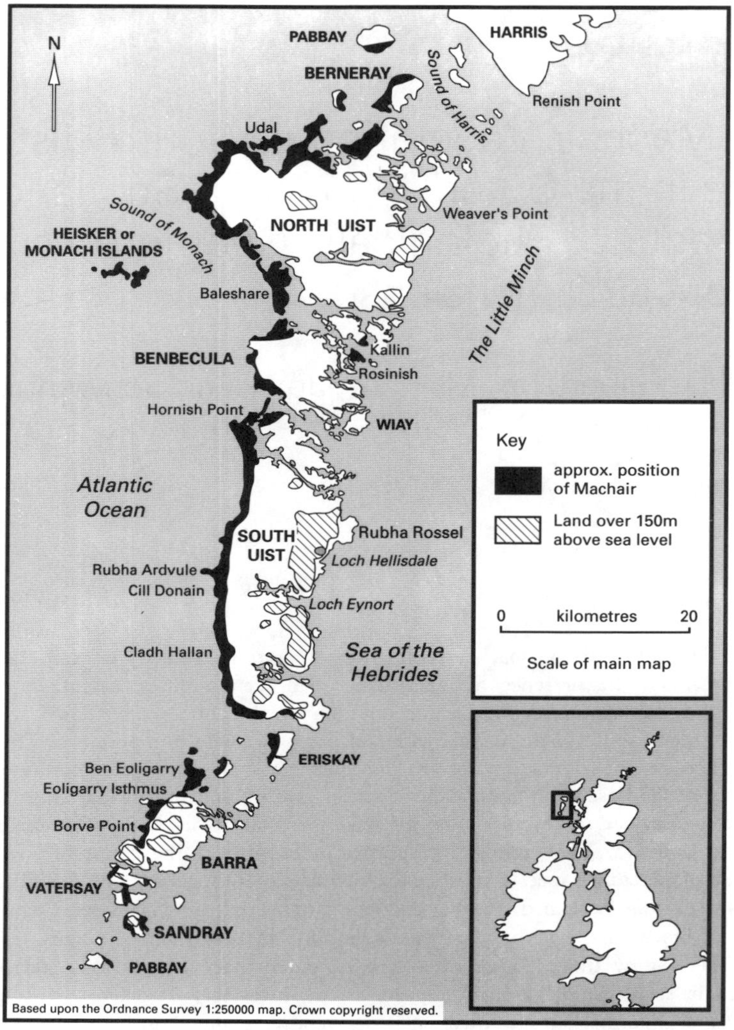


Figure 2.1 A location map showing South Uist and Barra in the Outer Hebrides of Scotland, the approximate distribution of machair and the principal sites mentioned in the text.

## INTRODUCTION

'Machair is a local, landscape term applied to dune pasture (often calcareous) subject to local cultivation and developed in humid and windy conditions in north and west Scotland' (Ranwell 1974; for a wider discussion, see Ritchie 1967, 1976, 1979; Ranwell 1974, 1977, 1980; Mather and Ritchie 1977; Angus 1991; Hudson 1991; and for a general account of sand dunes in Scotland see Robertson-Rintoul and Ritchie 1990). The machair and coastal dunes comprise important geomorphological and ecological resources in western Scotland, and numerous sites in the Outer Hebrides are scheduled as Sites of Special Scientific Interest (Nature Conservancy Council 1977, McKirdy and Foxwell 1985). This chapter seeks to develop a wider understanding of the origins of the vegetation and landscapes of Holocene coastal sand dunes and machair of the Outer Hebridean islands of South Uist and Barra (figure 2.1). It describes recent investigations of the interrelationships that occur between the vegetation and soils in the coastal sand dunes and machair landforms and, in addition, examines their relationships with various aspects of the machair stratigraphy, geomorphology, and human history of this region. In particular, the chapter introduces and discusses the significance of a previously unrecorded but important type of sedimentary stratification, which in the study area is invariably associated with the colourful and distinctive vegetation of the machair grassland. The antiquity of this relationship is explored using historical, archaeological and chronostratigraphic information.

In this account, particular emphasis is given not only to soils but also to sediments recycling in a dynamic, aeolian sedimentary environment. The chapter has three main parts. The first summarizes the complex and remarkably diverse literature that exists for the various aspects of the local machair environment. The second presents the results of a detailed investigation of the floristics of the machair and their environmental associations within the two study areas. The third describes the character of 'machair stratification' and its association with machair grassland and places this type of deposit within the wider chronological, geomorphic and human history of the machair.

## THE MACHAIR OF SOUTH UIST AND BARRA

The character of the machair in the study region is illustrated in plate 2.1. Yellow, white, silver and purple flowers grow on a deflation plain, with occasional areas of bare sand, behind the coastal dune cordon colonized by marram and lyme grass. The contemporary flora of the machair of the Outer Hebrides has been investigated by a number of researchers (e.g. Gimingham 1964; Dickinson et al. 1971; Randall 1973, 1974, 1976; Ranwell 1974, 1977; and Ratcliffe 1977; Currie 1979; Dickinson and Randall 1979; Pitkin et al. 1983; Boyd and Boyd 1990; Pankhurst and Mullin 1991). Most of these accounts contain descriptions and species lists for the flora, but there are few detailed studies of the plant community structure of machair vegetation, apart from the paper by Dickinson and Randall (1979). They describe the machair as consisting of a series of sub-habitats which are summarized in table 2.1. Variation in the floristic composition of machair vegetation is attributed to a number of factors. Soil type is an important control in this context. The sands of the machair consist of reworked fluvioglacial deposits washed inshore together with large quantities of comminuted shell sands (Ritchie 1979, Hudson 1991). The machair sands are alkaline in nature and the fresh blown sand has a pH of between 7.0 and 8.0. The pH falls with the gradual accumulation of organic matter and the leaching of soils. On the machair plains of Barra and South Uist,

**Plate 2.1** Machair grassland vegetation, in full flower in June 1990, on a deflation plain inland of the coastal dune cordon near Kildonan, South Uist (minibus for scale).

the pH normally falls to around 6.0. Vegetation and soil development are also reflected in the proportion of organic matter in the soils and in the ability of the soil to hold moisture. Most of the machair soils, with the exception of the dune slacks, are freely drained. Both Glentworth (1979) and Boyd and Boyd (1990) describe these soils as aeolian sand with regosols. Many authors have noted that shell sand may be blown across machair vegetation each winter, with deeper burial being caused by sand blown by severe storms.

The ecological history of the machair has been little investigated. The recent series of palynological studies of machair sites by Edwards (personal communication) and Brayshay (1992) are as yet unpublished. The exception is the monograph of Whittington and Ritchie (1988), whose results concentrate upon the prehistoric period. Unpublished excavations by the University of Sheffield in late Iron Age deposits at Cill Donain (*O. S.* spelling Kildonan – figure 2.1) and a Bronze Age site at Cladh Hallan (figure 2.1) on South Uist have revealed the almost total dominance of barley cultivation in the arable economy and a pastoral economy focused principally upon sheep and cattle. These studies throw little direct light upon the origins and structure of the brilliantly coloured vegetation that is a hallmark of the machair in summer. The earliest written descriptions of the study area by Munro (1549) are of limited help in that they tend to identify such features as the greenness of the vegetation growing on the sand-covered gneissose bedrocks such as Ben Eoligarry on Barra (figures 2.1 and 2.2). This hill and isthmus were similarly

Table 2.1 Sub-habitats of machair described by Dickinson and Randall (1979)

| *Sub-habitat type* | *Dominant species* |
|---|---|
| 1. Dune types | |
| (a) Active dune front | *Ammophila arenaria* |
| | *Atriplex hastata* |
| | *Honkenya peploides* |
| (b) Moribund dune front | *Festuca rubra* |
| | *Plantago lanceolata* |
| | *Galium verum* |
| | *Lotus corniculatus* |
| | *Trifolium repens* |
| (c) Dune back | *Ammophila arenaria* |
| | *Galium verum* |
| | *Plantago lanceolata* |
| | *Trifolium repens* |
| | *Senecio jacobaea* |
| | *Rhytidiadelphus squarrosus* |
| (d) Sand hill | *Festuca rubra* |
| | *Ammophila arenaria* |
| | *Galium verum* |
| | *Plantago lanceolata* |
| | *Trifolium repens* |
| | *Rhytidiadelphus squarrosus* |
| 2. Grassland types | |
| (a) Machair grassland | *Festuca rubra* |
| | *Poa pratensis* |
| | *Ammophila arenaria* |
| | *Trifolium repens* |
| | *Plantago lanceolata* |
| | *Bellis perennis* |
| | *Lotus corniculatus* |
| | *Rhytidiadelphus squarrosus* |
| (b) Dune slack | *Carex flacca* |
| | *Poa pratensis* |
| | *Festuca rubra* |
| | *Potentilla anserina* |
| | *Hydrocotyle vulgaris* |
| | *Rhytidiadelphus squarrosus* |

described by Rory McNeill, writing at about 1613–22, with greater reference being paid to the Eoligarry Isthmus (see Campbell 1936 and figure 2.2). The records of the visits of the Reverend Dr John Walker (see McKay 1980) give extensive accounts of machair and include descriptions of the distinctive and important yellow flowering *Galium verum*, the widespread cultivation of bere (barley) and oats on the machair, the recent introduction of the potato, the existence of major areas of sand drift, and the occurrence of storms which, on

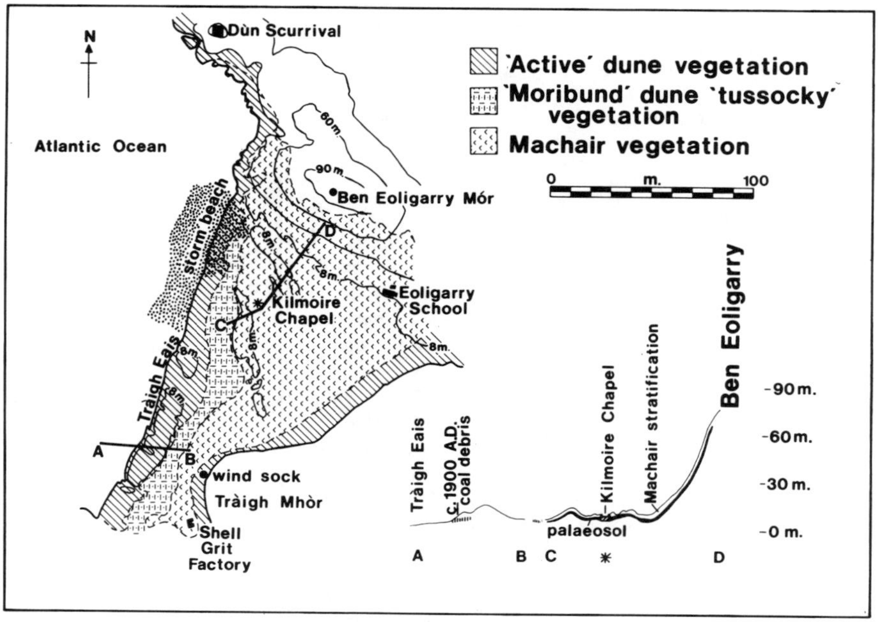


Figure 2.2 The topography, vegetation, and sample sites on the Eoligarry Isthmus on the island of Barra in the Outer Hebrides, and a generalized stratigraphy of the machair deposits on the Eoligarry Isthmus and Ben Eoligarry. A blanket of deposits exhibiting 'machair stratification' covers all deposits except those of the modern coastal dune cordon, the formation and deposition of which is attributed to the twentieth century on the evidence of coal debris located near their base attributed to shipwrecks coming ashore at Tràigh Eais in the period 1890–1900. The upper palaeosol is dated by a combination of archaeological and historical evidence to the period from the end of the sixteenth century to the end of the nineteenth century AD.

19 February 1749, breached the Eoligarry isthmus, dividing Barra into two islands. It is also from about this time that the song '*Smeorach Chlann Domhnaill*' ('The Mavis of Clan Donald'), probably composed in the 1750s by John MaCodrum (living at Paible, North Uist), became known, with its description of '... a beautiful land, a land of the plain, the land of the smiling coloured flowers, the land of mares and stallions and kine.... The land most lovely to be seen...' (see Matheson 1938, Angus and Eliott 1992) – this song might be a recognition of the colourful machair vegetation of summer.

Human activity on the machair is and obviously has been widespread, particularly on the machair plains. Such activity has a long and complex history, and is likely to have been a major influence upon the character of machair vegetation. The history of the Outer Hebridean machair has most recently been summarized by Angus and Elliott (1992, and references therein). Nowadays, the various machair grasslands are grazed extensively by sheep and cattle, and in many areas grazing and burrowing by rabbits is intense. Caird (1979) and Boyd and Boyd (1990) describe how parts of the machair are used for cultivation as 'lazy beds'. This may involve the use of a shallow plough and the cultivation of patches of ground for potatoes, and some cereals – typically a mixture of oats and rye – aided by the incorporation of dung and kelp into the soil. Elsewhere, it is possible to see the patches of ground displaying the typical parallel troughs and ridges which are the result of spade cultivation, nowadays associated with the growing of potatoes. The potato was introduced to the Outer Hebrides in 1743 and to Barra in 1752 (see Walker 1764 & 1771). In the past, the machair and the foreshore and coastal foredune ridge have been excavated for shell sand to aid the cultivation of peat soils further inland. The phosphate- and nitrate-rich archaeological midden sites in the machair have been 'quarried' for similar reasons. Disturbance of the foreshore and coastal dunes is also likely to have occurred in the kelp boom of the early nineteenth century (Macdonald 1811, Angus and Elliott 1992), as well as from the harvesting of marram (*Ammophila arenaria*) for use in thatch, mats, ropes, and bedding (see Beveridge 1911, Campbell 1958, Angus and Elliott 1992), and the digging of the roots of lady's bedstraw (*Galium verum*) which were used for dyeing. All of these activities came to be restricted by landowners seeking to curb the unwanted erosion and deposition of sand in these coastal lands. Ritchie (1979) emphasized the importance of artificial drainage in the nineteenth century as a potential cause of sand drift Such drainage affected many of the long chains of lochs which ran down to parts of the machair. Wet sand is not particularly susceptible to wind blow, but when dry, in the wind-dominated climate of the Outer Hebrides (see Boyd and Boyd 1990), these sands must have been readily deflated. These human agencies may also have combined with natural climatological and oceanographic fluctuations to cause accelerated deflation, deposition and a general movement of sand on the machair lands. In their recent review of the documentary history of the erosion of the machair, Angus and Elliott (1992) conclude that the magnitude of erosion that occurred between the sixteenth and nineteenth centuries AD was greater than that which has been noted in the twentieth century. They repeat Ritchie's (1979) observation that the machair had largely stabilized by the latter part of the nineteenth century.

The dominant view of the longer-term geomorphic history of Uist coastal sands is Ritchie's (1979) widely known, complex, multistage model of the evolution of machair landforms, which was derived from his research in the present study area. Ritchie (1967, 1979, 1985) and Whittington and Ritchie (1988) suggested that, in these Western Isles, the sand dune systems initially became established as relatively large and tall sand dunes at about 7000–8000 years BP in a broadly similar location to that of the present shoreline. The machair landforms (broadly defined) are seen as the product of this vast influx of outwash and shell sands, and their subsequent re-working by wind. The large dunes progressively blew out to produce the present common topographic combination of a 'high machair', immediately inland of the present shoreline, and lower relief further inland. Lochs, marshlands and flatter areas were impounded, leading to machair lochs inland of the present shoreline. The coastal sand hills of the 'high machair' are then thought to have eroded episodically, with the deflated sand being transported and re-deposited further inland. Re-deposition (broadly) occurred in two types of situation. 'Layer-cakes' made up of sheets of re-deposited sands accumulated further inland. More stable episodes were marked by the development of fossil soils, upon which archaeological settlements developed – from Neolithic to recent historic in age (Ritchie 1979, Whittington and Ritchie 1988). Further deflation and other forms of erosion then produced a process of scarp retreat which has resulted in the presence of a steep west-facing escarpment fronting a second type of 'high machair' inland of the machair lochs. Episodic deposition of sand from the coastwards 'initial' high dunes as well as any 'intermediate' dunes, into impounded lochs or wet marshy ground, produced the flatter sand sheets of the 'low machair' which were rendered relatively resistant to further deflation by the winter water table reaching the land surface. Within this sequence of events, fossil soils, often containing evidence of human activities, developed in more stable periods. Together with the winter water table in the machair, these soils served as stable base levels, which inhibited dune deflation to the lower parts of the system (plate 2.1 and Ritchie 1967, 1976, 1979). Most recently, a fringe of modern sand dunes, characterized by the familiar dune species *Ammophila arenaria*, with patches of *Elymus farctus* in some places, has developed to give the final topographic component of these coastal dunes; for example, the modern coastal fore dunes along the shorelines of the Eoligarry isthmus on Barra (figure 2.2).

The further evaluation of these geomorphological models on Barra and South Uist requires a more detailed understanding both of depositional environments and chronology than is currently available. At present, the most detailed, published chronological information derives from the studies of Crawford and Switsur (1977), who presented a chronology of sand dune development and human activity using historical, archaeological and radiocarbon dating evidence for the Udal Peninsula on the north coast of North Uist, and from the research of Whittington and Ritchie (1988) at Rosinish and Kallin (figure 2.1). Unfortunately, the extent to which these interpretations, or dune development chronologies, can be extrapolated south to the present study areas is not clear at present. On South Uist and Barra, dating evidence is weaker. There are substantial uncertainties about the antiquity of the deposits and palaeosols that are mappable in the coastal sands of the machair. Nor are relationships clear between such deposits and the relatively few archaeological sites in the study area which have been excavated within them (see, for example, Young and Richardson 1956–60; MacLaren 1974; Ritchie 1976, 1979; Barber 1987).

## THE VEGETATION OF THE MACHAIR AND ITS DETERMINANTS

The nature of machair vegetation and its environmental associations was established for the purposes of this study by the collection of floristic and soil data from the following areas:

- The west coast of Barra at the sheep- and cattle-grazed Borve Headland; the area immediately to its north at Allasdale where sheep-grazing is dominant; and on the Eoligarry Isthmus which is currently infested with rabbits and has a history of rabbit trapping lasting nearly one hundred years. The southern part of the Eoligarry isthmus is sheep-grazed, whereas the northern sector, including Ben Eoligarry has a long history of exclusive cattle grazing (figures 2.1 and 2.2).

- On South Uist in a zone west of Upper Loch Kildonan, adjacent to excavation sites of Cill Donain, immediately south of the headland of Rubha Ardvule (plate 2.1 and figure 2.1). This area has a significant rabbit population and is grazed extensively by sheep and some cattle. In many places the ground surface displays clear evidence of recent and extensive patch-cultivation by plough and spade.

Sampling in these areas of varying land use on each island was deliberately biased with quadrats being located in a zone across the machair wherever there was a perceptible change in the vegetation. This sampling method enabled as much as possible of the variability in floristics to be recorded. In each quadrat, percentage cover data for all species of higher plants, bryophytes, liverworts and lichens were collected. Recording was carried out in June each summer in the period 1988–92. Identifications were based on Clapham et al. (1989), Stace (1991), Watson (1978), Smith (1978) and Duncan (1970). Further information on species and their habitats was taken from Pankhurst and Mullin (1991). Bare sand, bare peat/soil, bare rock, litter and standing water were also recorded as percentage cover. Data were collected at each quadrat on sand/soil/peat depth, soil pH, percentage soil moisture content (gravimetric method) and percentage soil organic matter (assessed by loss-on-ignition); 147 quadrats were recorded on Barra and South Uist, with 80 quadrats on Barra and 67 on South Uist. The quadrats contained 176 species.

In order to define the main plant communities and to show their variation, the floristic data were first classified using Two-way Indicator Species Analysis (TWINSPAN Hill 1979a). Detrended Correspondence Analysis (DECORANA) (Hill 1979b, Hill and Gauch 1980) and Canonical Correspondence Analysis (CANOCO) (ter Braak 1986, 1987, 1988a,b) were used to show the relationships between the community types and to obtain information on environmental gradients and their correlations with floristic variation (Jongman et al. 1987, Kent and Coker 1992).

### *Plant community variation*

Eight main plant communities (termed A–H here) emerged from the TWINSPAN classification. These are summarized in table 2.2. The relationships between these community groupings are portrayed in figure 2.3, which is derived from the Detrended Correspondence Analysis.

*Group A.* These are quadrats from the front of the mobile dunes closest to the sea, and correspond to the active dune group of Dickinson and Randall (1979). Bare sand is dominant, with only two to four colonizing species. They are found to the extreme right of figure 2.3.

**Table 2.2** Major plant communities of the machair sampled on Barra and South Uist resulting from Two-way Indicator Species Analysis (TWINSPAN)

| *Group* | *Number of quadrats* | *Characterizing species (>50% constant by group)* | *Associated species (25%–50% constant by group)* |
|---|---|---|---|
| A | 6 | Bare sand<br>*Elymus farctus*<br>*Atriplex hastata*<br>*Cakile maritima*<br>*Plantago lanceolata*<br>*Taraxacum officinale* | *Galium verum*<br>*Leontodon autumnalis*<br>*Ammophila arenaria* |
| B | 17 | *Ammophila arenaria*<br>Bare sand<br>*Ranunculus acris* | *Festuca rubra*<br>*Achillea millefolium*<br>*Senecio jacobaea*<br>*Poa subcaerulea*<br>*Taraxacum officinale*<br>*Brachythecium albicans* |
| C | 38 | *Festuca rubra*<br>*Trifolium repens*<br>*Achillea millefolium*<br>*Galium verum*<br>*Plantago lanceolata*<br>*Ranunculus acris*<br>Bare sand<br>*Carex arenaria*<br>*Lotus corniculatus*<br>*Poa subcaerulea*<br>*Senecio jacobaea* | *Ammophila arenaria*<br>*Cerastium glomeratum*<br>*Bellis perennis*<br>*Brachythecium albicans*<br>*Taraxacum officinale*<br>*Agrostis stolonifera*<br>*Koeleria macrantha* |
| D | 35 | *Plantago lanceolata*<br>*Festuca rubra*<br>*Trifolium repens*<br>*Galium verum*<br>*Achillea millefolium*<br>*Lotus corniculatus*<br>*Bellis perennis*<br>*Poa subcaerulea*<br>*Ranunculus bulbosus*<br>*Euphrasia officinalis*<br>*Senecio jacobaea*<br>*Centaurea nigra* | *Luzula campestris*<br>*Agrostis capillaris*<br>*Ammophila arenaria*<br>*Ranunculus acris*<br>*Leontodon autumnalis*<br>*Poa trivialis*<br>*Plantago coronopus*<br>*Rhytidiadelphus squarrosus*<br>*Taraxacum officinale*<br>*Lolium perenne*<br>*Cerastium glomeratum*<br>*Rhinanthus minor*<br>*Prunella vulgaris* |
| E | 15 | *Trifolium repens*<br>*Fesruca rubra*<br>*Ranunculus acris*<br>*Potentilla anserina*<br>*Holcus lanatus*<br>*Agrostis stolonifera*<br>*Bellis perennis*<br>*Plantago lanceolata* | *Anthoxanthum odoratum*<br>*Poa subcaerulea*<br>*Vicia cracca*<br>*Carex flacca*<br>*Rhinanthus minor*<br>*Rumex acetosa*<br>*Equisetum arvense*<br>*Dactylorhiza majalis*<br>*Poa pratensis*<br>*Ophioglossum vulgatum* |

Table 2.2 *(continued)*

| *Group* | *Number of quadrats* | *Characterizing species (>50% constant by group)* | *Associated species (25%–50% constant by group)* |
|---|---|---|---|
| F | 8 | *Potentilla anserina*<br>*Anthoxanthum odoratum*<br>*Carex echinata*<br>*Carex nigra*<br>*Hieracium* sp.<br>*Luzula campestris*<br>*Agrostis capillaris*<br>*Hylocomium splendens*<br>*Molinia caerulea*<br>*Rhytidiadelphus squarrosus*<br>*Rumex acetosa*<br>*Trifolium repens* | *Dicranum scoparium*<br>*Epilobium palustre*<br>*Eriophorum angustifolium*<br>*Iris pseudacorus*<br>*Juncus effusus*<br>*Potentilla palustris*<br>*Agrostis stolonifera*<br>*Calluna vulgaris*<br>*Dactylorhiza majalis*<br>*Equisetum arvense*<br>*Festuca rubra*<br>*Holcus lanatus*<br>*Peltigera rufescens*<br>*Plagiothecium undulatum*<br>*Pleurozium schreberi*<br>*Poa pratensis*<br>*Sphagnum recurvum*<br>*Thuidium tamariscinum*<br>*Viola riviniana* |
| G | 10 | *Agrostis stolonifera*<br>*Caliergon cuspidatum*<br>*Ranunculus flammula*<br>*Eleocharis palustris*<br>*Juncus articulatus* | *Caltha palustris*<br>*Ranunculus repens*<br>Bare sand<br>*Carex arenaria*<br>*Mentha aquatica*<br>*Potentilla anserina* |
| H | 18 | *Rumex acetosa*<br>*Iris pseudacorus*<br>*Caltha palustris*<br>*Potentilla anserina*<br>*Cardamine pratensis* | *Epilobium palustre*<br>*Holcus lanatus*<br>*Agrostis stolonifera*<br>*Caliergon cuspidatum*<br>*Carex nigra*<br>*Dactylorhiza majalis*<br>*Equisetum palustre*<br>*Galium palustre*<br>*Angelica sylvestris*<br>*Equisetum arvense*<br>*Equisetum fluviatile*<br>*Filipendula ulmaria*<br>*Hydrocotyle vulgaris*<br>*Juncus effusus*<br>*Mentha aquatica*<br>*Phragmites australis*<br>*Stellaria palustris*<br>*Vicia cracca* |

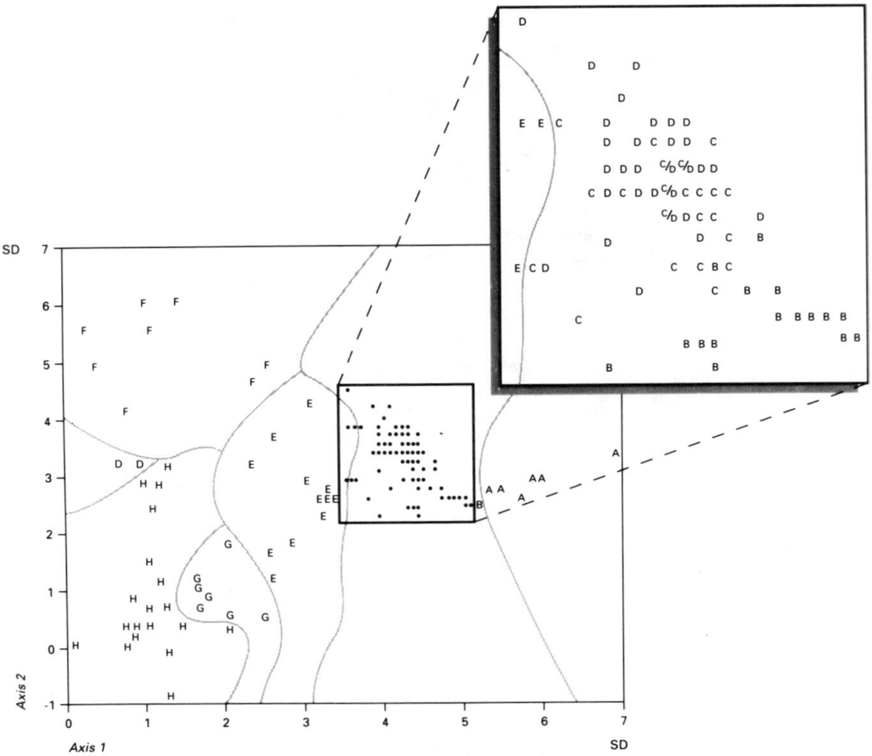


Figure 2.3 A quadrat ordination diagram derived from Detrended Correspondence Analysis of the 147 machair quadrats from South Uist and Barra. TWINSPAN groups (plant community types A–H – see table 2.2) have been superimposed upon the ordination. The area enlarged in the insert corresponds broadly to the machair grasslands (Groups C and D).

*Group B.* Quadrats in this group are from the tops and backs of active dunes and from blow-outs. There is still a significant proportion of bare sand but vegetation cover and species numbers are significantly increased in comparison to Group A. Hence, they are slightly to the left of Group A on the ordination diagram (figure 2.3).

*Groups C and D.* Taken together, these form a large group of 73 quadrats (approximately one half of the total data set) and they correspond to the machair plain communities – the 'machair grassland' of Dickinson and Randall (1979) (and table 2.1 here). *Festuca rubra, Trifolium repens, Galium verum, Achillea millefolium, Plantago lanceolata, Lotus corniculatus* and *Poa subcaerulea* are the characterizing dominant species. The subdivision into Groups C and D is related to increasing richness of associated species (D is richer than C) and a gradient of increased soil moisture and organic matter. The quadrats in Group D, thus, represent more mature machair grassland. Table 2.3 shows the associated environmental variables, with higher soil moisture and organic matter

Table 2.3 Environmental characteristics of machair TWINSPAN groups

| *Group* | *Number of quadrats* | | *pH* | *% soil moisture* | *% organic matter* |
|---|---|---|---|---|---|
| A | 6 | Mean | 7.84 | 2.6 | 0.82 |
| | | Max. | 8.00 | 8.7 | 0.90 |
| | | Min. | 7.60 | 0.4 | 0.66 |
| B | 17 | Mean | 6.96 | 8.5 | 3.23 |
| | | Max. | 8.20 | 64.7 | 15.81 |
| | | Min. | 6.50 | 1.2 | 0.65 |
| C | 38 | Mean | 7.01 | 16.1 | 6.04 |
| | | Max. | 7.80 | 59.4 | 17.39 |
| | | Min. | 6.40 | 0.8 | 0.81 |
| D | 35 | Mean | 6.78 | 34.7 | 11.01 |
| | | Max. | 7.40 | 138.0 | 22.95 |
| | | Min. | 6.40 | 6.6 | 3.82 |
| E | 15 | Mean | 6.77 | 132.8 | 20.79 |
| | | Max. | 7.60 | 504.9 | 61.53 |
| | | Min. | 5.40 | 3.2 | 3.05 |
| F | 8 | Mean | 5.18 | 589.0 | 52.29 |
| | | Max. | 7.60 | 1638.0 | 90.54 |
| | | Min. | 5.20 | 60.0 | 23.50 |
| G | 10 | Mean | 6.61 | 124.4 | 15.18 |
| | | Max. | 7.60 | 448.1 | 60.75 |
| | | Min. | 5.20 | 12.4 | 1.05 |
| H | 18 | Mean | 6.46 | 519.2 | 44.43 |
| | | Max. | 7.60 | 1546.1 | 78.94 |
| | | Min. | 5.20 | 128.1 | 9.34 |

values and lower pH in Group D than in Group C. The insert on the quadrat ordination diagram (figure 2.3) also shows how these machair grassland quadrats are distributed as a closely related set of samples, but with a local continuum within them. However, the 'tightness' of their distribution within the overall ordination shows that floristically they are a very clearly defined group.

*Group E.* This small but more widespread group occurs just to the left of Groups C and D on the ordination diagram (figure 2.3). The same dominant machair grassland species are found as for Groups C and D (table 2.2), but there are also a few more neutral/acid species such as *Holcus lanatus*, *Agrostis stolonifera* and *Rumex acetosa*, together with a damper element occurring in depressions in the machair and on the margins of dune slacks including *Carex flacca*, *Poa pratensis* and *Equisetum arvense*. This interpretation is confirmed by the soil moisture data (table 2.3).

*Group F.* This small group of diverse quadrats is characterized by still more acid and wet conditions (table 2.3). Organic matter values are also much higher. These sites are in hollows, often on the inner edge of the machair plain at its junction with the 'blacklands' – where the alkaline shell sands abut on to the acidic gneiss bedrocks and its overlying peat cover. The lowest pH values occur in this group. In many places these sites are indicated by the presence of *Iris pseudacorus*. This plant community type is mostly found in South Uist rather than Barra. The quadrats are spread out in the upper left section of the ordination diagram (figure 2.3).

*Group G.* These are also wet-hollow and wet-slack communities but they are not as wet or species-rich as the quadrats in Group H. The ordination diagram shows them to be transitional between Groups E and H.

*Group H.* The wet-slack and machair loch margin communities are found in this group. Many of these quadrats contain the indicator species *Iris pseudacorus*, *Caltha palustris* and *Cardamine pratensis*. There is a wide range of associated species. Soil moisture and organic matter values are relatively high, although not as high as Group F. The pH values indicate that these are communities based on machair shell sand rather than more acidic siliceous parent materials.

### *Discussion*

These analyses of vegetation and environmental data appear to suggest the importance of 'natural', as opposed to human agencies, as the principal factors in determining the nature of floristic variation in the areas of machair studied. This point is emphasized in table 2.4, which shows the close intercorrelation of environmental variables with ordination axes, which are summaries of floristic variation. The first axis of the floristic data is very highly correlated with both soil moisture and organic matter and this represents the primary environmental gradient across the machair. Interestingly, pH emerges as significantly correlated with both axes but is more important on the second. Thus, it is a secondary gradient and is more highly correlated with the second axis of floristic variation.

In terms of human impact on the vegetation, it is the machair grassland (the common Groups C and D, and to a lesser extent Groups B and E) which are of greatest interest. These quadrat data, deliberately sampled to obtain maximum

**Table 2.4** Intercorrelations of environmental variables with ordination axes from canonical correspondence analysis (CANOCO)

| *Variable* | *Ordination axes* | | | |
|---|---|---|---|---|
| | *1* | *2* | *3* | *4* |
| % soil moisture | −0.89 | 0.25 | 0.09 | 0.01 |
| % organic matter | −0.88 | −0.03 | 0.19 | 0.05 |
| pH | 0.48 | −0.70 | 0.15 | 0.04 |

diversity at a number of sites with notably different land usage, illustrate that these quadrats are surprisingly uniform, as seen in the expanded upper part of the ordination diagram. Hence, human disturbance does not immediately appear to be directly related to floristic variation in these data. This is not to underplay the role of human activities for these vegetation types; human activity may often 'lie behind' the environmental variables, by causing blow-outs, building or re-claiming sand dunes, lowering or raising water tables, or adding dung or seaweed to cultivated patches.

The remainder of this account focuses upon the dominant Groups C and D which are shown above to contain immediately recognizable and colourful machair species of the machair plain – the machair grassland of Dickinson and Randall (1979) – *Galium verum*, *Lotus corniculatus*, *Ranunculus acris*, *R. bulbosus*, *Bellis perennis*, *Taraxacum officinale*, *Euphrasia officinalis* and *Prunella vulgaris* (see table 2.2). The other groups are associated with particular types of habitat, such as sand dunes (Groups A and B) and lochs, marshes and wetlands (Groups G and H) and/or ecotonal situations (Groups E and F).

## THE GEOMORPHIC AND STRATIGRAPHIC ASSOCIATIONS OF MACHAIR GRASSLANDS

Stratigraphic evidence has been obtained by observations of eroding faces, field mapping, and as part of the (unpublished) University of Sheffield archaeological excavations by Ms Jo Heron at Tràigh Eais on the Eoligarry Isthmus of Barra, on the South Uist machair at Cladh Hallan by Mr Edward Moth, and at Cill Donain by Dr Marek Zvelebil (figures 2.1, 2.2 and 2.4); all of which have been assisted by information obtained from trial pits and boreholes. The type of microstratification described below is easily missed in exposures which have been 'trowelled', or 'stirred' when obtaining borehole samples – it is recorded primarily here from observations on faces which have been exposed to the weather.

At the study sites, the Holocene deposits that make up the machair were seen to be made up of six distinct types of sediments – typically associated with a particular type of landform – which could be both mapped and given stratigraphic significance. These deposits are: modern sand dunes; late- to mid-Holocene sand dunes and sand sheets; beach conglomerates; palaeosols; midden and other types of archaeological deposits and buildings; and the present blanket of sand which exhibits machair stratification. In brief, the most important of these in the present context is the blanket of sand displaying machair sedimentary stratification (plate 2.1). This has, with the exception of the deposits of the modern dune cordon (figures 2.2 and 2.4), been shown by widespread mapping studies to bury all older deposits, and to extend almost to the summit of Ben Eoligarry (figure 2.2).

### *'Machair stratification'*

Extensive field mapping at the study sites has demonstrated that this type of stratification occurs beneath thin soil A horizons and is always associated with

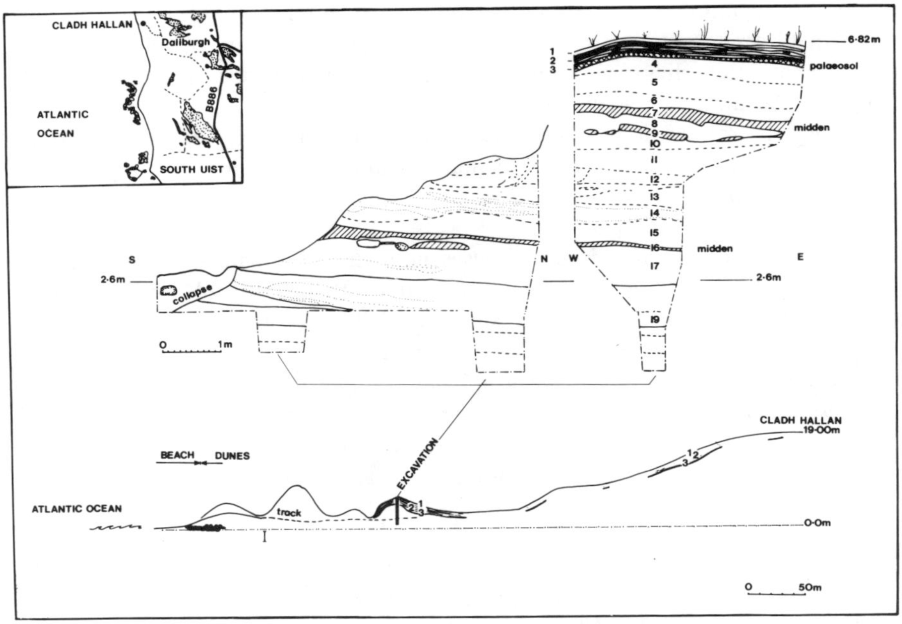


Figure 2.4 The stratigraphy of machair deposits at Cladh Hallan, South Uist, located in figure 2.1. The soil A horizon is shown as study Unit 1, the blanket deposit exhibiting machair stratification is Unit 2, and the uppermost palaeosol is Unit 3. The upper midden shown as Unit 7 is dated to the Bronze Age.

machair grassland vegetation (Groups C and D). This deposit is characterized by a type of stratification – informally termed here 'machair stratification' – that does not seem to have been recognized previously. Deposits exhibiting this type of stratification are illustrated in plate 2.2, and figures 2.4 and 2.5. Typically these deposits are 0.5–1.5 m thick (sometimes up to 2.5 m thick) and 'blanket' the machair landscape inland of the modern coastal dune cordon (figures 2.2 and 2.4). Micromorphological examination shows these deposits to consist of many superimposed thin couplets of (1) white/grey to brown comminuted shell sand (sometimes graded) 1–3 mm thick with a sharp lower boundary and (2) grey to black peaty shell sand, not visibly graded, variously 1–4 mm thick, with a bioturbated and/or diffuse lower boundary (figure 2.5). These couplets are essentially horizontal or gently undulating and are typically more or less parallel to both the modern ground surface and an underlying palaeosol, even on hillslopes of 10–15°, such as the south face of Ben Eoligarry (figure 2.2). The couplets may be traceable for distances of up to 10 m, but at other sites may be disrupted by burrows and other minor dislocations. Sometimes the couplets thin and coalesce with underlying couplets. This deposit almost invariably rests upon a palaeosol (plate 2.2).

Further details have been revealed by petrographic studies of thin sections through both these deposits and the recognizable palaeosols that occur beneath them. No one slide will reveal all the important micromorphological features

**Plate 2.2** Machair stratification between the surface soil and a palaeosol, and overlying dune sands in the dune sequence in the Eoligarry isthmus, Barra. The front of a skull of a horse protrudes from the lower dune sands in the middle of the photograph. The surface of the palaeosol contains rabbit traps dateable to the period approximately AD 1890–1900.

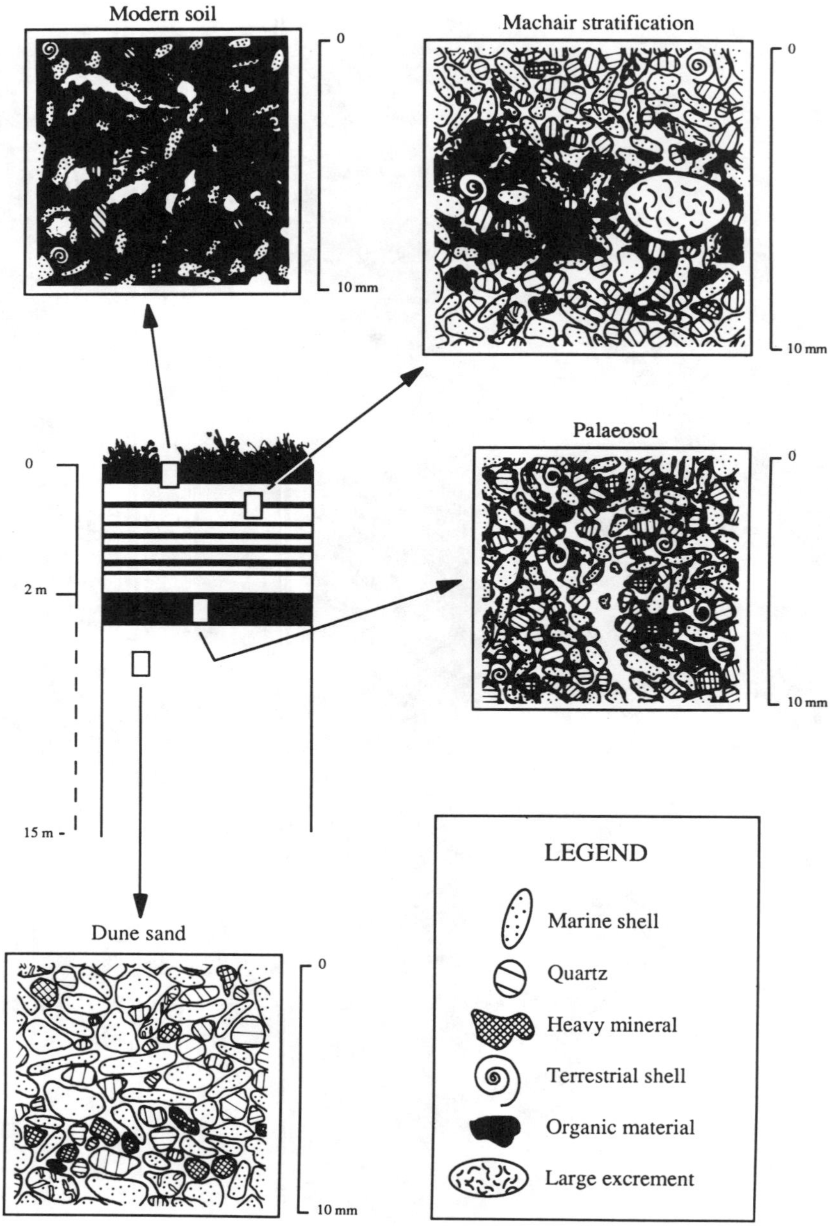


Figure 2.5 A generalized representation of the major micromorphological features present in machair soils and sediments.

present. In consequence, figure 2.5 is a generalized description, which groups together the features seen in an extensive survey of both the machair stratification and the palaeosols in these dunes. The paler lamellae are seen to consist of comminuted, often platy, shell debris (predominantly of thicker-shelled marine species) in a wide variety of different stages of abrasional 'rounding' and chemical etching/weathering. Some of these shell fragments have attached to their outer surfaces detrital, well-macerated organic materials, whereas other shell fragments contain such organic debris within borings into the shell made by fungi or algae. Grains of quartz and heavy minerals occur throughout these shell sands. The paler layers also contain passage features associated with the activities of soil invertebrates, as well as disruptions caused by root channels. These features have often collapsed, but are readily recognized by rearrangements of platy shell debris. The darker lamellae also contain detrital shell debris, quartz and heavy mineral grains. However, the most characteristic feature of them is the presence of a layer of loosely scattered, relatively fresh, amorphous plant material and fungal debris which is sometimes present as large organomineral excrements and aggregates. Some of this material appears to be detrital, while some appears to be *in situ*. Passage channels occur similarly within these organic lamellae.

The properties of both these types of lamellae differ notably from those exhibited by mature soils, recognizable palaeosols, and the sands of underlying and modern coastal sand dunes in the study areas. For example, the unpublished investigations by Schwenninger show the coastal aeolian dune deposits to be characterized by a rather uniform degree of weathering of the surfaces of the marine shell debris that compose them; there is also an absence of aggregates of amorphous organic matter and excrements. Thin sections of mature modern soils and palaeosols (e.g. figure 2.5) contain large quantities of calcitic fine materials (e.g. micrite) embedded within a matrix of organic matter and excrements – a feature largely absent in the machair stratification. Debris from thin-shelled terrestrial molluscs is noticeably more abundant in both these types of soils. These micromorphological studies demonstrate that the thinly laminated machair stratification corresponds with neither mobile sand dunes nor with moderately well-developed modern and fossil soils. Similarly, it is evident that these laminae are not textural in origin. They are demonstrably not the equivalent of graded bedding. Neither, given the comparative sizes of the organic aggregates involved, can they be the result of the infiltration or translocation of 'fines' down profile (see, for discussion, Robinson and Rich 1960, Dijkerman et al. 1967, Bond 1976, Kemp and McIntosh 1989). Heavy minerals are distributed quite differently to the lamellae of shell debris and organic materials, showing that these deposits are not the result of the fairly common phenomenon of preferential concentration of heavy minerals at an accumulating sand dune or beach sediment surface. Overall, there is an absence of features which would suggest the presence of pedogenic horizonation.

Several of the properties of the machair stratification quite clearly point to derivation and re-working. For example, the diverse weathering states of the shell debris suggest that material is being derived from a range of variously weathered shell sand sources within the sand dunes of the area. The occasional presence of organic debris adhering to the sides and occupying corroded tubes in shell debris in the paler shell sand lamellae suggests that this particular material was previously incorporated in biologically active soil surfaces. Similarly, the darker organic-rich lamellae do not show a dense continuous matrix of organic matter and excrement as evidenced in the palaeosols and mature modern machair soil. The organic matter is present in what sometimes appears to be detrital aggregates brought in from sources elsewhere – that is, eroding nearby soil surfaces. Furthermore, the presence of root channels and animal burrows emphasizes that these materials are likely to have served as soils, supporting their own plant and animal communities. This analysis points strongly to machair stratification being the result of the deposition of the following:

- thin layers of sand from adjacent eroding sand dunes and/or other exposed shell sand surfaces such as rabbit burrows and, occasionally, more organically rich topsoils; and

- the deposition of detrital organic material from topsoils and burrowing invertebrates as well as the *in situ* decay products of biological materials which collected on the developing land surface.

The former process is evident and frequently commented upon after storms in the area – the sand being spread widely in the complex airflows that occur behind the main dunes. The sand is derived from exposed dune faces, soil surfaces and blow-outs in the area. The inferred deposition of organic debris by deflation of topsoil and the local decay of plant and animal tissues is less easy to observe by casual observation. Machair stratification appears, therefore, to be the result, in whole or large part, of the repetitive re-working and accumulation of materials from sand dune complexes in which open sand and topsoils are eroding.

This interpretation offers an immediate explanation for several of the major features of the vegetation of the machair grasslands and the sediments that underlie them – that is, the vegetation in Groups C and D identified previously. The evident uniformity of the machair grasslands can be reasonably suggested to be the result of a number of factors:

- Frequent shallow burial during winter months and following storms at all times of the year. Provided that the burial is not too deep, the vegetation, which is largely made up of perennial species with some biennials and annuals, will regrow through the sand in the following months. The perennial species which make up the machair grassland (tables 2.1 and 2.2) are assumed to be adapted to this burial process. However, these hypotheses need to be tested.

- At regular intervals, the depth of deflated sand overlying the machair grassland may become too thick for perennial species to grow through. If this occurs, the existing vegetation and any underlying organic horizons will become trapped within the sand and start to form one of the layers of machair stratification.

- On and within the sand which has buried the existing vegetation, the machair species will re-establish by re-working from the seed bank, within the blown sand and from vegetative fragments which allow vegetative reproduction. Most of these species will be perennials and biennials. Annuals are not a large proportion of machair species. Further research is required on the seed banks, the relationships between the species composition of seed banks and overlying vegetation, and the mechanisms of recolonization of blown sand within the machair grassland (for discussion, see Leck et al. 1990, Fenner 1992, Warr et al. 1993).

- To these can be added the similar types of near-surface re-working of seed and sediment which are likely to be brought about, or accelerated, by what appear to be the rather different forms of land use that occur on the wind-dominated environment of the machair; that is, patch-cultivation by plough or spade, and animal grazing, defecation, burrowing and bioturbation.

The wider environmental trends influencing the machair vegetation identified in the above vegetation analysis (soil organic matter, moisture, and pH) can be postulated to relate to the consequences of wider spatial interactions across the machair between the geomorphic processes of deflation, transportation and deposition, and those of biological colonization and soil development, as well as those other broader-scale impacts upon the machair surface that stem from human activities, or climatic and oceanographic fluctuations. Sites closer to sand sources are likely to experience truncated soil development through burial in sand, whereas sites further from sand sources – that is, at the landward limits of the machair – are likely to encounter a more constant environment in which a soil A horizon can develop, and with this higher levels of soil organic matter, soil moisture and lower soil pH.

### *The stratigraphic position of machair stratification*

Stratigraphic evidence may also have considerable importance for understanding the antiquity of the machair. To date, all field mapping in the region suggests an almost invariable association between machair grasslands and machair stratification, as defined here. Detailed mapping and examination of the stratigraphy of the machair sands on the islands of Vatersay and Sandray as well as on Barra (especially the Eoligarry Isthmus – figures 2.1 and 2.2) and at Cladh Hallan and Cill Donain on South Uist (Gilbertson unpublished, and see figures 2.1 and 2.4) have shown that machair stratification always extends as a blanket from behind modern sand dunes to the machair inland, and that it nearly always overlies a distinct palaeosol (plate 2.2). This mapping also shows that the deposit occurs over both the *high-* and *low-*machair of Ritchie (1976, 1979). The only known exception to this 'rule' is one location on the Eoligarry Isthmus where machair stratification occurs to a depth of 10 cm below the uppermost palaeosol. In the present study, this type of stratification has never been detected lower in the sedimentary succession of these machair sands. For example, the 12 m thick sand and dune sequences at Cladh Hallan beneath the uppermost palaeosol, completely lack this type of stratification (figure 2.4, lower part). There is one possible exception: that is the reference to occasional 'thin layers (including organic material)', recorded in the machair at Sidhean Rosinish and Rosinish *VII* on Benbecula by Whittington and Ritchie (1988, their figure 24) which might be a similar type of stratification. However, these authors' detailed examinations have shown that these thin organic bands accumulated in dune slacks, and thus have different origins and topographic associations to machair stratification (Whittington, personal communication).

The associations between machair stratification, machair grassland, an uppermost palaeosol and the uppermost part of the stratigraphic sequence suggest that machair grassland, as defined here in its modern widespread distribution, is a comparatively young feature, especially when seen in terms of the 7000–8000 year history of the machair sand dunes and sheets identified by Ritchie (1979, 1985) and Whittington and Ritchie (1988). The antiquities of the uppermost palaeosols in these sequences from South Uist and Barra have been determined by a variety of methods. AMS dating of *in situ* seeds from the near-surface palaeosol below the machair stratification at Cladh Hallan gave a date of 2969 ± 75 years BP (OxA-3352) (figure 2.4). The antiquities of the uppermost palaeosols at the Cill Donain sites located in figure 2.1 are not known, but they are younger than the AMS dates on seeds of 3710 ± 80 BP (OxA-3353) and 3560 ± 80 BP (OxA-3354) at Cill Donain I; and 1115 ± 70 BP (OxA-3355) and 1670 ± 75 BP (OxA-3356) for Cill Donain III (for details of these uncalibrated dates, see Hedges et al. 1992). All these Cill Donain AMS dates are from sites some 2–4 m below the base of the uppermost palaeosol and hence the base of the machair stratification at these sites. The antiquity of one uppermost palaeosol – that in the Eoligarry Isthmus – can be identified with precision from a combination of archaeological and historical evidence. The contact surface between the base of the machair stratification and the uppermost palaeosol is occupied by large number of rabbit traps, attributable to two rabbit-catchers who lived in the derelict remains of a chapel of Kilmoire (located on figure 2.2) between approximately 1890 and 1905 (Mackinnon 1991, personal communication). The unpublished excavations by Heron show that this chapel was constructed after the upper palaeosol had begun to develop. The chapel was known to exist in 1846 (Buchanan 1942). It was present in 1695 when its position was precisely located by Martin (1695), and is similarly described in the story of Rorie McNeill, written sometime in the period 1613–22 (MacFarlane, in Campbell 1936). However, the report of the earlier visit of the Cleric, Dean Donald Munro, to the Eoligarry Isthmus in 1549, which noted important landscape details of this area, did not record the presence of the chapel. The upper palaeosol therefore appears to have begun to develop on open mobile sand sheets from the period AD 1549 to AD 1613–22 and to have become buried in deposits with machair stratification from the beginning of the twentieth century. Two further conclusions about the antiquity of the present machair stratification could be drawn from these new archaeological and stratigraphic data. Either the machair stratification above the uppermost palaeosol started to form immediately after the dated palaeosols developed, in which case the lowest components exhibiting this stratification are strongly time-transgressive through the study region; or more probably, the processes of re-working responsible for this type of stratification (aided and abetted by severe storms) resulted, over time, in a working-down to whatever stable surface was present. Typically this was a palaeosol and/or an archaeological feature. The wider geomorphic importance and strength of palaeosols and archaeological deposits in sand dunes have been noted previously by Ritchie (1966, 1979), Gilbertson (1981, 1983), and Gilbertson and Foale (1977). The latter explanation accounts for the association of the base of deposits exhibiting machair stratification with the surface of palaeosols, and predicts that it is the strength and not the age of palaeosols that is significant in this context. This interpretation also suggests that the antiquity of a palaeosol may be not be readily predictable from its position in an exposure in the machair.

### *Land use and machair stratification*

The present limited historical and stratigraphic evidence does not suggest a simple general correlation between types of land use and the presence of deposits exhibiting machair stratification. The exposures of the Late Iron Age at Cill Donain III revealed abundant evidence of a farming economy essentially similar to that which prevails today, but there is no sign of coeval machair stratification. This is also the case for Late Neolithic – early Bronze age deposits at Cill Donain 1, and for the post-medieval deposits associated with the Kilmoire chapel at Eoligarry. The present data are inadequate to establish whether or not there are any relationships between periods of widespread accelerated sand drift and the presence of machair stratification. The one area for which there is information is the Eoligarry Isthmus. The historical evidence described above demonstrates that the sixteenth to nineteenth centuries AD, suspected by Angus and Eliott (1992) and Ritchie (1979) to have been characterized by a greater degree of erosion, were associated at this isthmus with a period of widespread soil stability with the formation of a significant palaeosol. Machair stratification at this isthmus was contemporary with the formation and erosion of the large coastal fore dunes that back the beach of Tràigh Eais – processes which can also be dated to the end of the nineteenth century by reference to the coal debris found near the base of the dunes and known to be associated with a particular shipwreck (figure 2.2 and MacKinnon, personal communication). At present, the simplest approach would be to anticipate that too much, as well as too little, sand mobility might be inimical to the formation of machair stratification. It may be dependent upon the coincidence of a particular set of events, such as those associated with the local formation and decay of large, adjacent sand dunes at Tràigh Eais.

## DISCUSSION AND CONCLUSIONS

The present study of the floristics of the machair vegetation at four sites on Barra and South Uist has identified similar types of vegetation to those found previously further north on South Uist by Dickinson and Randall (1979). The principal type of vegetation has been shown to be distinctive machair grassland, which dominates the typical machair inland of the coastal dune cordon, and away from wetland or ecotonal situations. Floristic variation has been shown in the present survey to be correlated with soil moisture and organic matter. Variations in soil pH are seen to be of secondary importance in the machair grassland. The prime feature to emerge from the analysis is the great uniformity of the vegetation data. This uniformity has been explained by two sets of interacting processes:

- as a product of the widespread deflation, re-working and re-deposition of both the soils and near-surface sediments of the machair and dune systems by animals, land-use practices, and wind along this storm-prone, Atlantic shoreline; and

- as a result of heavy selection pressure in favour of perennial species that comprise most of the machair grassland and are capable of growing up through the resulting sand layer.

If burial of the vegetation is too deep, then the existing vegetation and any underlying organic material become 'fossilized' within the sand, giving rise to the organic layers that characterize machair stratification. This hypothesis could be tested by the establishment of controlled experiments to bury machair grassland to varying depths and then observing patterns of re-growth and re-colonization.

The nature of the re-working of the machair sands has now been established by micromorphological studies. A type of stratification, here termed 'machair stratification' has been identified and found to be invariably associated with machair grassland at the study sites. The types of land use practised in the storm-prone area are likely to encourage similar forms of re-working of the seed bank. The overall pattern of machair vegetation and soil properties appear, therefore, to reflect the complex interactions between geomorphic and biological processes and human activities. This analysis suggests that the features of the machair will tend to occur in particular sets of human, geographic, geological and biological circumstances in which the appropriate materials, environmental processes and geographical situations are to be found. The stratigraphic analysis presented suggests that the current wide distribution in the study area of machair stratification (and by association machair grassland) may be a relatively young feature of the machair. On the Eoligarry isthmus, it dates from the beginning of the present century. The key to understanding its stratigraphic distribution, invariably overlying an uppermost palaeosol, again appears to relate to re-working. In this case, the depth of deflation, and hence re-deposition, is controlled by the erosional resistance of this uppermost palaeosol, or an archaeological feature offering similar erosional resistance. The recognition of the role of re-working may be a useful key to understanding the biogeographic as well as the geomorphic landscapes of the machair grasslands of the study area.

ACKNOWLEDGEMENTS

The authors are grateful to their colleagues from the Sheffield Environment and Archaeology Research Campaign in the Hebrides (SEARCH) for their help in the field and access to excavations on Barra and South Uist; to the SERC-supported AMS dating programme at Oxford that supplied the AMS radiocarbon dates; to Dr R. A. Kemp and Professor J. Rose of the Department of Geography at Royal Holloway College, University of London for help and support with the micromorphological studies; to Brian Rogers of the Department of Geography of the University of Plymouth for cartographic assistance; to the British Ecological Society for Small Ecological Project Grant no. 860; to Historic Scotland and Loganair for financial support, assistance and sponsorship which helped to support the field programme or excavations; to Mr Patrick Ashmore, Drs Paul Buckland, Mary Elliott, Jon Sadler and Graeme Whittington for comments, advice and information; to Scottish Natural Heritage, the Scottish Department of Agriculture and Fisheries, and local grazing committees and landowners who gave permission to carry out the work described; and to the South Uist Historical Society and Miss Mary Mackinnon of Castlebay, Barra for giving freely of their knowledge of the local history of the study areas. Jean-Luc Schwenninger acknowledges the provision of research studentship bursaries from the Department of Archaeology and Prehistory, the University of Sheffield, NATO, and the E.C., Barbara Brayshay acknowledges the award of an SERC APPEAL Research Studentship.

REFERENCES

Angus, S. (1991) 'The use of the term machair in Gaelic and English', *Hebridean Naturalist* 7: 16–22

Angus, S. and Eliott, M. M. (1992) 'Erosion in Scottish machair with particular reference to the Outer Hebrides', in Carter, R. W. G., Curtis, T. G. and Sheehy-Skeffington, M. J. (eds) *Coastal Dunes: Geomorphology, Ecology and Management for Conservation* (Balkema, Rotterdam) pp. 93–112

Barber, J. (1987) *The Western Isles* (John Donald, Edinburgh)

Beveridge, E. (1911) *North Uist: its Archaeology and Topography, with Notes upon the Early History of the Outer Hebrides* (William Brown, Edinburgh)

Bond, W. J. (1976) 'Illuvial land formation in a laboratory column of sand', *Journal of the Soil Science Society of America* 50: 265–7

Boyd, J. M. and Boyd, I. L. (1990) *The Hebrides* (Collins New Naturalist, London)

Brayshay, B. A. (1992) The vegetation and vegetational history of Barra and South Uist in the Outer Hebrides of Scotland. Unpublished Ph.D. dissertation, University of Sheffield

Buchanan, D. (1942) *Reflections on the Isle of Barra* (Sales and Co., London)

Caird, J. B. (1979) 'Land use in the Uists since 1800', *Proceedings of the Royal Society of Edinburgh* 77B: 505–26

Campbell, J. L. (ed.) (1936) *The Book of Barra – Being an Account of the Island of Barra in the Outer Hebrides Written by Various Authors at Various Times, Together with Unpublished Letters and Other Matters Relating to the Island* (Routledge, London)

Campbell, J. L. (ed.) (1958) *Gaelic words and expressions from South Uist and Eriskay, Collected by the Rev. Fr. Allan McDonald of Eriskay (1859–1905)* (Dublin Institute for Advanced Studies, Dublin)

Clapham, A. R., Tutin, T. G. and Moore, D. M. (1989) *Flora of the British Isles*, 3rd edn (Cambridge University Press, Cambridge)

Crawford, I. A. and Switsur, V. R. (1977) 'Sandscaping and C14: the Udal, N. Uist', *Antiquity* LI: 124–36

Currie, A. (1979) 'The vegetation of the Outer Hebrides', *Proceedings of the Royal Society of Edinburgh* 77B: 219–65

Dickinson, G. and Randall, R. E. (1979) 'An interpretation of machair vegetation', *Proceedings of the Royal Society of Edinburgh* 77B: 267–78

Dickinson, G., Mitchell, J. and Tivy, J. (1971) 'The application of phytosociological techniques to the geographical study of vegetation', *Scottish Geographical Magazine* 87: 83–102

Dijkerman, J. C., Cline, M. G. and Olson, G. W. (1967) 'Properties and genesis of textural subsoil lamellae', *Soil Science* 104: 7–16

Duncan, U. K. (1970) *Introduction to British Lichens* (Buncle, Arbroath)

Fenner, M. (ed.) (1992) *Seeds: the Ecology of Regeneration in Plant Communities* (CAB International, Wallingford)

Gilbertson, D. D. (1981) 'The impact of past and present land use on a major coastal barrier system', *Applied Geography* 1(2): 97–119

Gilbertson, D. D. (1983) 'The impacts of off-road vehicles in the Coorong Dune and Lake Complex of South Australia', in Webb, R. H. and Wilshire, H. G. (eds) *The Environmental Effects of Off-road Vehicle Use: Impacts and Management in Arid Regions* (Springer-Verlag, Berlin) pp. 487–506

Gilbertson, D. D. and Foale, M. (eds) (1977) *The Southern Coorong and Lower Younghusband Peninsula of South Australia* (Nature Conservation Society of South Australia, Adelaide)

Gimingham, C. H. (1964) 'Maritime and sub-maritime communities', in Burnett, J. H. (ed.) *The Vegetation of Scotland* (Oliver and Boyd, Edinburgh) pp. 67–142

Glentworth, R. (1979) 'Observations on the soils of the Outer Hebrides', *Proceedings of the Royal Society of Edinburgh* 77B: 123–137

Hedges, R. E. M., Housley, R. A., Bronk, C. R. and van Klinken, G. J. (1992), 'Radiocarbon dates from the Oxford AMS system: *Archaeometry* datelist 15', *Archaeometry* 34(2): 337–57

Hill, M. O. (1979a) *TWINSPAN – a FORTRAN Program for Arranging Multivariate Data in an Ordered Two-way Table by Classification of the Individuals and the Attributes* (Ecology and Systematics, Cornell University, Ithaca, New York)

Hill, M. O. (1979b) *DECORANA – a FORTRAN Program for Detrended Correspondence Analysis and Reciprocal Averaging* (Ecology and Systematics, Cornell University, Ithaca, New York)

Hill, M. O. and Gauch, H. G. (1980) 'Detrended correspondence analysis, an improved ordination technique', *Vegetatio* 42: 47–58

Hudson, G. (1991) 'The geomorphology and soils of the Outer Hebrides', in Pankhurst, R. J. and Mullin, J. M. (eds) *Flora of the Outer Hebrides*, (Natural History Museum Publications, London) pp. 19–27

Jongman, R. G. H., Ter Braak, C. J. G. and van Tongeren, O. F. R. (1987) *Data Analysis in Community and Landscape Ecology* (PUDOC, Wageningen, The Netherlands)

Kemp, R. A. and McIntosh, P. D. (1989) 'Genesis of a banded soil in Southland, New Zealand', *Geoderma* 45: 65–81

Kent, M. and Coker, P. (1992) *Vegetation Description and Analysis: a Practical Approach* (Belhaven Press, London/CRC Press, Boca Raton)

Leck, M. A., Parker, V. T. and Simpson, R. L. (eds) (1990) *Ecology of Soil Seed Banks* (Academic Press, London)

Macdonald, J. (1811) *General view of the Hebrides or Western Isles of Scotland: with Observations on the Means of their Improvement, Together with a Separate Account of the*

*Principal Islands; Comprehending their Resources, Fisheries, Manufactures, Manners and Agriculture* (Published for the Board of Agriculture by Sir Richard Philips, London, and Silvester Doig and Andrew Stirling, Edinburgh)

Macfarlane, W. *Geographical Collections Relating to Scotland* (includes the re-writing of an original manuscript attributed to Rorie McNeill between approximately 1613 and 1622), in Campbell (1936), op. cit., 41–7

Maclaren, A. (1974) 'A Norse house on Drimore Machair, South Uist', *Glasgow Archaeological Journal* 3: 9–18

Martin, M. (1695) *A Description of the Western Isles of Scotland* (Bell London; McKay's reprint, Stirling). Reprinted for Barra in Campbell (1936), op. cit., 49–58

Mather, A. S. and Ritchie, W. (1977) *The Beaches of the Highlands and Islands of Scotland* (Countryside Commission for Scotland, Redgorton, Perth)

Matheson, W. (ed.) (1938) *The Songs of John MacCodrum, Bard to Sir James Macdonald of Sleat* (Oliver and Boyd, Edinburgh, for the Scottish Gaelic Texts Society)

Mckay, M. (ed.) (1980) *The Reverend Dr John Walker's Report on the Hebrides of 7764 and 1771* (John Donald, Edinburgh)

Mckirdy, A. P. and Foxwell, E. A. (1985) 'Sand dunes: their coast protection function and the Geological Conservation Review', in Doody, P. (ed.) *Sand Dunes and their Management: Focus on Nature Conservation* No. 13. (Nature Conservancy Council, Peterborough) pp. 11–19

Munro, Sir Donald (1549) *A Description of the Western Isles of Scotland called Hybrides (c.1549)* (Edinburgh: Eneas MacKay's reprint 1934, Stirling). Reprinted for Barra in Campbell (1936), op. cit., 31–36

Nature Conservancy Council (1977) *Outer Hebrides: Sites of Geological and Geomorphological Interest* (Nature Conservancy Council, Newbury)

Pankhurst, R. J. and Mullin, J. M. (1991) *The Flora of the Outer Hebrides* (Natural History Museum Publications, London)

Pitkin, P., Barter, G., Curry P., Mackintosh, A., Orange, A. and Urquhart, U. (1983) 'A botanical review of the SSSIs of the north and west coasts of the Uists and nearby islands', Unpublished report, Scottish Field Unit, Nature Conservancy Council, Edinburgh

Randall, R. E. (1973) 'Calcium carbonate in dune soils: evidence for geomorphic change', *Area* 5(4): 308–10

Randall, R. E. (1974) 'Aspects of machair ecology on the Monach Isles', in Ranwell, D. S. (ed.) *Sand Dune Machair 1*, (NERC, Institute of Terrestrial Ecology, Norwich) pp. 15–18

Randall, R. E. (1976) 'Machair zonation of the Monach Isles N. N. R., Outer Hebrides', *Transactions of the Botanical Society of Edinburgh* 42: 441–62

Ranwell, D. S. (1974) 'Machair in relation to the British sand dune series (summary)', in Ranwell, D. S. (ed.). *Sand Dune Machair 1*, (NERC Institute of Terrestrial Ecology, Norwich), pp. 4–5

Ranwell, D. S. (ed.) (1977) *Sand Dune Machair 2*, (NERC, Institute of Terrestrial Ecology, Cambridge)

Ranwell, D. S. (ed.) (1980) *Sand Dune Machair 3*, Report of a meeting in the Outer Hebrides (Institute of Terrestrial Ecology, Cambridge)

Ratcliffe, D. A. (1977) *A Nature Conservation Review*, Vol. 2 (Cambridge University Press, Cambridge)

Ritchie, W. (1966) The physiography of the machair of South Uist, unpublished Ph.D. dissertation, University of Glasgow

Ritchie, W. (1967) 'The machair of South Uist', *Scottish Geographical Magazine* 83: 161–73

Ritchie, W. (1976) 'The meaning and definition of machair', *Transactions and Proceedings of the Botanical Society of Edinburgh* 42: 431–40

Ritchie, W. (1979) 'Machair development and chronology in the Uists and adjacent islands', *Proceedings of the Royal Society of Edinburgh* 77B: 107–22

Ritchie, W. (1985) 'Inter-tidal and sub-tidal organic deposits and sea level changes in the Uists, Outer Hebrides', *Scottish Journal of Geology* 21: 161–76

Robertson-Rintoul M. and Ritchie, W. (1990) 'The geomorphology of coastal sand dunes in Scotland: a review', *Catena* (Supplement) 18: 41–9

Robinson, G. H. and Rich, C. I. (1960) 'Characteristics of the multiple yellowish-red bands common to certain soils in the southeastern United States', *Soil Society of America, Proceedings* 24: 226–30

Smith, A. J. E. (1978) *The Moss Flora of Britain and Ireland* (Cambridge University Press, Cambridge)

Stace, C. (1991) *New Flora of the British Isles* (Cambridge University Press, Cambridge)

ter Braak, C. J. F. (1986) 'Canonical correspondence analysis: a new eigenvector technique for multivariate direct gradient analysis', *Ecology* 67: 1167–79

ter Braak, C. J. F. (1987) 'The analysis of vegetation–environment relationships by canonical correspondence analysis', *Vegetatio* 69: 69–77

ter Braak, J. J. F. (1988a) 'CANOCO – an extension of DECORANA to analyse species–environment relationships', *Vegetatio* 75: 159–60

ter Braak, C. J. F. (1988b) CANOCO – *a FORTRAN program for canonical community ordination by [partial] [detrended] [canonical] correspondence analysis, principal components analysis and redundancy analysis (version 2.1)*, (Agricultural Mathematics Group, Ministry of Agriculture and Fisheries, The Netherlands, Wageningen)

Walker, J. 1764 & 1771 *Report on the Hebrides* (reprinted by John Donald Publishers, Edinburgh, ed. M. Mackay, 1980)

Warr, S. J., Thompson, K. and Kent, M. (1993) 'Seed banks as a neglected area of biogeographic research: a review of literature and sampling techniques' *Progress in Physical Geography* 17(3): 329–47

Watson, E. V. (1978) *British Mosses and Liverworts*, 2nd edn (Cambridge University Press, Cambridge)

Whittington, G. and Ritchie, W. (1988) *Flandrian Environmental Evolution of Northeast Benbecula and Southern Grimsay, Outer Hebrides, Scotland*, O'Dell Memorial Monograph 21, 44 pp., University of Aberdeen

Young, A. and Richardson, K. M. (1956–1960) 'A Cheardach Mhor, Drimore', *Proceedings of the Society of Antiquarians of Scotland* 93: 135–73

3

# A Lake-sediment Record of Changing Land-use Intensity on the Isle of Lewis, Scotland

Jane A. Entwistle, H. F. Lamb, P. W. Abrahams and R. A. Dodgshon

ABSTRACT

In the absence of documentary evidence about agricultural practice in north-west Scotland before the late eighteenth century, a palaeoenvironmental approach to determining changes in medieval land-use intensity is attempted here. The recent sediments of Loch Stiapavat, at the northern tip of the Isle of Lewis, show stratigraphic variation in the abundance of catchment-derived chemical elements, in magnetic parameters and in some pollen taxa which suggest that agricultural exploitation of the lake catchment increased markedly at about AD 1700. The tentative chronology is based on a comparison of lead and zinc profiles (indicating atmospheric fall-out from industrial sources) with dated profiles elsewhere in Scotland. The problem of dating lake sediments, especially where erosional input is high, is a major obstacle to palaeoecological studies of the medieval period.

## INTRODUCTION

Farming in the western Highlands and Islands of Scotland, prior to crofting and the clearances, had a considerable impact on the environment (Dodgshon 1988, Dodgshon and Olsson 1988). Two particular features of farming were critical in determining the nature and scale of this impact. The first was the extent to which communities made use of labour-intensive techniques of cultivation based on tools such as the spade and caschrom, and ridge-forms such as the lazy-bed (Dodgshon 1992). These enabled cultivation to be extended over ground that in normal circumstances would have been regarded as too marginal. In order to make the land productive, lazy-beds were constructed to provide adequate rooting depth and free drainage (Glentworth 1979). The second was the extent to which communities exploited the extensive resources of their non-arable land to expand the output of their more limited reserves of arable. In addition to the nutrients transferred via stock grazing routines, communities invested vast quantities of labour into the harvesting of seaweed, shell-sand, turf, soil and peat. Together with the organic matter from old turf dykes and thatch, these various manures were added to the arable and used to build up lazy-beds on ground where soil was otherwise too thin or impoverished to sustain cultivation. Such transfers had both negative and positive effects on the environment, with some areas denuded or skinned and other areas enriched to the point of acquiring plaggen soils.

Documentary sources enable the farming practices underlying this anthropogenic impact to be defined in some detail for the late eighteenth and nineteenth centuries. Unfortunately, the absence of such documentation prior to the eighteenth century prevents us from establishing a deeper perspective either on the history of the farm practices involved or on the wider question of how the relationship between society and environment may have altered over time. However, some speculation may be allowed. First, the overall scale of human impact on the environment is likely to have varied in step with fluctuations in local population, although we have no way of knowing how closely the population history of the area fits standard models of British or European population history over the last millennium or so. Second, eighteenth century sources make it clear that hand cultivation and high levels of nutrient transfer were so demanding on labour as to be uneconomic or impractical except during times of labour abundance (Dodgshon 1992). Thus, these practices are likely to have flourished only during times of population expansion, with incentives and means moving in step with each other. Third, climatic variations may have redrawn the geography of production, releasing previously marginal land for cultivation in favourable phases but ruling them out in colder, wetter phases. Such climatic shifts may have worked in harness with population changes, accentuating the phases of growth and contraction.

In the absence of documentary sources, these ideas on the long-term relationship between farming and the environment may be examined through the record of environmental history, preserved as stratigraphic variation in the composition of sediments which have accumulated in the many lochans of the region. The approach assumes that much of the variation in organic and inorganic material influx to the lake results from changing land-use intensity over time (e.g. Pennington et al. 1976, Gaillard et al. 1991, Edwards and Whittington 1993). An increase in catchment-wide erosion may be inferred from enhanced mineral content, higher concentrations of soil-derived elements, and from magnetic mineral components in lake sediments (Mackereth 1966, Engstrom and Wright 1984, Higgit et al. 1991). Here, we use a multidisciplinary approach, involving sediment chemistry, particle size, mineral magnetics and pollen analysis to investigate the pre-documentary impact of changes in land-use close to the Butt (northern tip) of Lewis, in the district of Ness, Outer Hebrides, Scotland.

## DESCRIPTION OF THE STUDY AREA

Lewis is the largest and most densely populated of the Hebridean Islands. The word Lewis derives from the Gaelic word *leogach* or *loedus*, meaning 'marshy'

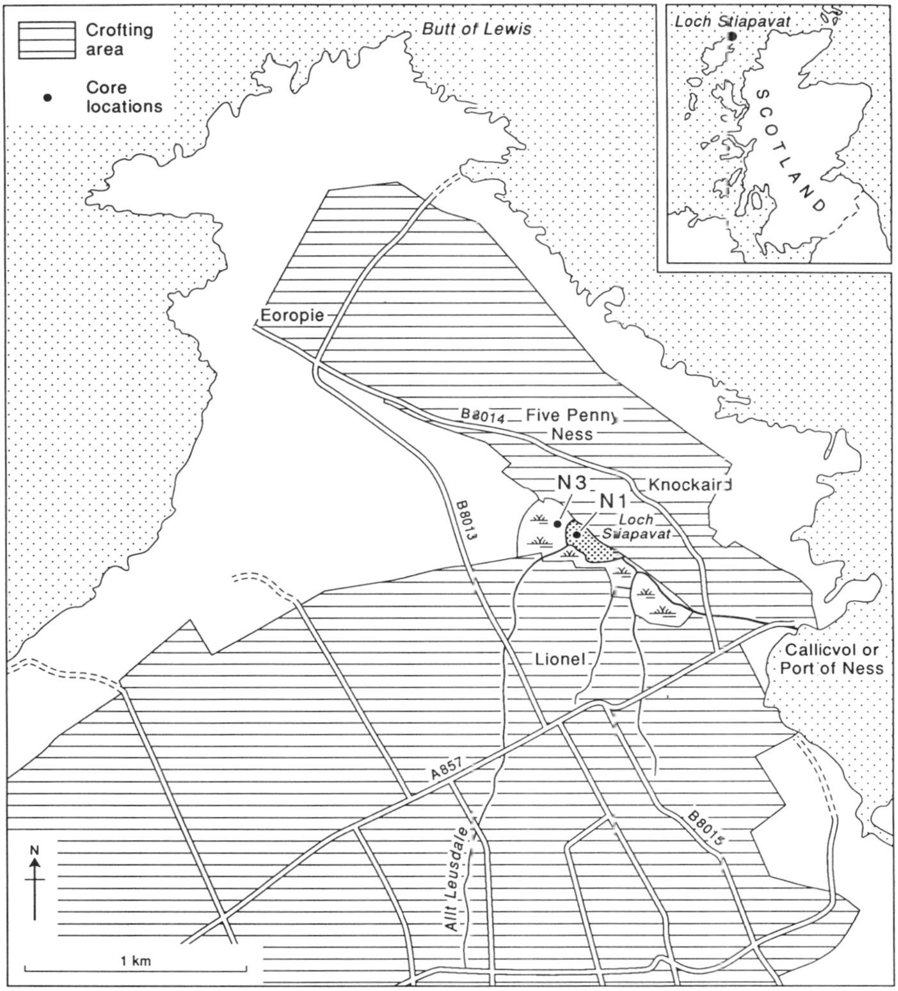


Figure 3.1 Loch Stiapavat and its catchment area (based on the 1 : 10 000 Ordinance Survey map).

or 'boggy' (Geddes 1955). As the interior of Lewis is largely covered by peat, settlements and agricultural land tend to be located around the rugged coastline (Glentworth 1979). The drainage is better in these areas, and shell-sand and seaweed are available for manure, both formerly very important in maintaining soil fertility (Fenton 1986). The Outer Hebrides are excessively windswept, with gales blowing on average one day in six at the Butt of Lewis (Birks and Madsen 1979). The Butt of Lewis is characterized by numerous cultivation rigs or lazy-beds, which dominate the landscape of this region. These rigs, up to 0.6 m high and 2 m broad, are a legacy of the former widespread cultivation in the area.

Loch Stiapavat is a small lake or lochan (2.8 ha) located 2.5 km south of the Butt of Lewis (grid reference NB 528643, altitude 15 m a.s.l.). Three townships border the loch (figure 3.1). The largest township is Five Penny Ness, lying to the north-west, although only a small proportion of the township land drains into the loch. The catchment is largely occupied by the townships of Knockaird and Lionel. Glacial tills, probably derived from mainland ice, lie to the south and north-west, blown sands to the west and Lewisian gneiss to the north-east (Peacock 1984). Soils of the western part of the catchment have been mapped as calcareous regosols, brown soils and gleys; most of the remainder is covered by non-calcareous gleys and humic gleys (Soil Survey of Scotland 1982). A sedge marsh encroaches on to the loch from its western margin.

## DOCUMENTED HISTORY

The earliest documentary references to the townships bordering the loch are provided by rental transcripts of 1718 and 1721 (*Report* 1902: app. 0, Macphail 1916: 314). Both list 'Five Pennys and Gearin' as having 88 tenants and Lionel as having only two tenants, although the latter are likely to have been tacksmen sub-letting land to others. Neither list mentions Knockaird. References to Gearin continued in accounts down to the 1760s. However, by the 1770s its place beside Five Penny Ness was taken by references to Knockaird (Scottish Record Office, hereafter SRO, GD46/1/267; GD46/1/212). Given that tenants listed in the 1760s as occupiers of Gearin reappear in the 1770s as tenants of Knockaird, the most likely explanation is that all or part of Gearin was detached from Five Penny Ness and established as the separate township of Knockaird, with 11 tenants. The first survey of the area was drawn up in 1809–11, but was not published in map form until 1817 (National Library of Scotland, A. Gibbs, Map of Lewis 1817). The survey confirms the importance of arable within the local farming economy. Indeed by Hebridean standards, the fields around Loch Stiapavat formed a well-cultivated landscape. The precise extent of arable for the period 1809–11 is shown in figure 3.2. At this time, the nucleus of Knockaird's settlement lay on the northern edge of its arable. In the case of Lionel, arable occupied the slopes (4–7°) running northwards and northwestwards down to the side of the loch, with its focus of settlement on the eastern edge of arable.

At the time of the 1809–11 survey these townships were still farmed as open-field or runrig communities, despite an estate declaration in 1795 binding all tenants and tacksmen to abandon the 'pernicious practice' of dividing their townships into runrig 'or common of arable' and, instead, to divide their land into separate holdings (SRO, GD/1/277). In fact, townships did not start to convert their runrig into crofts until after 1811. The impact of change on the townships around Loch Stiapavat is visible from the First Ordnance survey map of the area drawn up in 1853 (National Library of Scotland, First Edition O.S. 6 inch, Lewis 1853). By then, landholding in all three townships around the loch had been reorganized into crofts, but they had not yet been enclosed. The settlements had been relocated into a standard linear form along the newly aligned roads running through the townships.

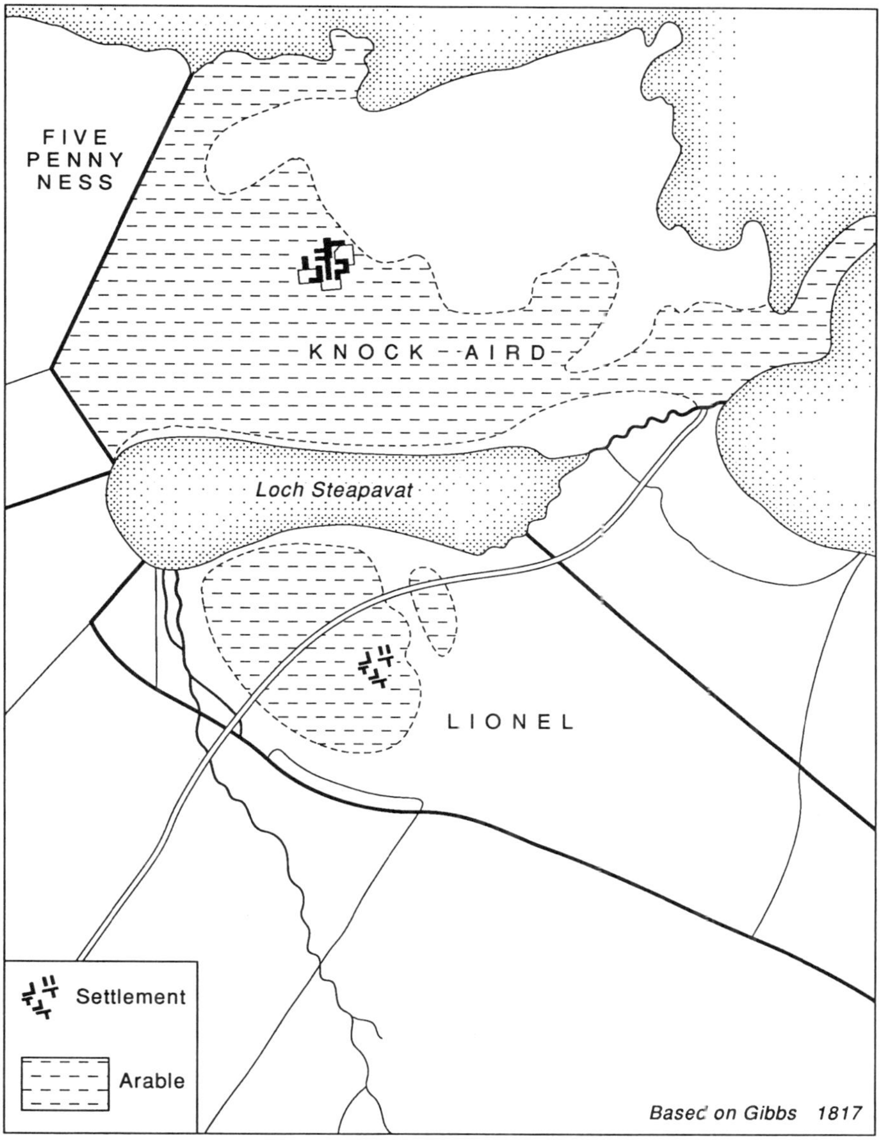


Figure 3.2 The area around Loch Stiapavat occupied by arable (after Gibbs 1817), based on a survey carried out in 1809–11.

METHODS

A 120 cm core (Ness-1) of the uppermost sediments of Loch Stiapavat was obtained from a water depth of 1 m using a 60 mm diameter Perspex tube equipped with a piston and extension rods. An additional 6 m core (Ness-3) was retrieved using a square-rod Livingstone piston corer from the littoral sedge marsh. The cores were logged by microscopic examination following the conventions of Troels-Smith (1955), sectioned into 1 cm slices and subsampled. Weight loss-on-ignition at 375 °C was determined as a measure of organic matter content (Ball 1964). The magnetic measurements reported here for the Ness-1 core (isothermal remanent magnetism, IRM, and saturated isothermal remanent magnetism, SIRM) were determined with a Molspin fluxgate magnetometer for a range of forward and reverse fields, 40–800 mT. Thompson and Oldfield (1986) provide fuller details of instrumentation and methods. In addition, a number of topsoil samples were collected from the slopes to the north and south of the loch and land adjacent to the inflowing stream Alt Leusdale. These samples were subjected to a range of magnetic measurement in an attempt to fingerprint the catchment topsoils and identify potential sources of sediment in the Ness-1 core.

Selected subsamples were treated with aliquots of $H_2O_2$ (100 volumes) to remove the organic fraction and wet sieved through a 63 $\mu$m sieve prior to particle-size analysis. The sand-sized fraction (> 63 $\mu$m) was negligible and thus discarded. Following dispersion in 50–70 ml of 0.005 per cent Calgon solution, samples were placed in an ultrasonic bath for 2–3 minutes prior to particle-size determination of the 1–63 $\mu$m size fraction (i.e. clay- and silt-sized) using a Sedigraph 5100 Analyser.

Sediment samples for chemical analysis were air-dried and powdered prior to a $HNO_3$–$HClO_4$ (4 : 1 v/v) digestion (Thompson and Wood 1982). Potassium (K) and sodium (Na) were determined by flame photometry, while calcium (Ca) and magnesium (Mg) were determined by atomic absorption spectrophotometry. A further 31 elements were determined by inductively coupled plasma mass spectrometry (ICP-MS), using a VG Elemental Plasma Quad II + instrument. Analytical precision (i.e. relative standard deviation, as defined by Thompson, 1983) of the 35 elements determined was better than 15 percent.

Pollen analyses were undertaken on 1 $cm^3$ subsamples from the Ness-1 core. Samples were processed by standard techniques involving initial treatment of the samples with HCl and NaOH, removal of silicate material by digestion with 40 per cent HF and digestion of cellulose by acetolysis (Faegri and Iversen 1989). Pollen was counted on a Zeiss Jenaval light microscope at × 500 magnification, with critical identifications made under oil immersion at × 1000 magnification. *Cerealia*-types were defined as grass grains greater than 40 $\mu$m in diameter (Innes and Shennan 1991). A minimum of 300 land pollen grains was counted at each level. Pollen data handling and presentation were undertaken using TILIA v1.09 and TILIAGRAPH v1.16 (Grimm 1991).

Radiocarbon dates were determined on bulk sediment samples by conventional liquid scintillation spectrometry. Infra-red stimulated luminescence (IRSL) was used to date two potassium–feldspar-rich sandy horizons the Ness-3 core (Duller 1992). Low concentrations of sand-sized potassium–feldspar grains precluded the application of this method to the sediments of core Ness-1.

Minimum variance cluster analysis was carried out using the multivariate statistical package MVSP (Kovach 1990) on a squared Euclidean distance dissimilarity matrix, derived from the data set of the 35 chemical elements

Table 3.1 Elemental groups based on minimum variance cluster analysis

| *Group I* | *Group II* | *Group III* | *Group IV* |
|---|---|---|---|
| Sc | La | Sr | Ni |
| V | Ce | Ca | Co |
| Cr | Pr | | Pb |
| Rb | Nd | | Ba |
| Cs | Sm | | Cu |
| Th | Eu | | Zn |
| Be | Y | | |
| K | Gd | | |
| Mg | Tb | | |
| Ga | Dy | | |
| Na | Ho | | |
| U | Er | | |
| | Tm | | |
| | Yb | | |
| | Lu | | |

determined on the core samples. This was undertaken to identify groups of elements which tend to vary down core in a similar manner. The data were standardized prior to cluster analysis so as to accord equal weighting to each variable (Johnston 1986).

## RESULTS: CORE STRATIGRAPHY

The sediments of the Ness-1 core are uniformly dark brown, organic-rich detrital muds, with no visible stratification. Microscopic examination indicates that diatoms are more abundant in the uppermost 14 cm. The uppermost 3 m of sediment core Ness-3 is largely comprised of organic-rich detrital muds with variable amounts of herbaceous detritus. Below this level the amount of inorganic material increases, with sand-rich units occurring at 450–466 cm, 507–510 and 550–590 cm respectively.

The elemental groupings produced by cluster analysis of the Ness-1 samples are given in table 3.1. Group I includes the alkali earth elements; Group II contains the rare earth elements (REE); Group III comprises Ca and strontium (Sr); while Group IV contains those elements that show a dramatic rise in concentration in the upper part of the core (e.g. lead, Pb and zinc, Zn). Stratigraphic profiles representative of each of the four groups are shown in figure 3.3.

On the basis of changes in organic matter content, modal particle size and elemental concentrations, the Ness-1 core may be divided into three units (figure 3.3). The basal unit (N1-1, 120–58 cm) has a mean loss-on-ignition (organic content) of 56 per cent, low elemental concentrations and a modal particle size generally coarser than 20 $\mu$m. In unit N1-2, (58–16 cm) the organic content of the sediment declines over 10 cm to a uniformly lower level (34 per cent), elemental concentrations are higher and, with the exception of one sample at

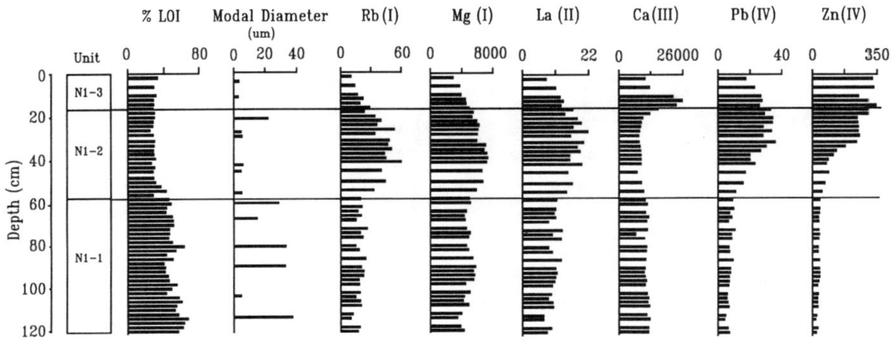


**Figure 3.3** Percentage loss-on-ignition (%LOI), modal particle size and concentration/depth profiles for elements typical of groups I–IV (identified by cluster analysis) in the Ness-1 sediment core: Group I represented by Rb and Mg; Group II represented by La; Group III represented by Ca; Group IV represented by Pb and Zn. Chemical concentrations in $\mu g\ g^{-1}$.

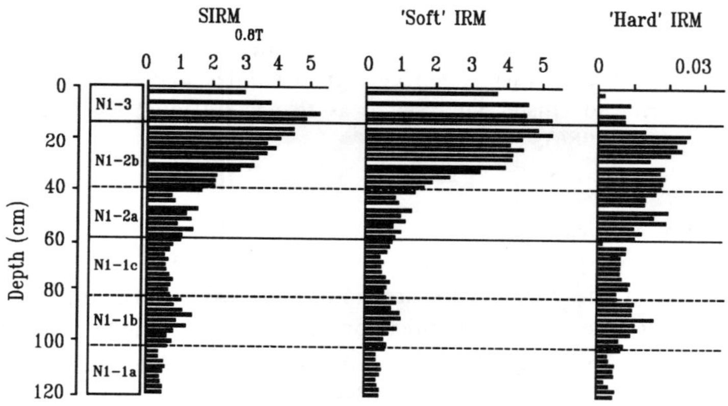


**Figure 3.4** SIRM, 'soft' IRM (SIRM − $IRM_{40}$) and 'hard' IRM (SIRM + $IRM_{-300}$) profiles for the Ness-1 sediment core. Magnetic measurements in $mAm^2\ kg^{-1}$

21 cm, the modal grain size decreases to about 5 $\mu$m. In unit N1-3 (16–0 cm) Group I and II elemental concentrations decline, while the organic matter content and the modal grain size remain unchanged. Maximum concentrations of 34 $\mu g\ g^{-1}$ Pb and 341 $\mu g\ g^{-1}$ Zn are observed at depths of 20 cm and 14 cm respectively. These concentrations compare to an average of 7 $\mu g\ g^{-1}$ Pb and 33 $\mu g\ g^{-1}$ Zn in unit N1-1. Zn concentrations remain high throughout unit N1-3, while concentrations of Pb decline in the uppermost unit. The Ca profile, representative of the Group III elements, shows elevated values between 16 and 10 cm (figure 3.3).

Mineral magnetic data from the core show a more complex stratigraphy (figure 3.4). Six magnetic units may be distinguished in the Ness-1 core, effectively subdividing the units defined above. Low values of SIRM, 'Soft' IRM (SIRM − $IRM_{-40}$) and 'hard' IRM (SIRM + $IRM_{-300}$) define units

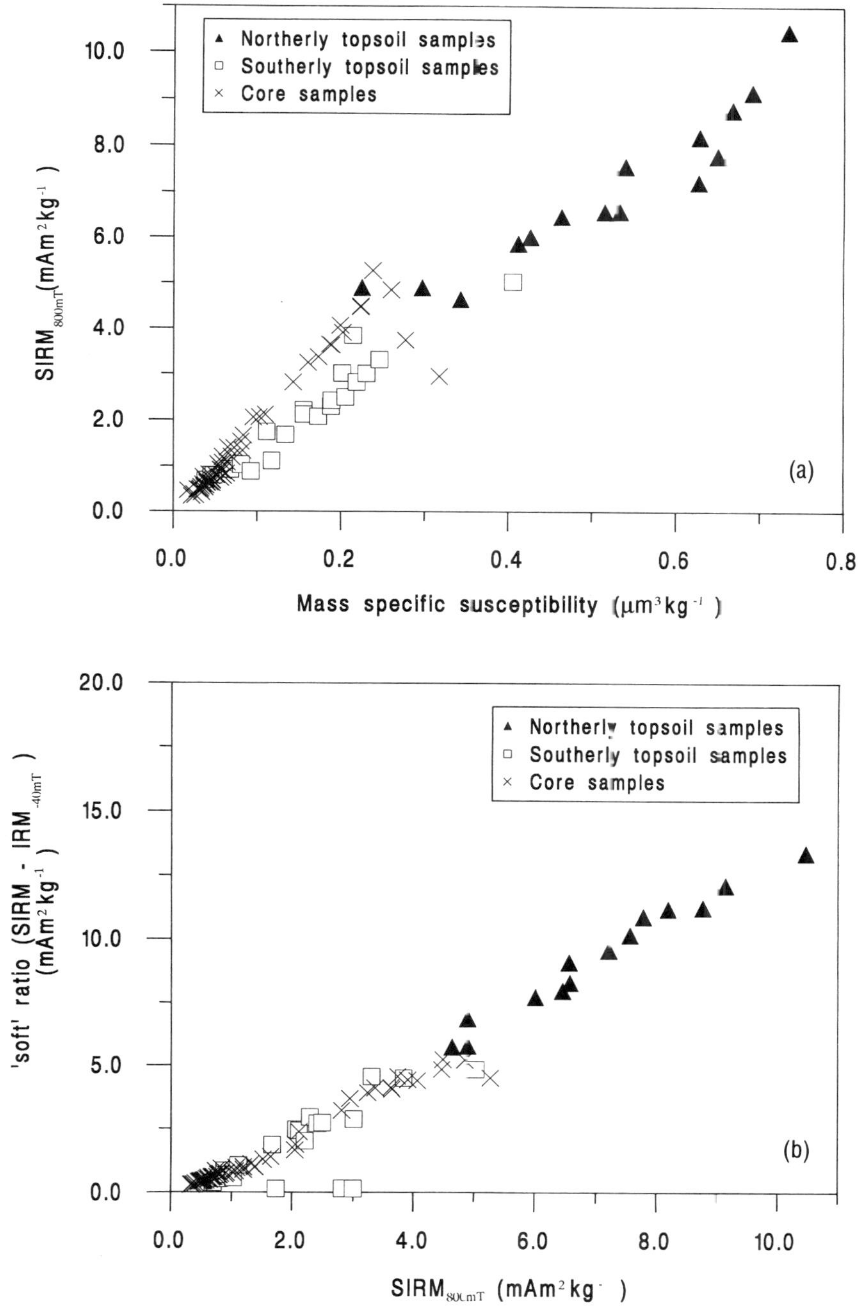


**Figure 3.5** The magnetic characterization of Loch Stiapavat sediments and catchment topsoil on the basis of (a) mass-specific susceptibility and SIRM, and (b) 'soft' component (SIRM $-$ IRM$_{-40\ mT}$) and SIRM.

**Table 3.2** Magnetic properties of topsoil samples and lake sediment samples from Loch Stiapavat and the surrounding catchment

| *Source* | *Mass-specific susceptibility* ($\chi_{lf}$) ($\mu m^3\ kg^{-1}$) | *SIRM* ($mAm^2\ kg^{-1}$) | *'Soft' component* ($SIRM - IRM_{-40}$) ($mAm^2\ kg^{-1}$) | *'Hard' component* ($SIRM + IRM_{-300}$) ($mAm^2\ kg^{-1}$) |
|---|---|---|---|---|
| Northerly topsoil samples | 0.23–0.73 | 4.64–10.46 | 5.72–13.37 | 0.42–1.32 |
| Southerly topsoil samples | 0.04–0.41 | 0.67– 3.84 | 0.14– 4.85 | 0.19–0.80 |
| Ness-1 sediment samples | 0.02–0.32 | 0.38– 5.27 | 0.31– 5.24 | 0.0 –0.26 |

***For a complete list see pp. 343–344***

***For a complete list see pp. 343–344***

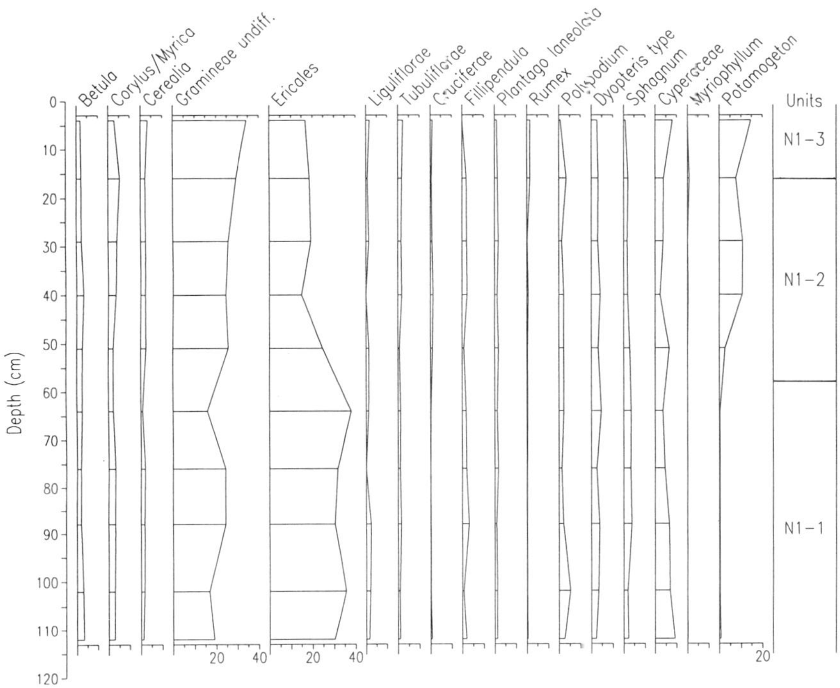


Figure 3.6 A pollen diagram for Loch Stiapavat. Data are expressed as a percentage of total land pollen, with the exception of aquatic taxa which are expressed as a percentage of the total pollen sum. The scale at base of the diagrams gives percentages for the bars.

N1-1a (120–105 cm) and N1-1c (85–58 cm). Intermediate values characterize units N1-1b (105–85 cm) and N1-2a (58–40 cm), although values for the top 2 levels in unit N1-2a are low. Above 40 cm (unit N1-2b), all three magnetic parameters increase strongly to peak values at slightly different levels between 20–10 cm before declining in the uppermost sediments (unit N1-3).

The magnetic parameters used for the characterization of the catchment topsoil and Ness-1 sediment samples are presented in table 3.2, along with the range of values found in this study. In addition, bivariate plots of selected magnetic parameters are presented in figure 3.5.

Selected pollen taxa are expressed as a percentage of the total land pollen (figure 3.6). The dominant pollen taxa present are Gramineae (grasses) and Ericales (heaths). Tree pollen grains are present but barely reach 5 per cent of the total. Cereal pollen grains are present in low abundances throughout the profile. Pollen of the aquatic macrophytes *Potamogeton* and *Myriophyllum*, expressed as a percentage of the total pollen sum, increase in unit N1-2.

**Table 3.3** Radiocarbon and Infra-Red Stimulated Luminescence (IRSL) ages. Radiocarbon ages are corrected to account for isotopic fractionation, following analysis of $^{13}C/^{12}C$ ratios, except where indicated (radiocarbon half-life taken as 5568 years)

| *Sediment core* | *Sample depth (cm)* | *Uncalibrated radicarbon age (yrs BP)* | *IRSL age (yrs BP)* | *Laboratory number* |
|---|---|---|---|---|
| Ness-1 | 60–61 | 3 070 ± 100 | – | Beta-58347 |
| | 119–120 | 3 020 ± 80 | – | Beta-58348 |
| Ness-3 | 100–104 | 2 670 ± 70 (uncorrected) | – | Beta-48871 |
| | 450–468 | | 25 000 ± 5 100 | N3Upper |
| | 550–568 | | 2 700 ± 200 | N3Lower |

INTERPRETATION

## *Chronology*

The luminescence behaviour of the Ness-3 samples indicates that the upper sand-rich unit contains a proportion of grains that were not well bleached at deposition and so the age for this sample is anomalously old (table 3.3). However, the age obtained for the lower sand-rich unit is believed to be a reasonable estimate of the time of deposition as the data imply that this sample was uniformly well bleached at deposition (Entwistle 1994). The radiocarbon ages on the Ness-1 core (one at 60 cm and the other at 119 cm depth; table 3.3) are both identical within one sigma errors. This suggests that at least one of these dates is incorrect. Given the evidence for erosional inwash of catchment material which may include old carbon, it is possible that the three radiocarbon ages are anomalously old (compare with Huttunen and Tolonen 1975, Jones et al. 1985, Pennington 1991). We therefore have less confidence in the radiocarbon dates than with the lower IRSL date and make a working assumption that this latter date is correct, providing an estimated sediment accumulation rate of approximately 2 mm $y^{-1}$. The sediment accumulation rate estimated from the Pb and Zn 'take-off' in the Ness-1 core (see below) also produces an approximate accumulation rate of 2 mm $y^{-1}$. Correlation of the Ness-1 and Ness-3 cores made on the basis of lithostratigraphy, organic matter, pollen and magnetic stratigraphy remains problematic and tentative. However, the similarity in accumulation rate is striking and in the absence of further chronological evidence is taken as support for our interpretation of the chronology.

As briefly mentioned above, an additional time marker is provided by the increased concentrations of Pb and Zn in the upper sediments of the Ness-1 core, which may be correlated with $^{210}$Pb-dated profiles of the same elements from other sites in western Scotland. Increasing amounts of these elements are assumed to have reached the lake sediments by atmospheric fall-out since the start of the Industrial Revolution (Murozumi et al. 1969, Price et al. 1978, Haworth and Lishman 1991). By comparison with previous Scottish palaeoeco-

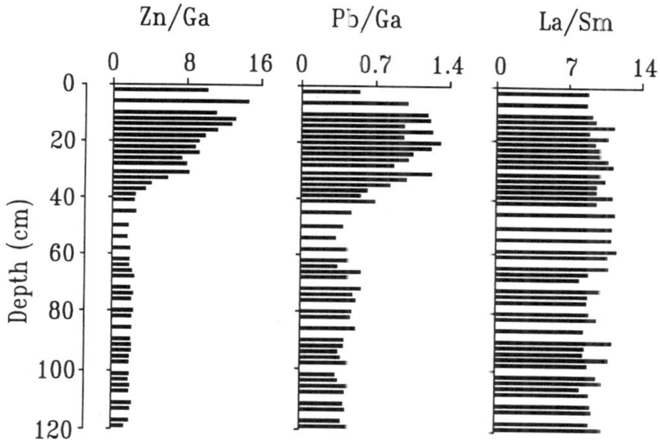


**Figure 3.7** Selected element ratios for the Ness-1 sediment core.

logical work (e.g. Farmer et al. 1980, Battarbee et al. 1985, Flower et al. 1988, Rippey 1990), it may be inferred that the increase in concentration of Pb and Zn towards the top of the core (figure 3.4) partly represents an increase in atmospheric pollution associated with the onset of large-scale British industrialization around AD 1800. However, the signal from atmospherically derived Pb and Zn may be complicated by the input of these elements into the loch from catchment sources. By normalizing Pb and Zn concentrations with respect to gallium (Ga), a lithophile element derived almost exclusively from catchment rock and soil sources (Mason 1966), it may be inferred that the increase in the Pb/Ga and Zn/Ga ratios at 35 cm (figure 3.7), rather than 58 cm, represents *c.*AD 1800. The increasing concentrations of Pb, Zn and other group IV elements between 58 cm and 35 cm may therefore be attributable to increased input via catchment derived inwash, whereas above 35 cm the elevated concentrations of the majority of group IV elements are due to an increase in the atmospherically derived input. Indeed the decrease in Pb concentrations in the top 10 cm of the core correlates well with improvements in atmospheric quality over recent years (Rippey 1990).

If the sediment at 35 cm depth is taken to represent AD *c.*1800, then, assuming a constant overall rate of sedimentation, the sediments have accumulated since this time at an average rate of about 2 mm $yr^{-1}$. In fact, the rate must have varied with changing intensities of catchment disturbance and erosion and changing patterns of sedimentation (Lehman 1975, Davis and Ford 1982, Edwards and Whittington 1993). This interpretation also assumes negligible post-depositional disturbance, bioturbation or diagenetic re-mobilization of Pb and Zn. Diagenetic mobilization is not totally ruled out as a contributor to the shape of the concentration profiles, as most of the Group IV elements, especially copper (Cu) and cobalt (Co), may be influenced by diagenesis in near-surface sediments (Ochsenbein et al. 1981; Finney and AnHuh 1989). However, in the absence of other chronological controls, we make the working assumption that the mean sediment accumulation rate has been about 2 mm $yr^{-1}$. Crudely estimated, the Ness-1 core may thus represent about 600 years of depositional history.

*Physical and geochemical evidence for catchment erosion*

The most straightforward evidence for increased erosional input to the lake is the decrease in organic content of the sediments above unit N1-1, together with the decreased modal particle size of the inorganic sediment fraction. This interpretation of the particle size changes follows Binford (1983), who concluded that increased catchment disturbance results in an increased inwash of very fine grained inorganic particles. In addition, elements of groups I and II (alkali and rare earth elements) show higher concentrations in unit N1-2. These are classified geochemically as lithophile elements (i.e. elements predominantly associated with silicate material; Mason 1966); numerous studies have linked increases in their concentrations and influx with increased catchment erosion and greater particulate input of minerogenic matter associated with human activity and land-use change within the catchment (Mackereth 1966; Ochsenbein et al. 1983, Jones et al. 1985). Exceptions may include Mg and V, the concentrations of which in lake sediments may be controlled by chemical, biological or anthropogenic factors (Bruland et al. 1974, Burden et al. 1986, Heathwaite and O'Sullivan 1991). However, in this study both Mg and V have concentration depth profiles similar to those of the other Group I and II elements, and so are inferred to be largely allogenic in origin (i.e. derived from outside the lake; Engstrom and Wright 1984). The majority of elements comprising Groups I and II are often considered to be conservative in lacustrine environments and, as such unlikely to be affected by post-depositional re-mobilization to any great extent (Kemp et al. 1976; Burden et al. 1986).

The elements of Group III (Ca and Sr) are also classified as lithophile elements (Mason 1966). However, in many lake sediments Ca has a large and variable authigenic component (i.e. formed within the lake; Engstrom and Wright 1984), largely due to biological activity, chemical sorption and precipitation (Mackereth 1966; Heathwaite and O'Sullivan 1991). Hence, this element cannot routinely be used as an indicator of catchment erosion. The subsurface maxima of Ca within unit N1-3 (figure 3.3) may be attributable to recent liming, although no historical documentation has been found to confirm this suggestion.

In the Ness-1 core an increase in the allogenic element concentrations, discernible for some elements (e.g. Mg) between 105 and 85 cm, and for all Group I and II elements above 58 cm (Fig 3), may represent an increase in particulate input, presumably as a result of greater catchment disturbance. However, such a conclusion is valid only if the chemical signature of the inwash is constant. Lithochemical uniformity is important, as increased concentrations of allogenic elements in the sedimentary profile could reflect a change in source area of the catchment inwash rather than increased influx. One technique for establishing source area consistency is to took at element ratios (e.g. Ca/Mg,

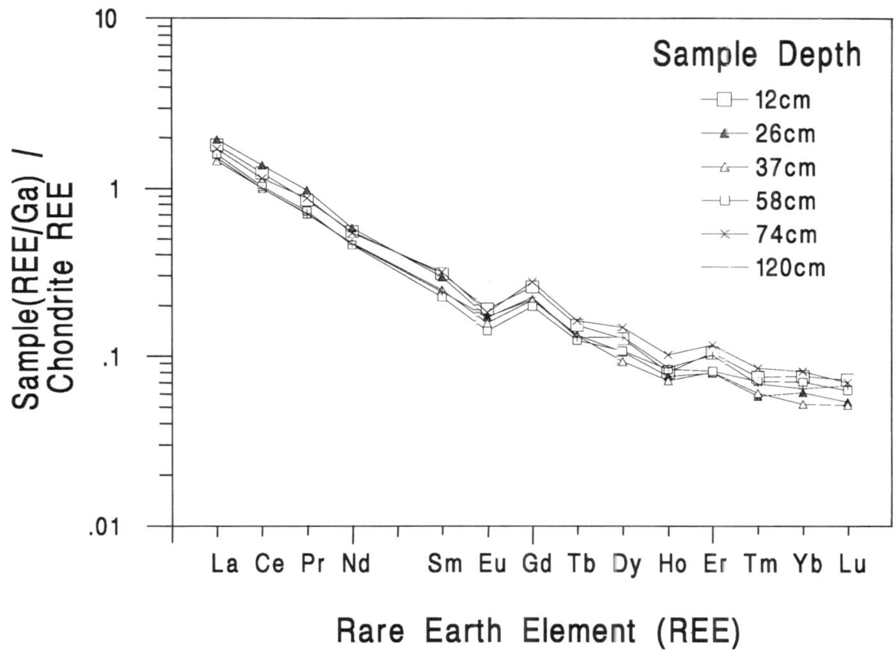


**Figure 3.8** Rare earth element spectra for randomly selected samples from the Ness-1 core. Chondrite data from Nakamura (1974).

Si/Al [silicon/aluminium], Jones and Bowser 1978; La/Sm [lanthanum/samarium], Williams 1988). The constant ratio of La/Sm throughout the Ness-1 sediment core (figure 3.7) suggests lithochemical consistency.

Further Support for lithochemical consistency comes from analysis of the REE (Group II). These elements have a limited mobility in the sedimentary environment, and among the lithophile elements are the most likely to represent a chemically inert record of allogenic input (Williams 1988). In this respect it is of interest that the cluster analysis separates this group of lithophile elements from those of Groups I and III. Differing lithologies can be distinguished through analysis of their REE spectra, as both the total concentration and the relative proportions of the REE vary considerably in common rock-forming minerals (Wedepohl, 1969). In order to provide a clear graphical presentation of the REE spectra at subsequent levels down the core (after Wedepohl 1969), the abundance of each REE was first normalized to Ga and this value was then divided by the abundance of the corresponding REE in a carbonaceous chondrite (Nakamura 1974). The final value was plotted against the appropriate REE (figure 3.8). The plot shows that each of the six samples, chosen at random levels down the core, have similar REE spectra. This suggests that the allogenic input is from a constant lithological source, and so increases in the concentrations of the Group I and II elements (e.g. at 58 cm) are attributable to increased inwash of catchment-derived particulate material, rather than inwash from a different source area or source material.

Apart from a single sample in unit N1–3 (figure 3.4), the modal particle size remains unchanged at about 5 μm. Following Binford (1983) this could suggest continued catchment disturbance up to the present (although biogenic silica, in the form of diatoms, may also contribute to the silt size fraction and thus complicate the interpretation; Edwards and Whittington 1993). However, the concentration of the Group I and II elements declines within this unit, suggesting that the intensity of catchment disturbance has decreased over recent years. An alternative interpretation of this decrease in elemental concentrations is that constant particulate input has been diluted by increased authigenic sediment accumulation, as suggested by the increased abundance of siliceous diatom frustules above 14 cm. This may also account for the observed increase in the concentration of Ba (a Group IV element) within the upper sediment layers, as Ba is strongly associated with biogenic silica (Henshaw 1978).

### *Magnetic mineralogy*

Stratigraphic variations in the concentration of ferrimagnetic minerals in lake sediments have been used by numerous workers to infer changing intensities of catchment disturbance and soil erosion (Thompson et al. 1975, Dearing and Flower 1982, Sandgren and Fredskild 1991). The magnetic parameter SIRM is sensitive to the concentration of magnetic minerals in the sediment, while magnetic remanence (IRM), grown in magnetic fields of varying magnitude, may be used to characterize the sediment mineralogy. 'Soft' IRM (SIRM-$IRM_{-40}$) is a useful indicator of the relative proportions of soft ferrimagnetic minerals (e.g. magnetite) in a sample, while the 'hard' IRM (SIRM + $IRM_{-300}$ is indicative of the abundance of antiferromagnets (magnetically 'hard' minerals, e.g. haematite; Oldfield 1991). Variations of the magnetic parameters may be due to other factors in addition to changes in the concentration of ferrimagnetic/antiferromagnetic minerals. These include authigenic mineral formation (e.g. greigite; Snowball and Thompson 1988, Hilton 1990, Williams 1991), microbial production of magnetite (Moskowitz et al. 1989, Stolz et al. 1990), and changes in magnetic grain size (Maher 1988). However, the three phases of intermediate and high SIRM in the Ness-1 core (units N1-1b, N1-2a and N1-2b; figure 3.4) are believed to predominantly reflect an increase in the influx or inorganic material to the loch, since the physical and geochemical evidence from these units suggests an increase in catchment disturbance. Stratigraphic variation in magnetic properties or the Ness-1 core includes changes in the relative abundance of magnetically 'soft' ferrimagnetic minerals and hard antiferromagnets. Above 35 cm (*c.*AD 1800) there is a proportionally greater increase in the concentration of ferrimagnetic minerals than in the antiferromagnetic component, despite the chemical ratios suggesting lithological consistency. An atmospherically derived magnetic component may partly be the cause of this, and may contribute to the high SIRM values above 35 cm (Hunt et al. 1984, Oldfield and Richardson 1990). This explanation accords well with observed increases in the concentration of the group IV elements (such as Pb and Zn), and with the fact that the organic matter content remains unchanged. One might expect levels of organic matter to decline further if a catchment-derived minerogenic source was invoked to explain the peak in SIRM. The reduction in SIRM above 10 cm (within unit N1-3) probably reflects an increase in the concentration or non-magnetic diatom frustules in the sediment, rather than any significant change in the atmospheric influx of magnetic minerals. In the Ness-1 core, atmospheric contribution to the 'hard' IRM signal is less clear given the presence of a background signal from catchment-derived sources (compare with Oldfield and Richardson 1990).

Mineral magnetic analysis was also applied to investigate potential sediment sources. Numerous investigations have demonstrated linkages between catchment erosional sources and lake sediments on the basis of magnetic similarities between sources and core sediments (Dearing et al. 1981; Flower et al. 1984, 1989; Oldfield et al. 1985). Attempts to construct sediment linkages typically compare values of mass specific susceptibility ($\chi$), SIRM and a range of demagnetization parameters using bivariate plots (e.g. Stober and Thompson 1979; Oldfield 1983, Dearing 1992). Figure 3.5 and table 3.2 highlight the similarity between the Ness-1 sediments and the catchment topsoils located to the south of the loch and in the vicinity of the inflowing stream catchment. This overlap suggests that catchment topsoils, particularly those to the south of Loch Stiapavat, do in fact contribute to the inorganic elastic material within the Ness-1 core.

In summary, the subdued SIRM peaks of units N1-1b and N1-2a (figure 3.4) probably represent an increase in the concentration of magnetic minerals derived predominantly from catchment sources, while the greatly elevated SIRM values in unit N1-2b are believed to result from both catchment-derived sources and atmospheric loading of the sediments. The magnetic data complements the chemical results and provides a degree of 'fine tuning' to the interpretations. In this case, the magnetic data suggest a phase of soil erosion (N1-1b 105–85 cm, *c.*AD 1470–1570, assuming a sediment accumulation rate of 2 mm $yr^{-1}$) which is evident only in very slightly enhanced lithophile element concentrations (figure 3.3).

### *Catchment vegetation*

The area around Loch Stiapavat has been treeless throughout the period represented in core Ness-1, and probably for much longer, as a comparison with Little Loch Roag, 55 km to the south-west, suggests (Birks and Madsen 1979). The slight reduction in heath taxa in unit N1-2 (after *c.*AD 1700, given the assumed chronology) may be evidence for increased arable, but the pollen stratigraphy is more notable for the low frequencies of cereal-type pollen and weed taxa than for any positive evidence of arable farming. The high values of *Potamogeton*, in units N1-2 to N1-3 may be a response to nutrient enrichment of the lake as a result of runoff from a cultivated catchment.

## DISCUSSION: THE HISTORICAL CONTEXT OF ENVIRONMENTAL CHANGE

Taking as a working assumption that the Ness-1 sediment core covers about

600 years, then the data illuminate those changes which occurred from about AD 1400 onwards. Assuming further that population in the area broadly followed trends elsewhere, this would have been soon after a critical point of change, the point at which the population expansion of the medieval period had given way to population contraction and stagnation. Not only might the Black Death have thinned numbers and reduced agricultural activity, but the climatic optimum of the twelfth to early fourteenth centuries would have started to give way to the colder, stormier conditions and reduced arable range of the Little Ice Age. Arguably, the low levels of catchment disturbance evident in the basal unit of Ness-1 (120–58 cm, *c.*AD 1400–1700) need to be seen against this backcloth of contraction and stagnation. Within this overall phase though, the subdued increase in catchment-derived material influx evident in N1-1b (105–85 cm, possibly around AD 1470–1570 given the assumed chronology) may be taken as indicative of a slightly increased level of human activity on the site and, therefore, of a slight recovery of local population. If national climatic trends are a guide, any such recovery may have been aided by a slight amelioration of climate that has been detected (Lamb 1985) between the mid-fifteenth and mid-sixteenth centuries, before the harsher years of the late sixteenth and seventeenth centuries.

The increasing levels of catchment disturbance inferred above 58 cm (*c.*AD 1700) suggest that the Hebrides may have shared in the renewed increase in population that is evident for other parts of upland Britain by the end of the seventeenth century. The physical, chemical and mineral magnetic analyses suggest that this increased level of disturbance peaked between 40 cm and 20 cm in core Ness-1 (i.e. the late eighteenth and late nineteenth century). Such a conclusion accords closely with available population data for the Hebrides, which suggests a particularly strong increase in numbers from the mid- to late eighteenth century onwards and an eventual peak during the middle decades of the nineteenth century. The low levels of disturbance hinted at by the magnetic signal for the top levels of N1-2a (*c.*mid-eighteenth century) are difficult to interpret historically unless one sees them as a product of the short, temporary disruption to farming on the Seaforth Estate caused by the social dislocations of the 1745 Jacobite uprising. Arguably, the inferred increase in catchment-derived particulate inwash evident at 40 cm (the late eighteenth century) reflects not just the increase in population but the use of more intensive methods of cultivation based on the spade, caschrom and the lazy-bed. Finally, the reduction in SIRM and allogenic element concentrations evident above 20 cm, and especially above 15 cm, are believed to be a reflection of increased abundance of diatoms in the sediment (as determined by semi-quantitative microscopic examination following the method of Troels-Smith 1955) rather than a catchment-wide reduction in the level of disturbance.

This palaeoenvironmental approach, using a multitude of techniques, needs to be refined in a number of ways. First, there is a clear need to resolve the problem of dating sediments of this age range, especially where radiocarbon determinations are affected by older, inwashed organic matter, as is often the case when records of catchment erosion are investigated. In the context of lake sediments, a more exact dating of variations during the medieval period may be possible using tephrochronology and luminescence dating. Second, there is a need to extend the analysis back to obtain a comparative record of earlier phases of human activity, such as the possible peak of activity over the twelfth and thirteenth centuries. Third, it would be valuable to examine variations in key elements such as phosphorus and nitrogen, the flux of which may reveal long-term changes in the nature of farming practices. Finally, comparative analysis with other sites in the Hebrides would more clearly establish the regional dimension of change. Without documentary sources, a palaeoecological approach is the only available avenue to understanding longer-term changes in the relationship between farming and the environment in such marginal environments.

ACKNOWLEDGEMENTS

J. A. E. was funded by an Institute of Earth Studies studentship. The authors would like to thank Lisa Entwistle, Alex Cullum and Geoff Duller for assistance during fieldwork, and W. T. Perkins for supervision and guidance with the ICP-MS. This chapter is publication number 292 of the Institute of Earth Studies, University of Wales, Aberystwyth.

REFERENCES

Ball, D. F. (1964) 'Loss-on-ignition as an estimate of organic matter and organic carbon in non-calcareous soils', *Journal of Soil Science* 15(1): 84–92

Battarbee, R. V., Flower, R., Stevenson, A. C. and Rippey, B. (1985) 'Lake acidification in Galloway: a palaeoecological test of competing hypotheses', *Nature* 314: 350–2

Binford, M. W. (1983) 'Palaeolimnology of the Peten Lake District, Guatemala, I: erosion and deposition of inorganic sediment as interred from granulometry', *Hydrobiologia* 103: 199–203

Birks, H. J. B. and Madsen, B. J. (1979) 'Flandrian vegetational history of Little Loch Roag, Isle of Lewis, Scotland', *Journal of Ecology* 67: 825–42.

Bruland, K. W., Bertine, K., Koide, M. and Goldberg, E. D. (1974) 'History of metal pollution in southern California coastal zone', *Environmental Science and Technology* 8: 425–32

Burden, E. T., Norris, G. and McAndrews, J. H. (1986) 'Geochemical indicators in lake sediment of upland erosion caused by Indian and European farming, Awenda Provincial Park, Ontario'. *Canadian Journal of Earth Science* 23: 55–65

Davis, M. B. and Ford, M. S. (1982) 'Sediment focusing in Mirror Lake, New Hampshire', *Limnology the Oceanography* 27(1): 137–50

Dearing, J. (1992) 'Sediment yields and sources in a Welsh upland lake-catchment during the past 800 years', *Earth Surface Processes and Landforms* 17: 1–22

Dearing, J. and Flower, R. (1982) 'The magnetic susceptibility of sedimenting material trapped in Lough Neagh, Northern Ireland, and its erosional significance', *Limnology and Oceanography* 27: 969–75

Dearing, J., Elner, J. K. and Happey-Wood, C. M. (1981) 'Recent sediment influx and erosional processes in a Welsh upland lake catchment based on magnetic susceptibility measurements', *Quaternary Research* 16: 356–72

Dodgshon, R. A., (1988) 'The ecological basis of Highland peasant farming 1500–1800

AD', in Birks. H. H., Birks, H. J. B., Koland, P. E. and Nioe, D. (eds) *The Cultural Landscape: Past Present and Future* (Cambridge University Press, Cambridge) pp. 139–51

Dodgshon, R. A. (1992) 'Farming practice in the Western Highlands and Islands before crofting: a study in cultural inertia or opportunity costs?', *Rural History* 3: 173–89

Dodgshon, R. A. and Olsson, G. (1988) 'Productivity and nutrient use in eighteenth century Scottish highland townships', *Geographiska Annaler* 70B: 39–51

Duller, G. A. T. (1992) 'Luminescence chronology of raised marine terraces, southern North Island, New Zealand', unpublished Ph.D. thesis, University of Wales, Aberystwyth

Edwards, K. J. and Whittington, G. W. (1993) 'Aspects of the environmental and depositional history of a rock basin lake in eastern Scotland, UK, in McManus, J. and Duck, R. W. (eds) *Geomorphology and Sedimentology of Lakes and Reservoirs* (John Wiley, Chichester) pp. 155–80

Engstrom, D. R. and Wright, H. E. (1984) 'Chemical stratigraphy of lake sediments as a record of environmental change' in Haworth, E. Y. and Lund, J. W. G. (eds) *Lake Sediments and Environmental History* (Leicester University Press, Leicester) pp. 11–67

Entwistle, J. A. (1994) 'Farming, settlement and environment in the Hebrides circa 800 to the present day', unpublished Ph.D. thesis, University of Wales, Aberystwyth

Faegri, K. and Iversen, J. (1989) '*Textbook of Pollen Analysis*', 4th edn, revised by K. Faegri, P. E. Kaland and K. Krzwinski, (John Wiley, Chichester)

Farmer, J. G., Swan, D. S. and Baxter, M. S. (1980) 'Records and sources of metal pollutants in a dated Loch Lomond sediment core', *The Science of the Total Environment* 16: 131–47

Fenton, A. (1986) 'Seaweed', in *The Shape of the Past 2, Essays in Scottish Ethnology* (John Donald, Edinburgh) pp. 48–82

Finney, B. and AnHuh, C. (1989) 'High resolution sedimentary records of heavy metals from the Santa Monica and San Pedro basins, California', *Marine Pollution Bulletin* 20(4): 181–7

Flower, R. J., Dearing, J. and Nawas, R. (1984) 'Sediment supply and accumulation in a small Moroccan lake: an historical perspective', *Hydrobiologia* 112: 81–92

Flower, R. J., Stevenson, A. C., Dearing J., Foster, I. D. L., Arrey, A., Rippey, B., Wilson, J. P. F. and Appleby, P. G. (1989) 'Catchment disturbance inferred from paleolimnological studies of three contrasted sub-humid environments in Morocco', *Journal of Paleolimnology* 1: 293–322

Flower, R. J., Battarbee, R. W., Natkanski, J., Rippey, B. and Appleby, P. G. (1988) 'The recent acidification of a large Scottish loch located partly within a national nature reserve and a site of special scientific interest', *Journal of Applied Ecology* 25: 715–24

Gaillard, M.-J., Dearing, J. A., el-Daoushy, F., Enell, M. and Hakansson, H. (1991) 'A multidisciplinary study of the lake Bjäresjösjön (S. Sweden): land-use history, soil erosion, lake trophy and lake-level fluctuations during the last 3000 years', *Hydrobiologia* 214: 107–14

Geddes, A. (1955) *Isle of Lewis and Harris* (Edinburgh University Publications, Geography and Sociology, No. 2)

Gibbs, A. (1817) 'Maps of Lewis', Map Room, National Library of Scotland, Edinburgh

Glentworth, R. (1979) 'Observations on the soils of the Outer Hebrides', *Proceedings of the Royal Society of Edinburgh* 77B: 123–37

Grimm, E. C. (1991) *TILIA and TILIAGRAPH Software* (IIlinois State Museum)

Haworth, E. Y. and Lishman, J. P. (1991) 'Recent changes to upland tarns in the English Lake District', *Hydrobiologia* 214: 181–6

Heathwaite, A. L. and O'Sullivan, P. E. (1991) 'Sequential inorganic chemical analysis of a core from Stapton Ley, Devon, U.K.', *Hydrobiologia* 214: 125–35

Henshaw, P. C. (1978) 'The interactions of trace metals and magnetic minerals in sedimentary environments', unpublished Ph.D. thesis, University of Washington.

Higgit, S. R., Oldfield, F. and Appleby, P. G. (1991) 'The record of landuse change and soil erosion in late Holocene sediments of the Petit Lac d'Annecy, Eastern France', *The Holocene* 1(1): 14–28

Hilton, J. (1990) 'Greigite and the magnetic properties or sediment', *Limnology and Oceanography* 35(2): 509–20

Hunt, A., Jones, J. and Oldfield, F. (1984) 'Magnetic measurements and heavy metals in atmospheric particulates of anthropogenic origin', *The Science of the Total Environment* 33: 129–39

Huttunen, P. and Tolonen, K. (1975) 'Human influence in the history or Lake Lovojarvi', *Finskt Museum*, pp. 68–105

Innes, J. B. and Shennan, I. (1991) 'Palynology of archaeological and mire sediments from Dod, Borders, Region, Scotland', *Archaeological Journal* 148: 1–45

Johnston, R. J. (1986) *Multivariate Statistical Analysis in Geography* (John Wiley, Chichester)

Jones, B. F. and Bowser, C. J. (1978) 'The mineralogy and related chemistry of lake sediments', in Lerman, A. (ed.) *Lakes: Chemistry, Geology and Physics* (Springer-Verlag, New York) pp. 179–236

Jones, R., Benson-Evans, K. and Chambers, F. M. (1985) 'Human influence upon sedimentation in Llangorse Lake, Wales', *Earth Surface Processes and Landforms* 10: 227–35

Kemp, A. L., Thomas, R. L., Dell, C. I. and Jaquet, J.-M. (1976) 'Cultural impact on the geochemistry of sediments in Lake Erie', *Journal of the Fisheries Research Board of Canada* 33: 440–62

Kovach, W. (1990) Multivariate Statistical Package, Plus–Version 2.0

Lamb, H. H. (1985) 'Climate and landscape in the British Isles', in, Woodell, S. R. J. (ed.) *The English landscape, Past, Present and Future* (Oxford University Press, Oxford) pp. 148–67

Lehman, J. T. (1975) 'Reconstructing the rate of accumulation of lake sediment: the effect of sediment focusing', *Quaternary Research* 5: 541–50

Macphail, J. R. N. (ed.) (1916) *Highland Papers*, Vol. II, Scottish History Society, 2nd Series, Vol. 12 (Edinburgh, 1916)

Mackereth, F. J. H. (1966) 'Some chemical observations on post-glacial lake sediments' *Philosophical Transactions of the Royal Society, London* 250B: 165–213

Maher, B. (1988) 'Magnetic properties of some synthetic sub-micron magnetites', *Geophysical Journal* 94: 83–96

Mason, B. (1966) *Principles of Geochemistry*, 3rd edn (John Wiley, Chichester)

Moskowitz, B. M., Frankel, R. B., Bazylinski, D. A., Jannasch, H. W. and Lovley, D. R. (1989) 'A comparison of magnetite particles produced anaerobically by magnetotactic and dissimilatory iron-reducing bacteria', *Geophysical Research Letters* 16(7): 665–8

Murozumi, M., Chow, T. J. and Patterson, C. (1969) 'Chemical concentration of pollutant Ph aerosols, terrestrial dusts and sea salt in Greenland and Antarctic snow strata', *Geochemica et Cosmochimica Acta* 33: 1247–94

Nakamura, N. (1974) 'Determination of REE, Ba, Fe, Mg, Na and K in carbonaceous and ordinary chondrites', *Geochemica et Cosmochimica Acta* 8: 757–75

Ochsenbein, U., Davison. W., Hilton, J. and Haworth, E. Y. (1983) 'The geochemical record of major cations and trace elements in a productive lake', *Archive für Hydrobiologica* 98: 463–88

Oldfield, F. (1983) 'The role of magnetic studies in palaeohydrology', in Gregory, K. J. (ed.) *Background to Palaeohydrology* (John Wiley, Chichester) pp. 141–65

Oldfield, F. (1991) 'Environment magnetism – a personal perspective', *Quaternary Science Reviews* 10(1): 73–85

Oldfield, F. and Richardson, N. (1990) 'Lake sediment magnetism and atmospheric deposition', *Philosophical Transactions of the Royal Society, London B* 327: 325–30

Oldfield, F., Maher, B. A., Donoghue, J. and Pierce, J. (1985) 'Particle-size related, mineral magnetic source sediment linkages in the Rhode river catchment, Maryland, USA', *Journal of the Geological Society of London* 142: 1035–46

Peacock, D. (1984) 'Quaternary geology of the Outer Hebrides', *Report of the British Geological Survey* 16(2).

Pennington, W. (1981) 'Records of a lake's life: the sediments', *Hydrobiologia* 79: 197–219

Pennington, W. (1991) 'Palaeolimnology in the English Lakes – some questions and answers over 50 years', *Hydrobiologia* 214: 9–24

Pennington, W., Cambray, R. S., Eakins, J. D. and Harkness, D. D. (1976) 'Radionuclide dating of the recent sediments of Blelham Tarn', *Freshwater Biology* 6: 317–31

Price, N. B., Malcom, S. J. and Hamilton-Taylor, J. (1978) 'Environmental impact of Pb and Zn in recent sediments, in *Biogeochemistry of Estuarine Sediments*, Proceedings of UNESCO/SCOR Workshop, Melnux, Belgium, November–December 1976, UNESCO, pp. 207–215

*Report to the Secretary for Scotland by the Crofters Commission on the Social Condition of the People of Lewis in 1901, as Compared with Twenty Years Ago* (Glasgow, 1902)

Rippey, B. (1990) 'Sediment chemistry and atmospheric contamination', *Philosophical Transactions of the Royal Society, London B* 327: 311–17

Sandgren, S. and Fredskild, B. (1991) 'Magnetic measurements recording late Holocene man-induced erosion in South Greenland', *Boreas* 20: 315–31

Snowball, I. and Thompson, R. (1988) 'The occurrence of greigite in sediments from Loch Lomond', *Journal of Quaternary Science* 3(2): 121–5

Soil Survey of Scotland 1982 1 : 250 000 Sheet 2, The Outer Hebrides, Soil Survey of Scotland, The Macaulay Institute for Soil Research, Aberdeen

Stober, J. A. and Thompson, R. (1979) 'An investigation into the source of magnetic minerals in some Finnish lake sediment'. *Earth and Planetary Science Letters* 45: 464–74

Stolz, J., Lovley, D. and Haggerty, S. (1990) 'Biogenic magnetite and they magnetisation of sediments', *Journal of Geophysical Research* 95(B4): 4355–61

Thompson, M. (1983) 'Control procedures in geochemical analysis', in Howarth, R. J. (ed.) *Statistics and Data Analysis in Geochemical Prospecting* (Elsevier, Amsterdam), pp. 39–58

Thompson, M. and Wood, S. M. (1982) 'Atomic absorption methods applied to geochemistry', in Cantle, J. E. (ed.) *Atomic Absorption Spectrometry* (Elsevier, Amsterdam) pp. 261–84

Thompson, R. and Oldfield, F. (1986) *Environment Magnetism* (Allen and Unwin, London)

Thompson, R., Battarbee, R. W., O'Sullivan, P. E. and Oldfield, F. (1975) 'Magnetic susceptibility of lake sediments', *Limnology and Oceanography* 20(5): 687–98

Troels-Smith, J. (1955) 'Karakterisering af løse jordarter', *Danmarks Geologiske Undersøgelse*, IV Raekke, Bd 3, Nr. 10

Wedepohl, K. H. (ed.) (1969) *Handbook of Geochemistry* (Springer-Verlag, Berlin)

Williams, T. M. (1988) 'The geochemistry and magnetic mineralogy of Holocene strata in acid basins: case studies of granitic catchments in the Galloway and Rannochmoor regions of Scotland', unpublished Ph.D. thesis, University of Edinburgh.

Williams, T. M. (1991) 'Ferrimagnetic sulfide formation in recent sediments of Loch Ba, Scotland and implications for magnetostratigraphic interpretation', *Quaternary Research* 35: 208–21

# 4

# A Scottish Broad: Historical, Stratigraphic and Numerical Studies Associated with Polleniferous Deposits at Kilconquhar Loch

Graeme Whittington and Kevin J. Edwards

ABSTRACT

A folk-memory recorded in the first *Statistical Account of Scotland* (1790–1) refers to the site of Redmire in Fife as one in which peat cutting occurred, eventually leading to flooding and the formation of Kilconquhar Loch. This is supported by documentary evidence extending back to the sixteenth century, as well as stratigraphic evidence from the loch basin. Pollen data from lake sediments reveal rich suites of pollen with an abundance of agricultural crop taxa including barley (*Hordeum*), oats/wheat (*Avena/Triticum*) and hemp (*Cannabis sativa*), in addition to pollen types characteristic of weed communities. The pollen assemblages change relatively gradually and contain many cultural indicator types in fairly low quantities. In an effort to detect structure within the assemblages and to throw light upon changing land-use practices within the catchment area of Kilconquhar Loch, the data are subjected to numerical analyses (mainly detrended correspondence analysis). These reveal possible changes in crop regimes (including rotation, a move towards fallow and/or sown pasture), as well as possibilities such as land drainage and the development of weed communities.

## INTRODUCTION

In the East Neuk of Fife, eastern Scotland, lies Kilconquhar Loch (figure 4.1). It looks completely natural and, given its geomorphological context, might be considered to have its origins as a kettle-hole loch in the coastal plain. Indeed, as an SSSI, it was described as such by the Nature Conservancy Council (Ratcliffe 1977). In his parish report for the first *Statistical Account of Scotland*, the Rev. A. Small (1790–1: 294–5) wrote:

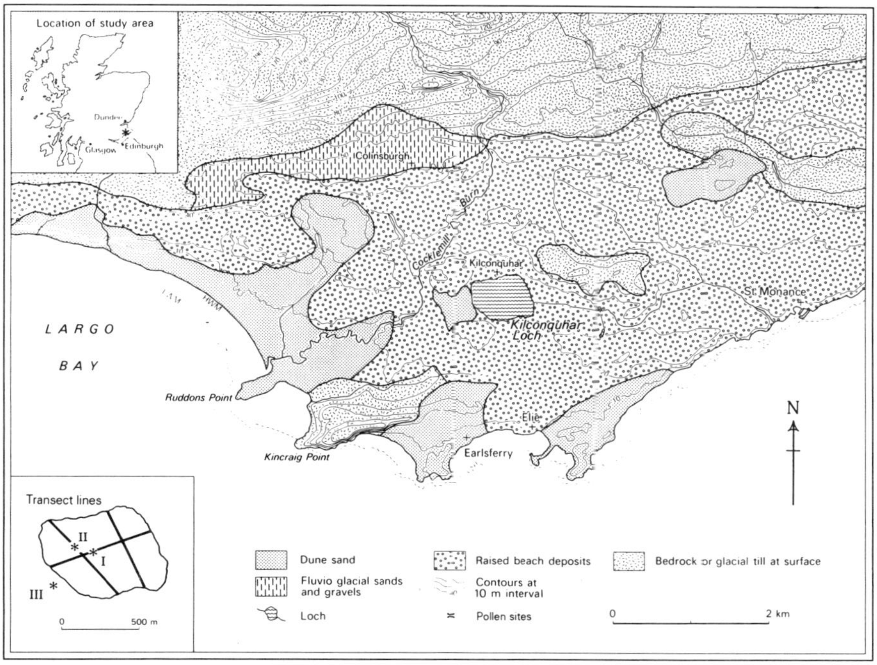


**Figure 4.1** The location of Kilconquhar Loch, the surrounding surface geology, coring transects and core sites.

> There is a beautiful piece of water on the south side of the village of Kilconquhar, commonly called Kilconquhar Loch, almost ¼ of a mile in length, and about ¼ of a mile in breadth. The Loch was originally called Redmire, from which much fuel was got, as peat and turf. It had a drain west-ward to the sea. There is a tradition here, that in the year 1624 or 1625, the drain was filled up with sand driven by a violent gust of wind from the sea, and that the water, thus stopped in its course, became a lake.

The survival of this folk-memory points to the possibility that Kilconquhar Loch originated as a flooded turbary, and would thus bear comparison with the Broads of East Anglia (Lambert et al. 1960). The date of possible inundation of 1624 or 1625 may, however, be somewhat late, given the evidence from a tombstone (now lost), in Kilconquhar churchyard, which recorded the drowning of a James Bellenden in 1593, while skating on the loch (Wood 1862).

Historically, then, the site of Kilconquhar Loch is of considerable interest and it was considered likely to repay documentary study. Furthermore, it was also possible that the methods of historical ecology, as represented by stratigraphic and palynological study, would throw light upon the origins of the loch and its later history.

## THE SITE

Kilconquhar Loch (National Grid Reference NO 488026) lies 1.5 km north of Elie at an altitude of 27 m OD (figure 4.1). It is a prominent landscape feature with a surface area of 0.38 $km^2$, but its average depth is only about 1 m and its catchment area is only 2 $km^2$. The water level is maintained by a weir at the outlet which drains eastward to the Firth of Forth at St Monance. The Loch is surrounded by raised beach, dune and blown-sand deposits which overlie tills, fluvioglacial sands and gravels, or calciferous sandstone bedrock of Carboniferous age. The gently rolling topography has arable husbandry as the predominant land use.

## HISTORICAL EVIDENCE

Kilconquhar is mentioned as early as 1548 in the *Register of the Great Seal* (Registrum Magni Sigilli Regum Scotorum, 1306–1659), a record of land transfer and resulting ownership. However, the closest reference to peat ex—ploitation is for 1592, alluding to the turbary of Firthmure in Balneill in the parish of Kilconquhar (RGS 2100: p. 716):

> dedit dictis personis dict. terras de Balneill, cum communi pastura et libertate, lucrandi lie feuall, faill et dovett in Firthmure

> gives the said person the said lands of Balneill, with common pasture and the right to win fuel, turf and divot in Firthmure

A similar reference is made in the 1603 *Register* to Firthmure and to another turbary, Schyremure, in the same parish (RGS 1405: 497):

> cum communis pastura et libertate lucrandi lie feuall, turvis, faill et dovettis in mora nuncupata Schyremure, in parochia de Kilconquhar

> with common pasture and the rights to win fuel, peat, turf and divot in the moor called Schyremure, in the parish of Kilconquhar

Nowhere in the records for Kilconquhar is mention made of peat winning at Redmire. This could be because Redmire was not subject to a change in ownership, or – perhaps more likely – because Firthmure and Schyremure were active turbaries, whereas by 1592, Redmire was flooded. This would fit with the date of 1593 given for the ice-skating accident and with a *Register* entry for 1598 for the adjacent lands of Carmurie in the Barony of Ardros and pertaining to a boundary (RGS 874: 284–6):

> begynnand at that part of the south syde of the gret loch callit of auld the Reidmyre quhair the gret mylneleid and watergang rynnis furth of that samyn loch unto the said mylne... et maximum partem dicti magni

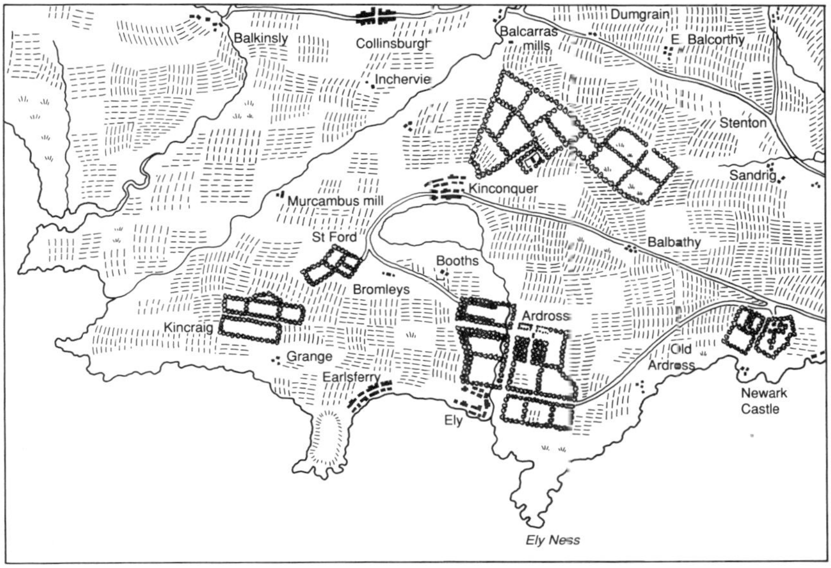


Figure 4.2 The Kilconqhar Loch area in General William Roy's map, *c.*1749 (by permission of the British Library).

> lacus terrumque super quibus inundavit et restagnavit (que paucis abhinc annis fuere terre aride et glebarum paludes *lie peitmyres*, quamvis tunc tecte dicto lie Reidmyre per restagnationem aquarum v i arenarum), cum libertate piscandii in dicto lacu

> beginning at that part of the south side of the great loch formerly called the Reidmyre where the great mill lead and waterway runs out of that same loch to the said mill... and the greater part of the said big lake and lands above which are inundated and flooded (which until a few years ago were dry land and meadows of the glebe and peat myres, until the said Reidmyre was flooded with water by the power of sand), with rights to fish in the said lake

The allusion to sand is powerful confirmation of the statement in the first *Statistical Account*, in which the Rev. Small referred to the loch's origins in the damming by wind-blown sand of a drainage route to the west. This area is certainly one of considerable volumes of sand. Maps by Blaeu (1654) and Moll (1725) show no stream outlet from the Loch, but are very limited in their detail, whereas Roy's Military Survey (*c.*1749) shows an exit to the south flowing to the Firth of Forth at Elie (figure 4.2). There was a water-mill at Elie in 1584, and, as seen above, there are references to the 'mylne dam and mylne leid' from 1598 (cf. RGS 874: 286; SRO, GD147/Box 6/Carmurie). Documentary references to the mill and its lead disappear in the late eighteenth century. Today, neither exists.

The mention of fishing rights in the *Register* entry for 1598, has an echo in 1617 (RGS 1550: 589):

> de terras et dominis de Kilconquhair, cum lacu et piscationibus ejusdem
>
> of the lands and domains of Kilconquhar, with its lake and piscatory

As was the case in the Norfolk Broads, the loch at Kilconquhar was clearly a recognized piscatory.

## STRATIGRAPHY

The documentary evidence alludes to a former mire, and to its demise as a peat cutting and subsequent flooding in response to a storm which blocked an outlet to the west. Stratigraphic evidence from the site and its environs is particularly instructive with regard to these points.

In figure 4.1 are shown the locations of three survey lines which were investigated stratigraphically with a Russian corer (Jowsey 1966). Essentially, three stratigraphic units are recognizable (see Whittington and Jarvis 1987 for further discussion). The basal deposits (numbered 4–12 in Figure 4.3) consist largely of sands and gravels with occasional layers of clay. A middle unit (number 3) consists of a compact layer of peat up to 150 cm in thickness and with sharp upper and lower contacts. An upper unit (numbered 1 and 2), also up to 150 cm in thickness, comprises fine and coarse silty organic lake muds with occasional molluscan shell fragments.

Given the supposed kettle-hole origin of the loch basin, the natural inference is that the basal unit had a late-glacial origin, and this is supported by pollen-analytical evidence (see next section). The peat deposit might, likewise, be assumed to have a postglacial origin, but the pollen evidence from core site KLI (figure 4.1) and an associated radiocarbon date of 10 340 ± 160 BP (I-15 071), suggests that it began to form in late-glacial times. The pollen spectra from the peat of core site KLII do, however, contain a record which extends into the postglacial period (Whittington and Jarvis 1987). This, along with irregularities in the levels of the peat surface and their lack of correspondence to the underlying topography, points strongly to the likelihood that the surface represents the remains of a cut-over raised bog. This is reinforced by the pollen assemblages recovered from the uppermost organic mud deposits. They contain a pollen flora dominated by cultural indicator taxa and provide such a sharp contrast with those in the underlying peat that a depositional hiatus must exist. The muds were obviously laid down subsequent to peat cutting and they reflect the floristic events in the vicinity of the flooded turbary which is now Kilconquhar Loch. The muds are calcareous and are subject to hard-water error; all the radiocarbon determinations from the muds in the KLI profile were falsely old.

The suppositions that the site was a former mire and that it ceased to exist as a result of peat cutting seem to be proven. In an effort to examine the third proposition, that-flooding was a response to a sand-blocked outlet to the

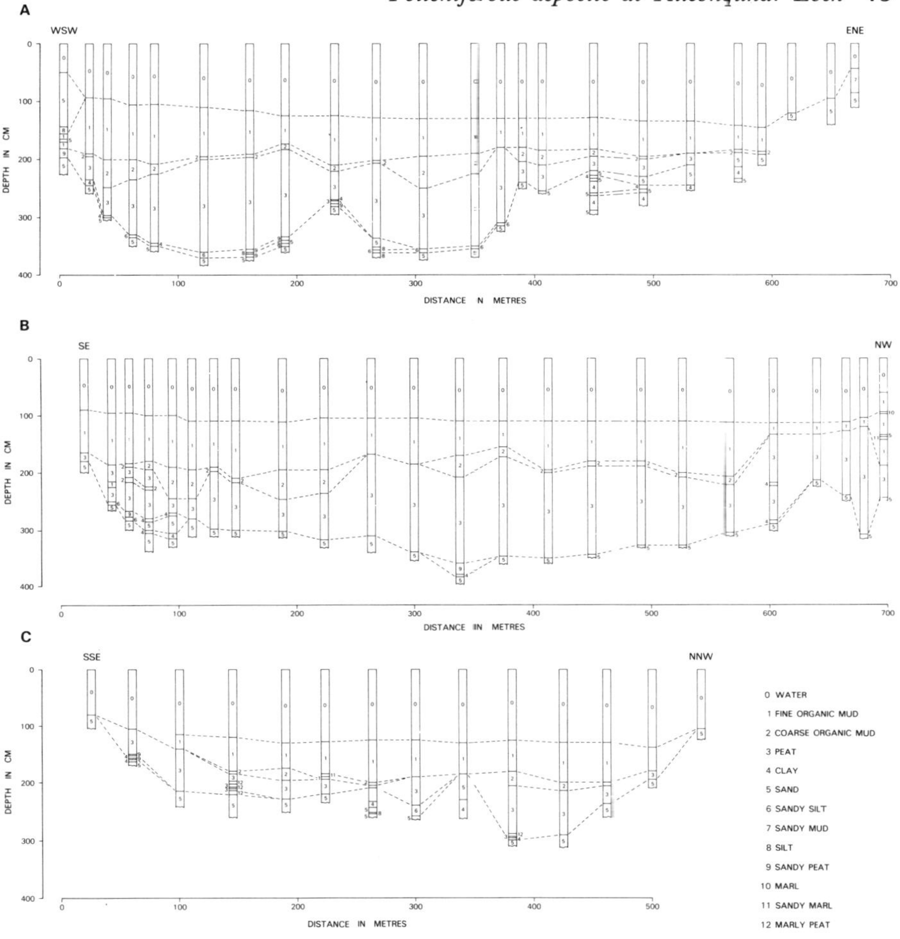


Figure 4.3 The stratigraphy of the sediments in Kilconquhar Loch. The transect lines are those shown in figure 4.1.

west, trial borings were made in the area of sand between Kilconquhar Loch and the Cocklemill Burn. This proved to be extremely difficult, since the blown marine sand deposits were largely inpenetrable beyond about 3 m. At one location, however, 50 m from the western edge of the Loch, the sand was penetrated at 3.1 m, where it gave way to 0.48 m of mud underlain by more sand. Pollen analysis of the mud (core KLIII, located in figure 4.1) showed it to be probably contemporaneous with the mud deposits in cores KLI and KLII, particularly with regard to a high *Cannabis* content. An extension of the loch, or the muds deposited in an outlet stream channel, clearly existed to the west of the loch. The organic deposits would, of necessity, have to represent a later stage in any damming process, with the deposition of the lower sand in KLIII reflecting the initial damming stage, followed by the flooding of Redmire. The uppermost sand layer could signify the later engulfing of either a residual stream outlet or an arm of the newly created loch.

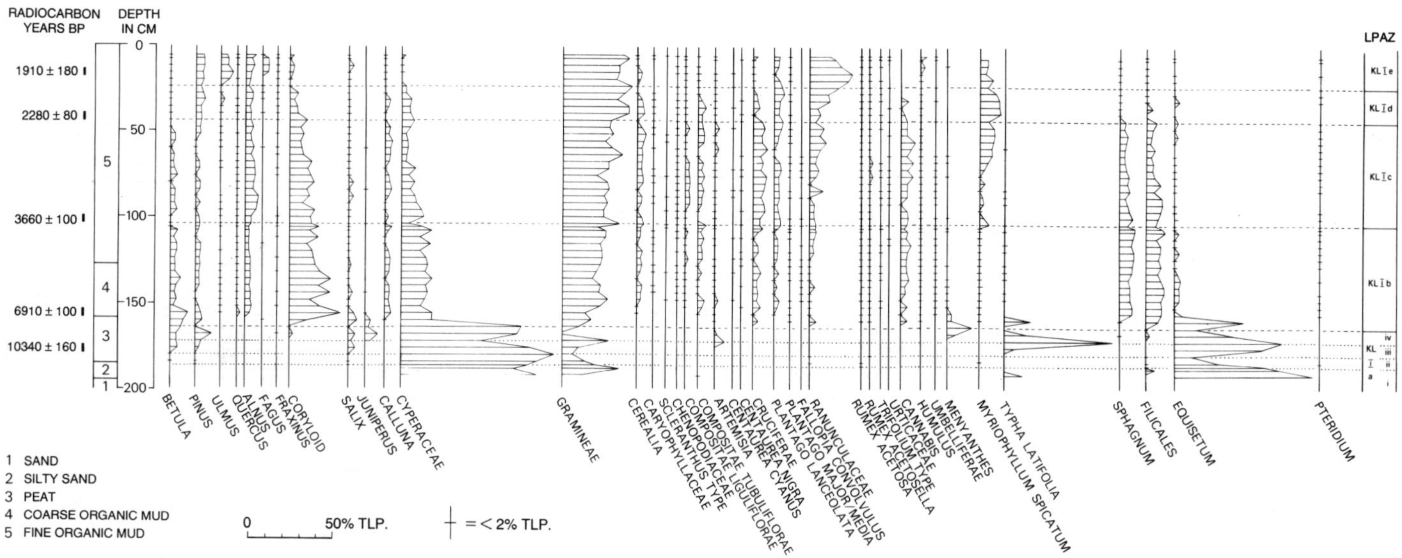


**Figure 4.4** The percentage pollen and spore diagram of selected taxa from core KLI.

The lower sand layer at site KLIII may have been deposited before 1593 (when James Bellenden drowned), but perhaps not long before this (the *Register* entry for 1598 talks of lands above the loch 'which until a few years ago were dry land' and that 'Reidmyre was flooded with water by the power of sand'). There was certainly a major storm in 1570 which led to an estimated 400 000 deaths around the coasts of the North Sea (Lamb 1982). The *Statistical Account* (Small 1790–1) recorded the tradition that 'sand driven by a violent gust of wind from the sea' blocked a western outlet from Redmire in 1624 or 1625. Is this reflected in the deposition of the upper sand layer? A vast storm in the Firth of Forth reportedly did enormous damage in 1625 (Cochrane, 1793; the storm may have been concurrent with a 'high sea flood' recorded in Hamburg at around the same time (H. H. Lamb, personal communication)).

## PALYNOLOGY

### *Methods and the presentation of results*

The deposits from the three cores obtained from the Kilconquhar Loch area were treated with standard NaOH, HF and acetolysis procedures (Faegri and Iversen 1989), including HCl, since the muds were calcareous. Residues were mounted in silicone oil and a pollen counting sum of 500 total land pollen (TLP) was employed. Identification was based on type slides. The high representation of Cannabaceae pollen made it desirable to distinguish between *Cannabis sativa* (hemp) and *Humulus lupulus* (hop). Within the limits of separation, this was achieved by measurement of pore protrusion (French and Moore 1986, Whittington and Gordon 1987). A conventional pollen diagram of selected taxa from core KLI is presented here (figure 4.4). Full and partial pollen data sets from Kilconquhar Loch, illustrating particular aspects of vegetational history, may be found elsewhere (Whittington and Jarvis 1987; Edwards and Whittington 1990). Pollen nomenclature largely follows Moore et al. (1991), and plant nomenclature follows Clapham et al. (1987). English names for pollen and spore taxa are shown in table 4.1.

### *Summary of vegetation history*

The basal sand deposits at KLI were largely bereft of pollen. The lowest local pollen assemblage zone KLIa (figure 4.4), occurs mainly in peat and is dominated by Cyperaceae and Gramineae. Other herbs are also present, including *Artemisia* and *Rumex* spp. and, together with *Juniperus communis*, are typical of Late Devensian and Early Holocene floras. Of the tree and shrub taxa, *Betula*, *Pinus* and *Salix* are worthy of note. The pollen of the aquatic taxa *Menyanthes trifoliata* and *Typha latifolia* type and the spores of *Equisetum* suggest that bog pools were also present.

A number of taxa close to the base of zone KLIb appear suddenly (e.g. *Alnus*, *Calluna vulgaris*, Cerealia type, Compositae tubuliflorae, Cruciferae, *Plantago lanceolata*, *Cannabis sativa*), expand markedly (e.g. Coryloid, Filicales and *Sphagnum*) or decline (e.g. *Betula*, *Pinus*, *Juniperus*, *Typha latifolia* type

**Table 4.1** Common name equivalents and abbreviations of pollen and spore taxa mentioned in the text and in diagrams

| *Latin name* | *Abbreviation*[a] | *English name* |
|---|---|---|
| *Alnus* | Al | alder |
| *Artemisia* | Ar | mugwort |
| *Avena/Triticum* | A/T | oats/wheat |
| *Betula* | Be | birch |
| *Calluna vulgaris* | Cal | heather |
| *Caltha* | Ct | marsh marigold |
| *Cannabis sativa* | Can | hemp |
| Caryophyllaceae | | pink family |
| *Centaurea cyanus* | | cornflower |
| *Centaurea nigra* | | knapweed |
| Cerealia undiff. | Cer | cereal type |
| Chenopodiaceae | | goosefoot family |
| *Chrysanthemum segetum* | | corn marigold |
| Coryloid | Co | hazel/bog myrtle |
| Compositae liguliflorae | Lig | dandelion sub-family |
| Compositae tubuliflorae | Tub | daisy sub-family |
| Cruciferae | Cru | cabbage family |
| Cyperaceae | Cy | sedge family |
| *Equisetum* | Eq | horsetail |
| *Fagus sylvatica* | Fa | beech |
| *Fallopia convolvulus* | | black bindweed |
| *Fraxinus excelsior* | | ash |
| Filicales | Fi | ferns |
| Gramineae | Po | grass family |
| *Hordeum* | Ho | barley |
| *Humulus lupulus* | Hu | hop |
| *Juniperus communis* | | juniper |
| *Linum* | | flax |
| *Menyanthes trifoliata* | | bogbean |
| *Myriophyllum spicatum* | | spiked water milfoil |
| *Pinus sylvestris* | Pi | Scots pine |
| *Plantago lanceolata* | Pl | ribwort plantain |
| *Plantago major/media* | | great/hoary plantain |
| *Pteridium aquilinum* | | bracken |
| *Quercus* | | oak |
| Ranunculaceae | Ra | buttercup family |
| *Rumex acetosa* | Rs | common sorrel |
| *Rumex acetosella* | Rl | sheep's sorrel |
| *Salix* | Sx | willow |
| *Scleranthus* type | | knawel |
| *Sinapis* | | cf. charlock |
| *Sphagnum* | Sp | sphagnum moss |
| *Trifolium* type | | clover type |
| *Typha latifolia* | | bulrush |
| *Ulmus* | Ul | elm |
| Umbelliferae | | carrot family |
| Urticaceae | | nettle family |

[a]Used in the DCA plots, figures 4.7a–e.

and *Equisetum*). This is indicative of a hiatus in the depositional record, which is reinforced by the sedimentary change from peat to lake mud.

The gradually changing nature of the pollen assemblages from KLIb onward is strongly suggestive of a slowly evolving agriculturally exploited landscape. Crop taxa such as *Cannabis sativa*, *Hordeum* and *Avena*/*Triticum* types (cf. Andersen 1979) are accompanied by a potential arable weed flora (e.g. Cruciferae, Chenopodiaceae, Urticaceae, *Fallopia convolvulus* and *Centaurea cyanus*). Apart from the dominant presence of Gramineae, taxa which can be frequent in pastureland and rough grazing are also common (e.g. *Plantago lanceolata*, Ranunculaceae, *Trifolium* type, *Rumex acetosa* type and *Pteridium aquilinum*). Taxa of the subfamilies Compositae tubuliflorae and Compositae liguliflorae are also present and may contain species found in either arable, pastoral or ruderal habitats. The pollen of *Humulus lupulus* is unlikely to represent hop cultivation, unless as an escape, and is more likely to be a constituent of a fen-carr community (cf. Stace 1991).

The rise of *Fagus* and *Ulmus* at the top of the profile in zone KLIe probably reflects the planting of woodland on large estates over the last two centuries.

### *Numerical analyses*

The pollen and spore assemblages which are coeval with lake mud deposition at Kilconquhar Loch contain many taxa which are characteristically associated with cultural activity. Many of these taxa are of low abundance; yet even these could conceivably preserve signs of changing land use within the pollen catchment area. In an attempt to detect ecological structure within the spectra, a data set from the longest organic mud profile, KLI, was subjected to numerical analyses, primarily involving ordination of the data.

Ordination, in which taxa are arranged in low-dimensional space such that taxa behaving in a similar manner are located close together, is a standard approach to the determination of structure in large ecological data sets (Whittaker 1978; Digby and Kempton 1987), including those derived from palynological analysis (Birks and Gordon 1985). Although a number of ordination techniques are available, detrended correspondence analysis (DCA; Hill 1979, Gauch 1982), is one of the most sophisticated and widely used (e.g. Almendinger 1992; Foster et al. 1992, Wiltshire and Edwards 1993) and was employed here. An initial filtering of the pollen and spore taxa from the mud section (0–156 cm) of core KLI was based on the requirement that a taxon should attain at least 2 per cent of a TLP + cryptogam counting sum at one level within the numerically analyzed section. Demonstrably aquatic taxa were excluded from the study as the focus here was upon terrestrial activity; it is recognized that nutrient changes consequent upon land-use change could have an effect upon aquatic species. The 2 per cent criterion was chosen so as to minimize possibly spurious correlations arising from taxa found in extremely low quantities. DCA was implemented by means of the appropriate option within the computer program TILIA (Grimm 1991). There was no down-weighting for low values, since it was considered unwise to bias further a data set in which taxa of different pollen productivity and dispersal are involved (cf. Turner 1986).

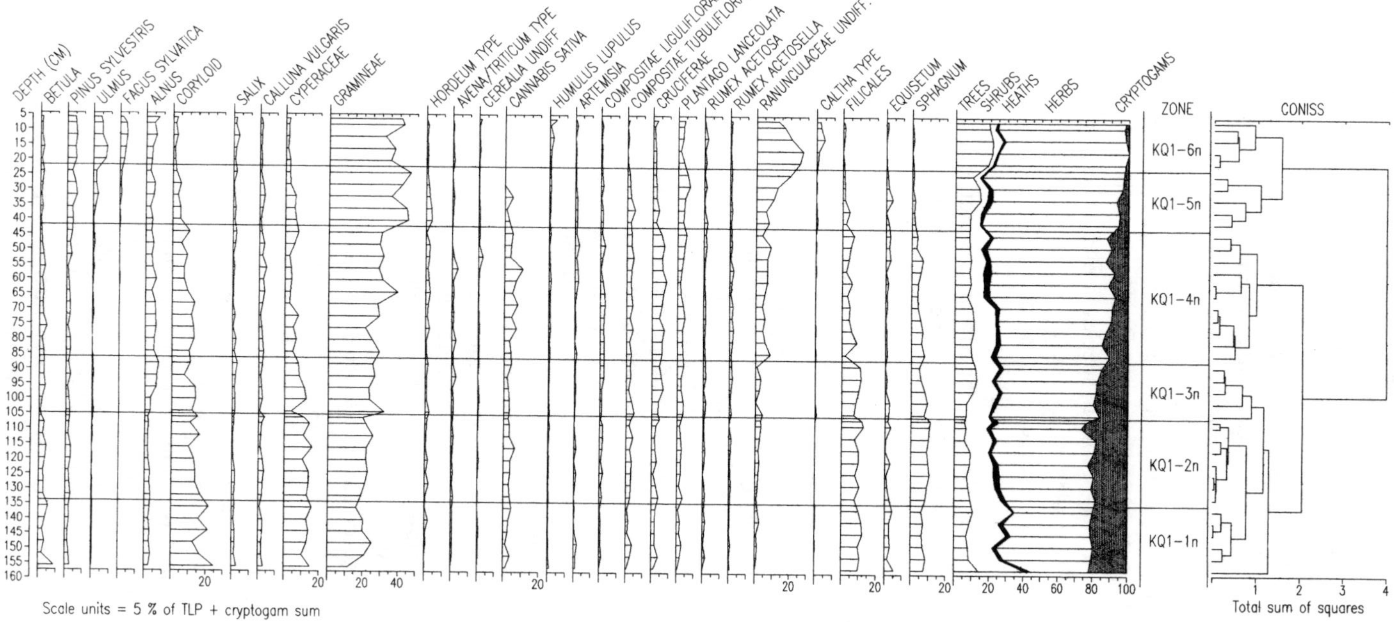


**Figure 4.5** The percentage pollen and spore diagram for the special numerical data set obtained from the mud section of core KLI.

A pollen diagram based on a total taxa sum for the complete (≥ 2 per cent) data set from the mud section of core KLI is shown in figure 4.5. The data set was partitioned into numerically determined pollen zones using a stratigraphically constrained cluster analysis (CONISS; Grimm 1987, 1991). The zones selected differ slightly from those determined from the fuller pollen and spore data set (cf. figure 4.4), and hence they are given different alpha-numeric titles (zones KQI-1n to KQI-6n in figure 4.5) so as to avoid any confusion.

DCA was performed on the spectra within the new zones, keeping to the criterion that any taxon had to have at least a 2 per cent representation in at least one level *within* that zone. This procedure is not dissimilar from that of Turner's (1986) applications of Principal Components Analysis, save for one important difference. The analyses at Kilconquhar Loch are based on the spectra within zones in the belief that zones are likely to show relatively stable ecological relationships (cf. Cushing 1967). Turner, however, examined pattern changes *across* individual anthropogenic zone boundaries in order to 'display the nature of a particular change in land use' without including 'noise from the changes at other zone boundaries' (Turner 1986: 222). Since noise and change are likely to be present, almost by definition, at zone boundaries, this was considered to be inimical to the establishment of stable associations, and Turner's approach was therefore not used for the Kilconquhar Loch data set.

### *Ordinations*

*Mean sample scores.* A DCA on the mean sample scores for each 'numerical' zone (with appropriate values extracted from a whole profile DCA on the core KQI data set) shows that the progressive change in the pollen and spore content of the assemblages through the profile is supported by the general linearity of the graphical plot (figure 4.6). The similarity in the axis 2 scores for each zone reflects the fact that the percentage variation (based on the sum of eigenvalues for the first four axes) explained by the second axis is only 5.36 per cent, compared to 89.90 per cent for axis 1.

KOI-1n. The DCA plot for the basal zone (figure 4.7a) shows the two crop taxa, *Hordeum* type and *Cannabis sativa*, at opposite ends of axis 2, suggesting either their growth in different parts of the pollen catchment area (for instance, on different soils or on different landholdings) or, less likely, that the hemp pollen derives from retting activities (see Whittington and Edwards 1989). Wetland taxa are not noticeably associated with the *Cannabis*, however, although this is not necessarily to be expected, as the pollen produced in the process of hemp retting is not likely to share production similarities with pollen from wetland plants. The graphical proximity of *Artemisia* and *Hordeum* type may signify that mugwort formed a weed in barley crops, and, likewise, Cruciferae (perhaps *Sinapis*) may have been a weed associate of hemp. Gramineae is also well separated from the arable crop taxa and may reflect fields under pasture grass and with Ranunculaceae rather than the normally ubiquitous *Plantago*

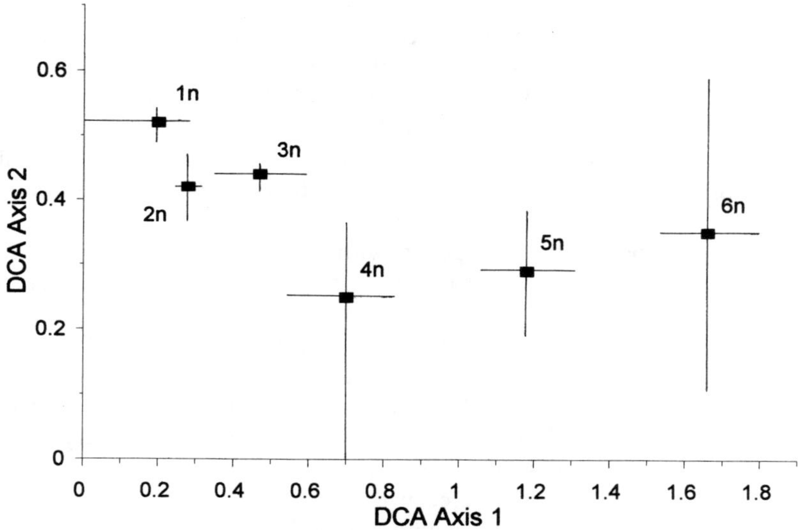


**Figure 4.6** The DCA plot of the mean sample scores for each pollen zone shown in figure 4.5. The range of scores within each zone for each axis is also shown.

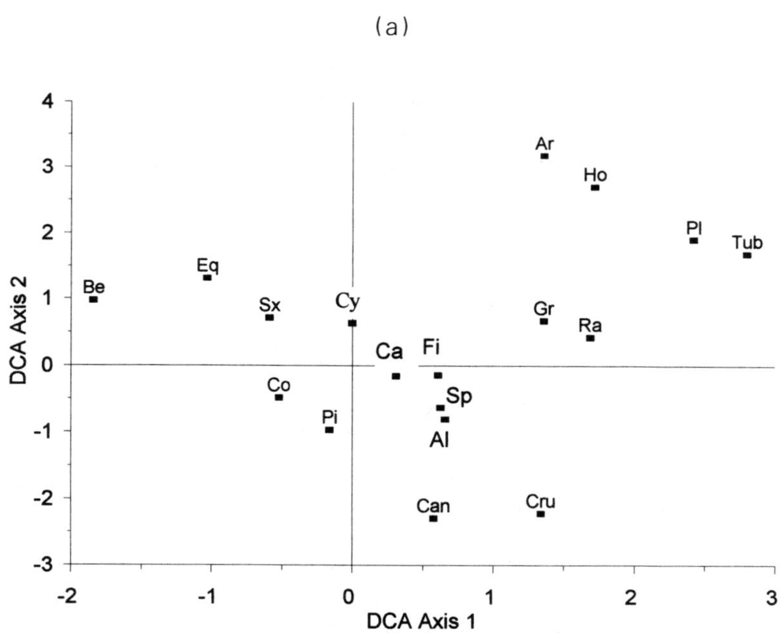


**Figure 4.7** DCA plots of taxa scores for axes 1 and 2 for the pollen zones shown in figure 4.5. Taxa abbreviations are listed in table 4.1. (a) KQ1–1n; (b) KQ1–2n; (c) KQ1–3n; (d) KQ1–4n; (e) KQ1–5n/6n.

(b)

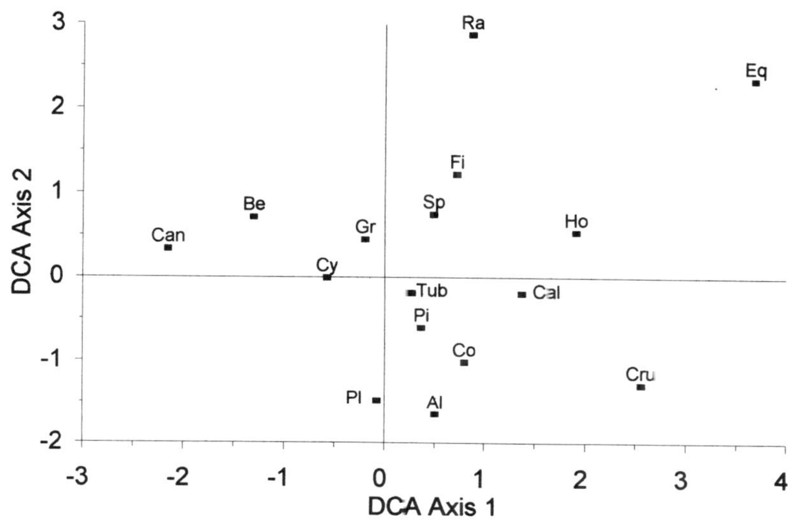


(c)

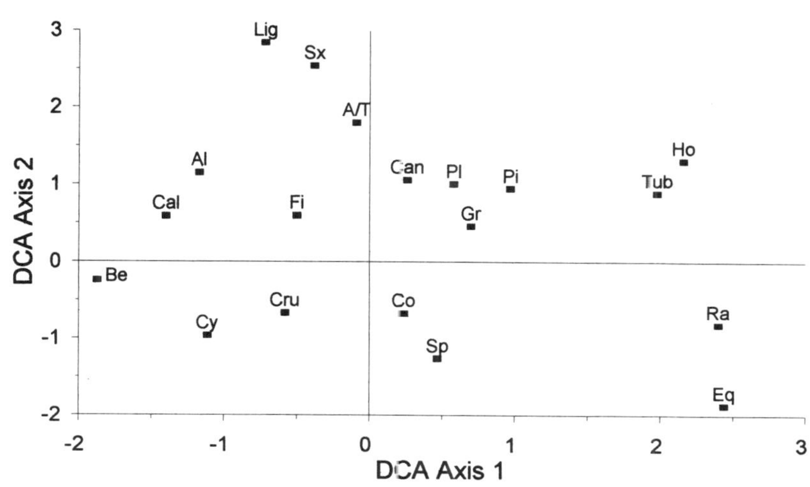


Figure 4.7 *(continued)*

(d)

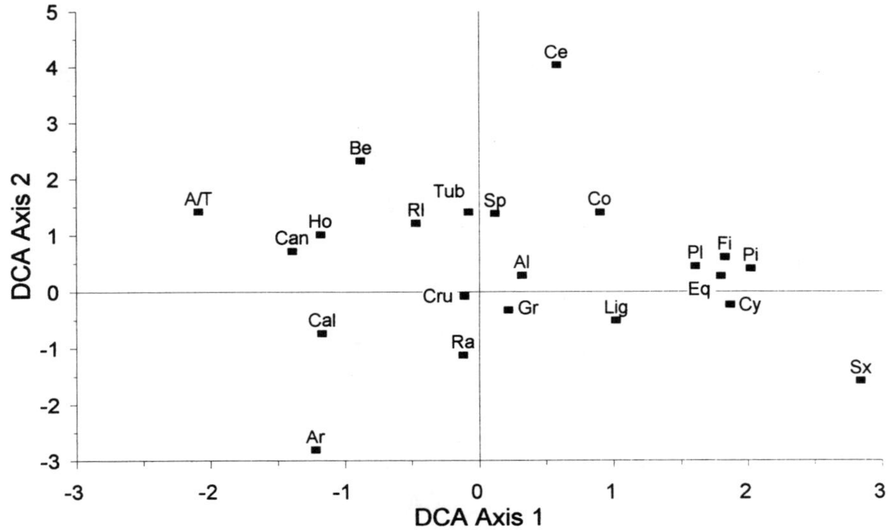


(e)

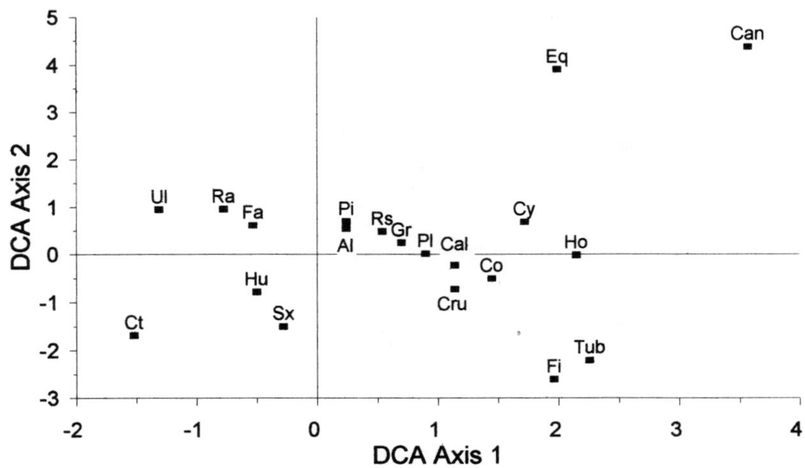


Figure 4.7 *(continued)*

*lanceolata* as an attendant weed. With regard to Gramineae, however, it must be noted that prior to the Improving Movement of the eighteenth century, sown pasture was not unknown but was relatively uncommon in Scotland (Franklin 1952). It is not possible to say what period is covered by the zone KQI-1n deposits, but the grass pollen may reflect grasses in meadows, or as components of ruderal communities, rather than pasture. The grouping of such taxa as *Equisetum*, *Salix*, Cyperaceae, *Sphagnum* and *Alnus* suggests their growth in damp areas, perhaps close to the loch-side.

KOI-2n. *Cannabis*, *Hordeum* and Gramineae continue to be well separated (figure 4.7b). The possible weed associations evident in zone KQI-1n are no longer present, perhaps indicating improved weed control programmes. Agricultural improvement with the drainage of damp areas may be shown by the striking disassociation of the damp habitat community (or communities) seen in the previous zone.

KOI-3n. The crop taxa *Hordeum* and *Cannabis* are now joined by *Avena/Triticum* type (figure 4.7c). The proximity of these last two and Gramineae may suggest that some crop rotation between oats/wheat and hemp was occurring, possibly with grass. The latter could in part be temporary ley as part of a rotation (cf. Franklin 1952: 157), or fallow (which, other than as part of outfield husbandry, was infrequent in many places until later in the eighteenth century (Handley 1953, Smith 1962), but which might explain the proximity of *Plantago lanceolata* to Gramineae in figure 4.7c). The expansion in the grass and plantain curves (figure 4.5) may indicate that grass sown for animal feed was becoming more important than arable agriculture, putting pressure on land previously devoted to hemp and oats/wheat, or that soil impoverishment was forcing crop rotation and/or fallow. The nearness of Compositae tubuliflorae to *Hordeum* type may indicate the infestation of barley crops with corn marigold (*Chrysanthemum segetum*). 'Guld' or 'gool' was the bane of medieval agriculturalists from at least the fifteenth century onwards in Scotland (Franklin 1952, Symon 1959). It was not only deemed a serious crime to allow its growth, but 'gool-riders' were empowered to fine recalcitrant farmers; it has been stated that the removal of corn marigold could increase corn yields ten-fold (Franklin 1952: 44). There is also an old rhyme (Symon 1959: 24):

> The guell, the Gordon and the hoodie craw
> Are the three warst faes Moray ever saw.
>
> The gool, The Gordon and the hooded crow
> Are the three worst foes Moray ever saw.

KOI-4n. *Cannabis*, *Hordeum* and *Avena/Triticum* are now closely associated graphically (figure 4.7d) and they were perhaps growing in a system of crop rotation. *Chrysanthemum* may have continued as a weed of barley fields, possibly accompanied by *Rumex acetosella* (cf. Van der Veen 1992). The eccentric position of the undifferentiated Cerealia may be due to its substantial representation in one spectrum only within the zone, or it may even be the case that the taxon comprises wild grasses growing in a community away from those of pasture grasses (in the area of wind-blown sand, for instance). The separation of the Gramineae from the other crop taxa may indicate that its extent within the pollen catchment area has expanded (as a crop or fallow), together with an associated weed flora which may have included Ranunculaceae, Compositae liguliflorae and even Cruciferae occurring as ruderals.

KOI-5n/6n. Zones KQI-5n and -6n were combined as statistical limitations within DCA, and the pattern of values for each zone did not permit full ordinations to be computed. *Hordeum* and *Cannabis* are now widely separated (figure 4.7e). The proximity of hemp to *Equisetum* may indicate retting activity, possibly using non-local hemp (Whittington and Edwards 1990) or, given the reduced pollen percentages for *Cannabis*, may even denote a change to the selective use of female plants to produce a different quality of fibre (Edwards and Whittington 1992). The disappearance of hemp from the middle of zone KQI-5n may signify the adoption of imported products or a wholesale switch to flax cultivation (*Linum* pollen is grossly under-represented in the fossil record). A poem, *The Preiching of the Swallow*, by schoolmaster R. Henryson, who lived in Dunfermline *c.*1420–90, and published in 1571, shows that hemp was being grown with flax in Fife:

> 'see ye yone Churll' (quod scho) 'beyond yone pleuch,
> Fast sawand hemp, and gude linget seid?
> Yone lint will grow in lytill time in deid
> And thairoff will yone Churll his Nettis mak
>
> 'see you the peasant,' (said she) 'beyond the plough,
> Fast sowing hemp, and good flax seed?
> That lint will grow in little time indeed.
> And thereof will the peasant make his nets'

The centroidal position of Gramineae may show the continued dominance of pasture grassland with *Plantago lanceolata* as a weed. The clustering of the suite of mire taxa (*Caltha*, *Salix* and *Humulus*) in the bottom left of the graph could conceivably reflect the emergence of a new wetland habitat, perhaps associated with drainage ditching. With regard to *Humulus*, such a position would appear to validate the procedure employed to separate its pollen from that of *Cannabis*. The graphical proximity of *Fagus* and *Ulmus* reflects the planting of beech and elm on the policies of large estates over the last two hundred years. The distribution of these tree types and those of other arboreal taxa (*Alnus*, *Pinus* and Coryloid) in figure 4.7e may show that all were widespread within the pollen catchment area.

## CONCLUSIONS

The conclusions arising from the work at Kilconquhar Loch may be stated as follows:

- Documentary and folk-memory evidence for a former turbary on the site of Kilconquhar Loch would seem to be validated by the stratigraphic investigations.
- The evidence for the damming of a western outlet from the Loch, and its being responsible for the flooding of the Redmire turbary, is more equivocal, but it is probably supported by the investigations reported here.
- The SSSI description of Kilconquhar Loch as a kettle-hole is probably correct, in that it was the ultimate origin of the depression in which the water body now stands, but it is incorrect regarding the proximal cause of the present-day loch.
- The chronology of organic mud deposition remains, for the moment, an intractable problem. On the strength of the documentary evidence, it would seem that the muds extend back to at least the late sixteenth century.
- Numerical studies of the pollen spectra reveal possible changes in crop regimes and land use throughout the period represented by the organic muds. There was a possible shift from crop growth in discrete areas to one where a rotation including grass leys or fallow became important. This change in practice may have been a reponse to the extension of pastoral activity, perhaps by means of sown grass, or to soil impoverishment. The development of crop–weed associations may also be evident. *Cannabis* cultivation in the locality may have given way to the retting of non-local hemp or to a change in hemp cultivation practices.
- The integration of documentary, stratigraphic, palynological and numerical methods has greatly enhanced our knowledge of the study area. The Norfolk Broads are not unique, and while Kilconquhar Loch represents a spectacular instance of the flooding of a medieval turbary in Scotland, it is surely not the sole British example beyond East Anglia.

## ACKNOWLEDGEMENTS

Work took place with the permission of A. S. McLeod of Nairn Estates and C. B. Sutherland of Kilconquhar Estate. We would like to thank Jack Jarvis and Colin Cameron for assistance with coring; Lois Wood for help with pollen counting; Kevin Burkhill, Janet Mykura and Graeme Sandeman for cartographic assistance; and the Russell Trust for the funding of radiocarbon dates. Mrs P. Bawcutt drew our attention to the poem by Henryson. Jon Sadler and Judy Turner are thanked for helpful discussions on numerical aspects of the research, and Paul Buckland commented on the text.

REFERENCES

Almendinger, J. C. (1992) 'The late Holocene history of prairie, brush-prairie and jack pine (*Pinus banksiana*) forest on outwash plains, north-central Minnesota, USA', *The Holocene* 2: 37–50

Andersen, S. T. (1979) 'Identification of wild grass and cereal pollen', *Danmarks Geologiske Undersøgelse Årbog 1978*: 69–92

Birks, H. J. B. and Gordon, A.D. (1985) *Numerical Methods in Quaternary Pollen Analysis* (Academic Press, London)

Cochrane, A. (1793) *Description of the Estate and Abbey of Culross* (Edinburgh)

Clapham, A. R., Tutin, T. G. and Moore, D. M. (1987) *Flora of the British Isles*, 3rd edn (Cambridge University Press, Cambridge)

Cushing, E. J. (1967) 'Late-Wisconsin pollen stratigraphy and the glacial sequence in Minnesota', in Cushing, E. J. and Wright, H. E., Jr (eds) *Quaternary Palaeoecology* (Yale University Press, New Haven) pp. 59–88

Digby, P. G. N. and Kempton, R. A. (1987) *Multivariate Analysis of Ecological Communities* (Chapman and Hall, London)

Edwards, K. J. and Whittington, G. (1990) 'Palynological evidence for the growing of *Cannabis sativa* L. (hemp) in medieval and historical Scotland', *Transactions, Institute of British Geographers* N.S. 15: 60–9

Edwards, K. J. and Whittington, G. (1992) 'Male and female plant selection in the cultivation of hemp, and variations in fossil *Cannabis* pollen representation', *The Holocene* 2: 85–7

Faegri, K. and Iversen, J. (1989) *Textbook of Pollen Analysis*, 4th edn, revised by K. Faegri, P. E. Kaland and K. Krzywinski (John Wiley, Chichester)

Foster, D. R., Zebryk, T., Schoonmaker, P. and Lezberg, A. (1992) Post-settlement history of human land-use and vegetation dynamics of a *Tsuga canadensis* (hemlock) woodlot in central New England. *Journal of Ecology* 80: 773–86

Franklin, T. B. (1952) *A History of Scottish Farming* (Thomas Nelson, London)

French, C. N. and Moore, P. D. (1986) 'Deforestation, *Cannabis* cultivation and Schwingmoor formation at Cors Llyn (Llyn Mire), Central Wales', *New Phytologist* 102: 469–82

Gauch, H. G. (1982) *Multivariate Analysis in Community Ecology* (Cambridge University Press, Cambridge)

Grimm, E. (1987) 'CONISS: a Fortran 77 program for stratigraphically constrained cluster analysis by the method of incremental sum of squares', *Computers & Geosciences* 13: 13–35

Grimm, E. C. (1991) *TILIA and TILIA-GRAPH* (Illinois State Museum, Springfield)

Handley, J. E. (1953) *Scottish Farming in the Eighteenth Century* (Faber and Faber, London)

Henryson, R. (1571) *The Morall Fabiliss of Esope* (Thomas Bassandyne, Edinburgh)

Hill, M. O. (1979) *DECORANA – a FORTRAN Program for Detrended Correspondence Analysis and Reciprocal Averaging* (Cornell University, Ithaca, New York)

Jowsey, P. C. (1966) 'An improved peat sampler', *New Phytologist* 65: 245–8

Lamb, H. H. (1982) *Climate History and the Modern World* (Methuen, London)

Lambert, J. M., Jennings, J. N., Smith, C. T., Green, C. and Hutchinson, J. N. (1960) *The Making of the Broads* (Royal Geographical Society Research Series No. 3, London)

Moore, P. D., Webb, J. A. and Collinson, M. E. (1991) *Pollen Analysis* (Blackwell Scientific Publications, Oxford)

RGS (Register of the Great Seal, *Registrum Magni Sigilli Regum Scotorum* (1306–1659)

Ratcliffe, D. (ed.) (1977) *A Nature Conservation Review, vol. 2 Site Accounts* (Cambridge University Press, Cambridge)

Small, A. (1790–91) 'The parish of Kilconquhar', *Statistical Account of Scotland* 19: 287–302

Smith, J. H. (1962) *The Gordon's Mill Farming Club 1758–1764* (Oliver and Boyd, Edinburgh; Aberdeen University Studies No. 145)

Stace, C. (1991) *New Flora of the British Isles* (Cambridge University Press, Cambridge)

Symon, J. A. (1959) *Scottish Farming Past and Present* (Oliver and Boyd, Edinburgh)

Turner, J. (1986) 'Principal components analyses of pollen data with special reference to anthropogenic indicators', in Behre, K.-E. (ed.) *Anthropogenic Indicators in Pollen Diagrams* (A. A. Balkema, Rotterdam) pp. 221–32

Van der Veen, M. (1992) *Crop Husbandry Regimes: an Archaeobotanical Study of Farming in Northern England 1000 BC – AD 500* (J. R. Collis Publications, Sheffield; Sheffield Archaeological Monographs 3)

Whittaker, R. H. (ed.) (1978) *Ordination of Plant Communities* (Junk, The Hague)

Whittington, G. and Edwards, K. J. (1989) 'Problems in the interpretation of Cannabaceae pollen in the stratigraphic record', *Pollen Spores* 31: 79–96

Whittington, G. and Edwards, K. J. (1990) 'The cultivation and utilization of hemp in Scotland' *Scottish Geographical Magazine* 106: 167–73

Whittington, G. and Gordon, A. D. (1987) 'The differentiation of the pollen of *Cannabis sativa* L. from that of *Humulus lupulus* L.', *Pollen et Spores* 29: 111–20

Whittington, G. and Jarvis, J. (1987) 'Kilconquhar Loch, Fife: an historical and palynological investigation. *Proceedings of the Society of Antiquaries of Scotland* 116 (1986): 413–28

Wiltshire, P. E. J. and Edwards, K. J. (1993) 'Mesolithic, early Neolithic and later prehistoric impacts on vegetation at a riverine site in Derbyshire, England', in Chambers, F. M. (ed.) *Climate Change and Human Impact on the Landscape* (Chapman and Hall, London), pp. 157–68

Wood, W. (1862) *The East Neuk of Fife* (Oliver and Boyd, Edinburgh)

# 5

# A Rising Cloud of Midges in the Scottish Highlands?

Alasdair Roberts

ABSTRACT

It is argued that the biting Highland midge (*Culicoides impunctatus*) has become much more common in Scotland since the middle of the nineteenth century. The starting-point is an argument from silence: that incomers to the West Highlands, from Pont in the sixteenth century through to Boswell, Pennant and MacCulloch, made no reference to midge attacks. Gaelic poetry is equally silent, although the peat smoke environment of black houses may have given the Highlanders a measure of protection from midge bites. The possibility of a cultural explanation in terms of changed Victorian perception is rejected. It is not argued that midges were absent from the area prior to *c.*1850 (comments by Burt and about the 'Prince in the heather' show otherwise) but rather that the midge population has increased due to changes in land use. One proposed explanation is that arable land ceased to be drained as a result of depopulation. To this factor is added the impoverishment of sheep-sick pastures by over-grazing and the subsequent spread of deer forests. The effect of coniferous forestry plantations in the present century is also considered in a broad-ranging, if speculative, ecological inquiry. The possibility is raised of confirmation by other kinds of evidence.

## INTRODUCTION

The biting Highland midge, *Culicoides impunctatus*, distinguishable from the *Chiromonidae* which occupy the same general territory but scarcely bite, is found predominantly in areas where annual rainfall exceeds 120 cm, from the Western Isles to the mid-point of Scotland's mountains. The females seek their blood meals between May and September when tourists are in the area, and the subject is often trivialized. In 1992 mites and bats were respectively claimed as parasites and predators on the midge, only to be light-heartedly dismissed in a House of Lords debate (Hansard, 10 November 1992). *Culicoides impunctatus* requires to be taken seriously as an aspect of the Highland environment particularly since, as will be argued, midges were conspicuously absent from the Scottish Highlands prior to the middle years of the nineteenth century. During the period of fashionable Highland tours which began with Pennant, Johnson and Boswell, visitors made no reference to midge attacks; thereafter midge swarms were recorded frequently to the point of caricature. The key environmental change would seem to be the fact that land which had either been under intensive cultivation or used carefully for the pasturing of cattle was cleared to create sheep-walks. The result, over several decades, was a deterioration to 'sheep-sick' pastures, where (arguably) midges were able to flourish in an abbreviated food chain without many predators. Climatic change may also be relevant: a very high midge population appears to have persisted into the modern era of global warming.

## THE ARGUMENT FROM SILENCE

Can the documentary evidence of history shed light on *Culicoides impunctatus*? A good deal of written material is available, much of it in book form, with Johnson (1773) and Boswell (1773) only the best known of many visitors to the Scottish Highlands and Islands who recorded their impressions for a wider public. All were eager to record distinctive features; none made any reference to midges. Only in the second half of the nineteenth century do midges begin to feature regularly in travellers' tales, when they quickly become proverbial to the point of caricature. The first *Punch* midge cartoon appeared in 1851. Early accounts ranged from the impressions of literary-minded diarists, artists and what would now be called travel-writers (Faugas de St Fond 1784, Murray 1796, Leyden 1800, Wordsworth 1803, Southey 1819, Bowman 1825) to the more precise records of naturalists and scientists (Pennant 1772, Walker 1772, 1819). Midges are nowhere mentioned. The approach taken in this literature survey was to include every published book in English which was readily available, and then check for midge references during the period May to September while the author was in the West Highlands or Western Isles.

It must be emphasized that the economic historian now heading Scotland's environmental history movement is cautious about accepting an argument from silence, observing that poison ivy, a problem for early settlers in North America, was never discussed by them: 'Maybe in a world of great discomfort midges and poison ivy did not rank high on the list of human troubles' (Smout 1992). The scientist–author of a best-selling book on Scottish midges has made a similar comment on the historical evidence to the effect that these same diarists were silent on 'bed bugs, lice, horse-flies, stable-flies and mosquitoes... . In an age of horse transport these travellers seem to complain about many things but never horse-borne insects' (Hendry 1992). Travellers did mention unidentified biting insects which were clearly not midges, however, and the poet James Hogg made specific reference to bed-bugs: 'I got the best bed, but it... was also inhabited by a number of little insects common enough in such places' (Hogg 1804: 81). Moreover, anyone who has experienced a Highland midge attack in the hour or two before dusk in conditions of minimal sunlight and wind will acknowledge it to be quite different from other kinds of insect bite. Accounts from the 1850s onwards testify to this, sometimes linking midges with the impossibility of sketching: 'I had no longer sat down than up rose millions of midges, which sent me reeling down the craggy steep, half mad' (Weld 1860: 265) or using illustration to distinguish between good and bad days for sketching on the moors (Bede 1861: 23). Nothing similar was recorded by artists of the previous century (cf. Gilpin 1776).

Midges were proverbial long before the period under discussion, but only for their smallness. In an early-sixteenth version of the Bible in Scots, Matthew xxiii, 24 is rendered 'Blind leiders, clengeand a myge bot suelliand a camele' (Oxford English Dictionary 33: iv, 423). Describing an evening on the banks of Loch Lomond, Robert Burns used midges as a poetic image of country dancing: 'We flew at Bab the Bowster... like midges sporting in the mottie sun' (Brown 1973: 9); apparently he was not troubled by the biting variety. The word 'midge' has been used metaphorically, applied to 'a diminutive person' (OED) in England as in Scotland, where Lowland Scots used midge as a synonym for tiny. Many proverbs testify to the Gael's awareness of physical environment, yet the *mheanbhchuileag*, or tiny fly, where mentioned at all in Gaelic lore, is once again proverbial for its smallness: 'The cow is only a good deal bigger than the midge' (Mackintosh 1819: 109). By way of contrast, Highland culture takes notice of the biting *creathlag*, or cleg: 'The pithless poison is the fly's bite that stains the skin with blood' (Mackintosh 1819: 161).

Early and late, there are examples of midges not being mentioned when other insects are. On Timothy Pont's sixteenth-century map of Edrachillis in northwest Scotland is written 'All heir ar black flies in this wood... seene souking me[n]s blood' (Stone 1989: 30). These black flies must be clegs, which again feature in a key passage by the eighteenth-century Gaelic poet Alexander MacDonald, Alasdair MacMaighstir Alasdair. When offering a comprehensive list of biting insects to justify his loathing for Eigneig in Moidart, he wrote of 'every beast that on me preyed, the wasp, the gadfly and the bee' (MacDonald and MacDonald 1924: 275). This is more than an argument from silence: midges would surely have featured alongside *creathlaggan* or clegs (to say nothing of wasps and bees) if they had preyed on the eighteenth-century inhabitants of Moidart as they do now. Then there is a 127-page disquisition on 'The Moors' by the Edinburgh professor, John Wilson, based on walking tours in the early nineteenth century. At one stage his theme is the teeming insect life concealed by heather and moss which goes unnoticed by the rambler: 'All the while you are supposing yourself alone! ... But the whole wilderness – as you choose to call it – is crawling with insect life' (Wilson 1865: 361). His mention of biting red ants draws attention to the absence of midges. The contrast with Charles Weld, writing a quarter of a century later, is striking: 'Talk of solitude on the moors! – why, every square yard contains a population of millions of these little harpies, that pump blood out of you with amazing savageness and insatiability' (Weld 1860: 80).

## TWO EARLY MIDGE RECORDS

The argument from silence can be strengthened by evidence, at first sight negative with regard to the argument being advanced, from the first half of the eighteenth century. There are two early records of midge swarms which establish that *C. impunctatus* was present in the Highlands. The first comes from Edward Burt, a military engineer who helped to build the first of the roads attributed to General Wade in the period around 1730. It follows a discussion of Fort Augustus as the eastern limit of the heavy rainfall belt: 'I have but one Thing more to take Notice of in relation to the Spot of which I have been so long speaking, and that is, I have been sometimes vexed with a little Plague (if I may use the Expression), but do not you think I am too grave upon the Subject; there are great Swarms of little Flies which the Natives call *Malhoulakins*: ... These are so very small, that, separately, they are but just perceptible and that is all; and, being of a blackish colour, when a number of them settle on the skin, they make it look as if it was dirty; there they bore with their little augers into the pores, and change the face from black to red' (Burt 1754: ii, 345). The same climatic conditions applied then as now: 'They are only troublesome (I should say intolerable) in Summer, when there is a profound Calm; for the least Breath of Wind immediately disperses them.' The description of midges as intolerable places in sharper relief the silence of later visitors who travelled the military roads.

The second record is from the summer of 1746. When Prince Charles Edward Stuart made his enforced tour of the Highlands and Islands after Culloden, he was bitten by *Culicoides impunctatus*. According to one of his companions, high on the side of Glenmoriston, 'the evening being very calm and warm, we greatly suffered by mitches, a species of little creatures troublesome and numerous in the highlands' (Forbes 1896: iii, 379). If midges were troublesome and numerous in the 1740s, why are they never mentioned in the 1770s and thereafter? It may be that the former period was unusual in terms of milder climate (to be discussed) which helped to create a contrast with the latter period of increased arable and cattle farming. These two records certainly strengthen the general claim of changing midge populations as a reflection of changing environment, including climate, and therefore to the hypothesis advanced by the title of this chapter.

## PEAT SMOKE A PROTECTION?

Edward Burt observed of his 'little Plague' that 'the only Refuge from them is the House, into which I never knew them to enter' (Burt 1754: ii, 345). Several of the later journals of Highland visitors make reference to the peat smoke which filled the homes of ordinary people. Leyden (1800: 34) and Southey (1819; 177) both commented on the dirty appearance of those who dwelt in blackhouses and the effect on the bleary eyes of old people. Bowman additionally implied a link between peat smoke and protection from midges. The blackhouse was:

> of an oblong shape, about six yards long by three wide, and formed wholly of turf, the walls not five feet high, and the roof very steep, particularly at one end, where it rose a little higher in a conical shape and ended in a smoke hole. A good deal of the smoke, however, was making its escape through the entrance. The interior was dark and gloomy, for there was neither window nor lattice... . The floor was of earth and very uneven; and at the farther end, under the aperture, a few peats surrounded by stones were burning upon it... . I understood from several persons during the tour that the Highlanders are so much attached to their ancient dwellings that they can hardly be persuaded to exchange them for the more comfortable modern cottages. They are certainly much warmer, because the smoke and heated air must fill every cranny before it can escape through the hole in the roof... . The empyreuma of the peat smoke may also operate favourably in counteracting contagion; for it is a singular though well ascertained fact that bugs are never found, even in the greatest filth, where peat is used (Bowman, 1825, 60)

Bowman learned from his informants that insects swarmed in houses where coal was substituted for peat. The deterrent effect on all insects suggests that Highlanders smoked brown by their turf fires might have been protected against the particular insects which rose from the peat bogs in summer.

On the other hand, blackhouses were carefully constructed to provide ventilation, so this aspect should not be exaggerated. During the summer of 1746 the Young Pretender would have been the beneficiary of whatever protection peat smoke had to offer since he was given the highest place (a dubious way of honouring royal status) in several blackhouses. This is perhaps the time to note, in general, that some people are more affected by midges than others, although opinion differs as to whether the greater sufferers tend to be male or female; and also that, far from it being possible to build up immunity to midges by living with them, it is visitors who sometimes enjoy some days (only) of immunity. Modern crofters are much hampered by them at harvest. At any rate, peat smoke might help to explain the absence of references to midges in Gaelic oral literature, but it would not account for the silence of visitors to the Highlands in the late eighteenth and early nineteenth centuries. The peat reek hypothesis may be welcomed by anyone who is inclined to doubt the reality of a rising midge population, but it is not being proposed here except as one possible factor among other more important ones.

## SCIENTISTS AND MIDGES

Scientists began to visit the Highlands in the seventeenth century, concerned to record aspects of the environment 'which have not hitherto been mention'd by the Learned', an attitude first expressed by the Skye minister Martin Martin who emphasized the interest attaching to 'every Fiber of each Plant, and the least Particle of the smallest Insect' (Martin 1716: x). The argument from silence is therefore at its strongest in relation to the increasingly precise investigations of naturalists such as Walker and Pennant in the 1760s through to the geologist MacCulloch in the early nineteenth century. Two more scientists may be used to underline the point. William MacGillivray spent a year in his native Harris in 1817 (Ralph 1993: 4–10). His diary, held at the University of Aberdeen, shows MacGillivray's interest in flora and fauna developing into his particular specialism of ornithology. He explored the moors on a daily basis during August and September; nightly he recorded, labelled and speculated upon every natural phenomenon, including insects: 'We were assailed by winged Bugs, of what species I do not know, some of which I caught upon my skin. They did not bite, however, or at least not severely, for I did not detect many marks... . In regard to the object which I had in view on this trip, I may say that with the single exception of not having found specimens of the Itch insect I have attained them in a greater or less degree' (MacGillivray 1818). 'The itch', or scabies, was common among the Highlanders. Elsewhere MacGillivray refers to bed bugs as 'light-armed cavalry', but once again there is silence on midges.

In the summer of 1850 the entomologist H. T. Stainton was exploring the moors of Arran in search of the larvae of small moths (Stainton 1864: 104–11). Neither midge larvae nor swarms were recorded. Since then midges have received surprisingly little attention from experts. The impetus given by two postwar government surveys (Cameron et al. 1946, Cameron 1948) soon petered out, and midges were omitted from subsequent studies (e.g. Turnock 1970, Tivy 1973). On the other hand, Frank Fraser Darling, the founding father of modern ecology, was in no doubt about the harmful effect of these insects on crofting and tourism, and exaggerated their presence to the point where 'every square yard of the Highland and Island moors has its midges' (Darling and Boyd 1964: 117). He did not hold out any hope of a solution, but a recent research initiative has brought midges back on to the scientific agenda.

Fieldwork currently being conducted on the west coast of Scotland by a University of Aberdeen team of entomologists and ecologists is focused on the role of pheromones, with a view to incorporating these chemicals into a control mechanism for midges. This research has already begun to yield precise data on the lifestyle of *C. impunctatus*. Through the electron microscope applied to freeze-dried midges, females have been shown to possess three times as many potential olfactory sensilla as tactile hairs on their antennae, reflecting their dependence on smell for the finding of blood meals (Blackwell et al. 1992a). The video camera has been used to establish individual flight patterns within midge swarms, as well as the size, shape and behaviour of swarms adjacent to breeding areas of boggy ground and *Sphagnum* moss (Blackwell et al. 1992b). Such familiar generalizations as the tendency of midges to swarm before dusk in conditions of low wind have thus been confirmed by observation of insects hitherto considered too small for the normal tracking and trapping techniques of entomology. Perhaps the most useful evidence here, for its possible link with climatic change, concerns the negative effect of cold weather on the midge's twice-yearly breeding cycle (Blackwell et al. 1992c).

## CLIMATIC AND OTHER CHANGES IN THE ENVIRONMENT

Situated on the Atlantic edge of Europe, the West Highlands are particularly sensitive to climatic change (Lowe 1993). A lowering of sea temperatures in the seventeenth century brought down the snow-line by 300–400 m (Lamb 1988; 156), yet midges now breed as high as 1000 m above sea level. In addition to the much smaller breeding area then available, cold winters would have reduced the survival rate of midge larvae. During the opening years of the eighteenth century a milder climate appeared which may help to explain the references to midges discussed above. Westerly winds became more prevalent during the eighteenth century, but it was easterly winds out of central Europe which brought the potato blight to the Scottish Highlands (Lamb 1988: 158). The warm humid weather which favoured the spores of *Phytophthora infestans* would also have been ideal for *Culicoides impunctatus*. Potato blight reached Scotland in 1846, and the same 'cool, rainy nights and moderately warm cloudy days' (Devine 1988: 33) persisted into the period when midges were starting to be reported.

One effect of cool summers and high winds at the end of the seventeenth century was further to deplete the woodlands of north-west Scotland (Lamb 1977: 466–7). By then, the original forest cover had been reduced to about 5 per cent of the country's land surface due to a combination of climate, pressure of rural population and external demand (Smout 1993). In the eighteenth century, however, the Clanranald territories of Moidart and Arisaig still had extensive woods, which extended inland to the borders of Lochaber. The remnant of this today comprises oak, birch, alder, ash, elm and native Scots pine, and shelters a range of oceanic flora and vegetation including mosses and liverworts. It is unlikely that midges would have flourished in this complex ecosystem, with their larvae a prey to other insects such as dragonflies. Environmental change came quickly in the late eighteenth century: 'The transformation of Moidart and Arisaig into a short-term productive source of cheap wool and meat for urbanising and industrializing Britain was rapid and devastating. Little or no heed was now given to the protection of woodlands, and the marginal uses to which it was put were of less consequence to the landlord. A regime of continuous muirburn and deforestation was introduced' (Cheape 1993).

While the loss of natural woodland cannot itself be a prime cause of rising midge numbers (most of the Highlands having lost forest cover by the sixteenth century) the well-managed Clanranald forest casts an interesting perspective on Moidart's apparent freedom from midges when Alasdair MacMaighstir Alasdair lived there. It is now commonplace among environmentalists to argue for mixed woodland in place of evergreen, and a midge dimension to that debate may be suggested here. Certainly, forestry workers are notoriously afflicted by midges (Hendry 1989: 41). While the forestry plantations themselves do not provide particularly good breeding grounds for midges (or much else, due to the acidic effect of spruce needles) the element of shelter beside clearings and at the periphery may be relevant. Finally on this aspect, bracken and heather are less contentious than woodland and forestry estates, but bracken also provides shelter and has spread during the period associated here with rising midge numbers As for heather, a deteriorating Highland soil cover may owe something to persistent muir burning in the interests of sheep.

Sheep must have a huge bearing on the question at issue. The Highland Clearances have been endlessly debated since the early nineteenth century, but only now are historians, geographers and others beginning to consider the ecological effects of substituting sheep-walks for farmtowns. In the the time of lowest midge incidence, as argued here, an over-populated Highland region was drained and cultivated as never before and the traditional export of black cattle to the Lowlands was supplemented, in certain areas, by a greatly expanded trade in whisky. Thus marginal land was brought into cultivation: former high pastures or shealings, made fertile by cattle, carne under the plough to produce barley. Then came the Clearances. To choose a single West Highland example, the environmental effects of depopulation and sheep-walks have been described in terms of 'a very minute investigation' for one Lochaber estate close to the time when midges appeared (it is argued) in greatly increased numbers (Somers 1848: 126–36). Six farmtowns, each one formerly supporting 10 or 12 families, were deserted with their furrows 'still visible amid the fern'; 40 000 sheep were clipped each year on on one sheep-walk alone. Those people who did remain were on the poorest land, their crops blighted by vapours emanating from the moss. Meanwhile, the possibilities of arable farming were demonstrated by a Fort William banker who 'commenced to drain, burn, lime, and cultivate his moss... . Finer corn, turnips, and potato crops are nowhere to be seen' (Somers 1848: 131).

Long before 'green' had an environmental connotation, the crofters' historian James Hunter argued that the old Highland agricultural system bequeathed to sheep farmers fertile areas of lowland ground, along with upland pasture which had previously been grazed only in summer by cattle, and that 'purely extractive' sheep farming became the norm from the 1850s (Hunter 1973: 204). An attempt has recently been made to achieve a balanced view on the old question of sheep-sick pastures which was first debated in the 1870s (Mather 1993). Doubt has been expressed as to the alleged destructive effect of sheep on woodland, there being so little of it left when sheep appeared in large numbers. It has proved difficult, over the years, to establish reliable indicators for sheep performance, but it does seem clear that lambing percentages were in decline in the latter part of last century in all areas of the West Highlands. This can be attributed to the deteriorating 'green land' of former shealing pastures rather than to heather 'black land'. The knowledge of experts on such a key aspect of land management as heather burning is still surprisingly poor, partly because of the interaction of different environmental effects. In particular, it is difficult to distinguish the impact of sheep from that of red deer. By the 1880s deer had become a more economic proposition than sheep by 50 per cent, and there were an estimated quarter of a million red deer in the Highlands by the middle years of this century (Smith 1993). Sheep and deer now frequently share the same hill ground. Whatever their share of responsibility for a deteriorating environment, and whatever their direct effect (if any) on the actual midge-breeding areas of undrained moss, the presence of so many carbon dioxide-breathing blood meals can hardly have a negative effect on the midge population.

## CONCLUSIONS

The title of this chapter ends with a question mark, and it may be necessary to accept the old Scottish verdict of 'not proven'. Could scientific evidence be used to prove that the incidence of *Culicoides impunctatus* has risen (and fallen) from one historical period to another? Pollen analysis has become a valuable tool and might prove useful here. Fossil larvae have proved susceptible to investigation in the case of insects rather larger than the midge and are helping to enlarge the understanding of archaeologists. In the meantime, it is encouraging to find that entomologists and ecologists are turning to the subject with renewed interest. At present, those engaged in fieldwork find it hard to imagine that the boggy ground of their investigations was ever different from today, and climatic change finds more favour with specialists who are nevertheless agreed in finding interest in a new perspective from history.

This is an exploratory essay, suggesting a multifactorial explanation. The necessity of explanation depends, of course, on the possibility of a rising midge population being conceded. Given the many factors at work on a precarious Highland environment in the nineteenth century (and since) it is surely harder to imagine that changed land use had no effect on midges, although the apparently midge-free period which started in the second half of the eighteenth century may also have been atypical. The early period is one in which climate is likely to have been the main factor in environmental change, whereas the latter period's changes were man-made. We may never be able to have certain knowledge of how midge populations fluctuated over past time, but it would not be difficult to monitor their presence or absence from different areas of the Highlands now, and relate that to some of the variables which have been under discussion. This amounts to a practical as well as academic proposal which would benefit all those summer visitors who are attracted to the outdoors experience of 'unspoiled wilderness'. There are unanswered questions, at any rate, surrounding *Culicoides impunctatus* which deserve to be considered by the interdisciplinary areas of environmental history, historical geography and biogeography.

## REFERENCES

Bede, C. (1861) *Glencraggen: or a Highland Home in Kintyre* (Longman, London)

Blackwell, A., Mordue (Luntz), A. J. and Mordue, W. (1992a) 'Morphology of the antennae of two species of biting midge: *Culicoides impunctatus* (Goetghebuer) and *Culicoides nubeculosis* (Meigen) (Diptera, Ceratopogonidae)', *Journal of Morphology* 213: 85–103

Blackwell, A., Mordue (Luntz), A. J., Young, M. R. and Mordue, W. (1992b) 'The swarming behaviour of the Scottish biting midge, *Culicoides impunctatus* (Diptera: Ceratopogonidae)', *Ecological Entomology* 17: 319–25

Blackwell, A., Mordue (Luntz), A. J., Young, M. R. and Mordue, W. (1992c) 'Bivoltinism, survival rates and reproductive characteristics of the Scottish biting midge, *Culicoides impunctatus* (Diptera: Ceratopoginae) in Scotland', *Bulletin of Entomological Research* 82: 299–306

Boswell, J. (1773) *Journal of a Tour of the Hebrides* (Macmillan, London)

Bowman, J. E. (1875) *The Highlands and Islands* (Sutton, London)

Brown, R. L. (1973 edn), *Robert Burns's Tours of the Highlands and Stirlingshire 1787* (Boydell, Ipswich)

Burt, E. (1754) *Letters from a Gentleman in the North of Scotland*, 2 vols (John Donald, Edinburgh, 1974)

Cameron, A. E., Downes, J. A., Morison, G. D. and Peacock, A. D. (1946) *A Survey of Scottish Midges* (HMSO, Edinburgh)

Cameron, A. E. (1948) *Second Report on the Control of Midges* (HMSO, Edinburgh)

Cheape, H. (1993) 'Woodlands on the Clanranald Estates: a case study', in Smout, T. C. (ed.) *Scotland Since Prehistory: Natural Change and Human Impact* (Scottish Cultural Press, Aberdeen) pp. 50–63

Darling, F. Fraser and Boyd, J. M. (1964) *The Highlands and Islands* (London)

Devine, T. M. (1988) *The Great Highland Famine* (John Donald, Edinburgh)

Faugas de St Fond, B. (1784) *A Journey through England and Scotland to the Hebrides* (Hopkins, London, 1907)

Forbes, R. (1896) *The Lyon in Mourning* (James Thin, Edinburgh, 1975, 3 vols)

Gilpin, W. (1776) *Observations on the Highlands of Scotland* (Richmond, London, 1973)

Hendry, G. (1989) *Midges in Scotland* (Aberdeen University Press, Aberdeen)

Hendry, G. (1992) personal correspondence, 2 June and 29 June 1992

Hogg, J. (1804) *A Journey through the Highlands and Western Isles* (Byways, Edinburgh, 1981)

Hunter, J. (1973) 'Sheep and deer: Highland sheep farming, 1850–1900', *Northern Scotland* i: 199–222

Johnson, S. (1773) *Journey to the Western Isles of Scotland* (Gardner, Paisley, 1907)

Lamb, H. H. (1977) *Climatic History and the Future* (Methuen, London)

Lamb, H. H. (1988) *Weather, Climate and Human Affairs* (Methuen, London)

Leyden, J. (1800) *Journal of a Tour in the Highlands and Western Islands of Scotland* (Blackwood, Edinburgh)

Lowe, J. J. (1993) 'Setting the scene: an overview of change', in Smout, T. C. (ed.) *Scotland Since Prehistory: Natural Change and Human Impact* (Scottish Cultural Press, Aberdeen) pp. 1–16

MacCulloch, J. (1819) *A Description of the Western Isles of Scotland, Including the Isle of Man: Comprising an Account of Their Geological Structure; with Remarks on Their Agriculture, Scenery and Antiquities*, 3 vols (Hurst, Robinson, London)

MacDonald, A. and MacDonald, A. (1924) *The Poems of Alexander Macdonald (Mac Mhaighstir Alasdair)* (Northern Counties, Inverness)

MacGillivray, W. (1818) *Journal of a Year's Residence in the Hebrides by William MacGillivray from 3rd August 1917 to 13th August 1818* (Aberdeen University Library: Special Collections, MS 1033)

Mackintosh, D. (1819) *Gaelic Proverbs and Familiar Phrases* (William Stewart, Edinburgh)

Martin, M. (1716) *A Description of the Western Islands of Scotland* (James Thin, Edinburgh, 1981)

Mather, A. S. (1993) 'The environmental effect of commercial sheep farming in the Scottish Highlands', in Smout, T. C. (ed.) *Scotland since Prehistory: Natural Change and Human Impact* (Scottish Cultural Press, Aberdeen) pp. 81–5

Murray, S. (1796) *A Companion and Useful Guide to the Beauties of Scotland* (Byways, Hawick, 1982)

*Oxford English Dictionary* (1933) (Clarendon Press, Oxford) vi 423, 'midge'

Pennant, T. (1772) *Tour in Scotland* (Benjamin White, London)

Ralph, R. (1993) *William MacGillivray* (HMSO, London)

Smith, J. (1993) 'The impact of changing deer numbers on the Scottish Highlands since *c.*1780', in Smout, T. C. (ed.) *Scotland since Prehistory: Natural Change and Human Impact* (Scottish Cultural Press, Aberdeen) 89–97

Smout, T. C. (1992), personal correspondence, 11 May 1992

Smout, T. C. (1993) 'Woodland history before 1800', in Smout, T. C. (ed.) *Scotland Since Prehistory: Natural Change and Human Impact* (Scottish Cultural Press, Aberdeen) pp. 40–9

Somers, R. (1848) *Letters from the Highlands on the Famine of 1846* (Melvens, Glasgow, 1977)

Southey, R. (1819) *Journal of a Tour in Scotland* (James Thin, Edinburgh, 1972)

Stainton, H. T. (1854) *The Entomologist's Companion* (Reeve, London)

Stone, J. C. (1989) *The Pont Manuscript Maps of Scotland: Sixteenth Century Origins of a Blaeu Atlas* (Map Collector Publications, Tring)

Tivy, J. (1973 edn) *The Organic Resources of Scotland: Their Nature and Evolution* (Oliver & Boyd, Edinburgh)

Turnock, D. (1970) *Patterns of Highland Development* (Macmillan, London)

Walker, J. (1772) *Report on the Hebrides of 1764 and 1771* (John Donald, Edinburgh, 1980)

Weld, C. R. (1860) *Two Months in the Highlands, Orcadia and Skye* (Longman, London)

Wilson, J. (1865) *Recreations of Christopher North* (Blackwood, Edinburgh)

Wordsworth, D. (1803) *Recollections of a Tour Made in Scotland* (James Thin, Edinburgh, 1974)

# PART II

# England and Wales

6

# Medieval land use, agriculture and environmental change on Lindisfarne (Holy Island), Northumbria

K. Walsh, D. O'Sullivan, R. Young, S. Crane and A. G. Brown

ABSTRACT

This chapter arises from a long-term, interdisciplinary programme of research into the history of settlement on Lindisfarne. It explores the relationships between documentary, archaeological and palaeoenvironmental data on the land use and agriculture of Lindisfarne from the ninth to the seventeenth century. The data employed include ecclesiastical documents, archaeological evidence, pollen analysis from a lough on the island and geomorphological evidence of changing soil conditions. It is considered vital to view land-use history in the specific context of individual settlements, and the extent to which these were economically interdependent or independent. The evidence from Lindisfarne suggests increasing economic diversification within estate structures in the ninth and tenth centuries, but also some fragmentation. The more recent documentary evidence can be viewed in the light of evidence indicating changing land use, at least partially forced by a changing coastal environment.

> ... tell us more about the island in question. Don't try and be too concise – give us a detailed account of it from every point of view, geographical, sociological, political, legal – in fact, tell us everything you think we'd like to know, which means everything we don't know already.
>
> Thomas Moore, *Utopia* (1516)

## INTRODUCTION

This chapter explores the use of both documentary and non-documentary evidence for landscape change and economy on the island of Lindisfarne in

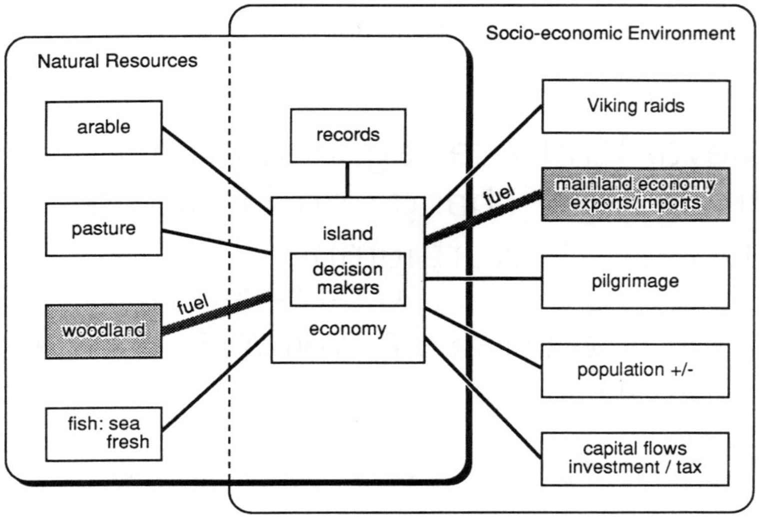


**Figure 6.1** A flow chart illustrating the major components of the Lindisfarne economy in the Middle Ages.

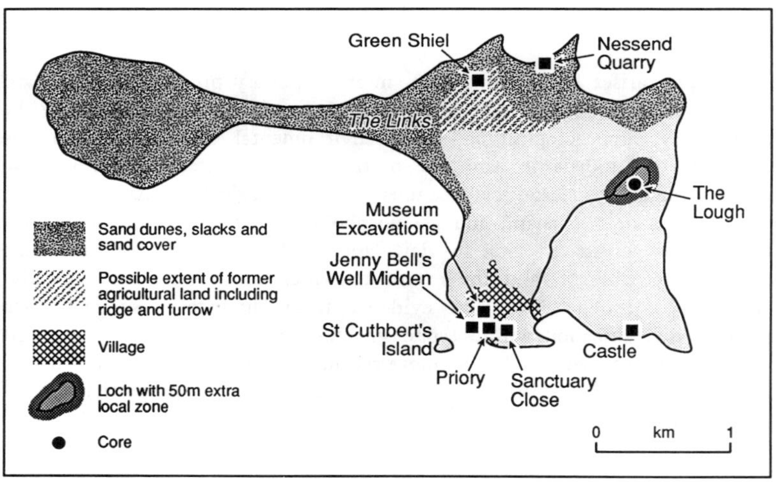


**Figure 6.2** A map of the island, showing the location of the dune system and archaeological sites.

the medieval period. An island has some advantages for this type of analysis due to its bounded nature in both landscape and socio-economic terms. Paradoxically this can help in the identification of the role of a settlement unit in the wider economy which, in this case, is that of the multiple estate. In any study of landscape change which attempts to combine documents and palaeoecological or geomorphological evidence, there is a danger of falling into the reinforcement syndrome, whereby weak palaeoecological/geomorphological data is supported by ambiguous documents, and vice-versa. This is most likely to occur when specific resources are being considered rather than the overall resource base of an economy. A key theme of this contribution is the nature of internal and external influences on the island environment, economy and community throughout the medieval period.

This chapter considers two medieval episodes, the eighth to tenth centuries and the twelfth to sixteenth centuries. In neither case is an attempt made to integrate the two very different types of evidence, using one source to reinforce the other. Our aim is rather to contrast the inferences of land-use change that can be made from the documents and from the archaeological and palaeo-ecological evidence. Figure 6.1 attempts to structure the main economic and social components of the Lindisfarne economy and especially those that would have had an impact on land use during the medieval period. Several sites have been excavated on the island as part of an archaeological project begun in 1983 (figure 6.2: Beavitt et al. 1990).

## THE EIGHTH TO TEN CENTURIES

### *Non-documentary sources: the evidence from Green Shiel and the Lough for land-use change*

The physical evidence for the Early Christian monastery on Lindisfarne is slight and there is no modern excavation on the site (O'Sullivan 1990). The most important source of evidence for the economy and land use of the island during the early medieval period comes from excavations and associated environmental studies in the north of the island around the farmstead site of Green Shiel which have been in progress since 1984 (O'Sullivan and Young 1991a,b).

*Green Shiel.* The Green Shiel settlement site is located close to the north shore of Lindisfarne, within what is now an extensive and relatively stable dune system which covers the whole of the northern area of the island, an area of about 1.5 km$^2$. The settlement (figure 6.3) comprises five buildings with associated yards, their occupation dated by 19 coin finds to the middle decades of the ninth century (*c.*AD 835–71). The animal bone assemblage is dominated by cattle; 45 per cent of the material from one building comes from calves under two years of age at death.

To the south of the settlement site an area of broad ridge and furrow is visible where the ridges and furrows on the modern land surface reflect those of the buried soil which is now covered by some 50–80 cm of blown sand. The existence of the settlement site and the buried soil alone indicate that this part of the island was at one time a viable farming environment, which at some point was lost to the developing dune system. Therefore much emphasis has been placed on the investigation of environmental change on this part of the island, especially the history of the dune system.

*The environment: the island without dunes.* Aerial photographs reveal that the dune system has only stabilized in its present form since the 1950s. It is

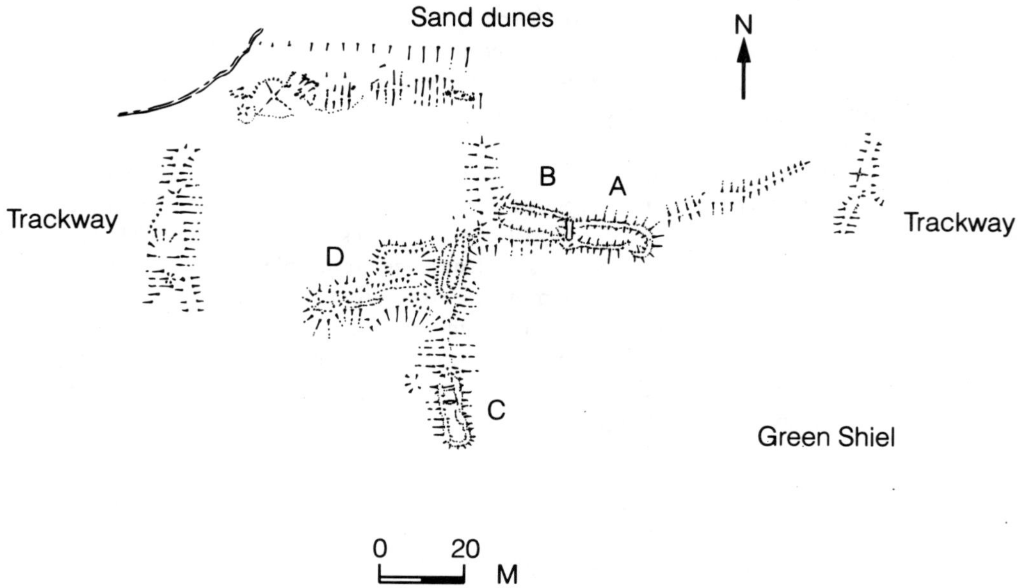


**Figure 6.3** The plan of the early medieval settlement at Green Shiel.

apparent from both air photographs and written accounts that the history of this system has been one dominated by periods of severe instability. Buildings within the system have been recurrently buried and exposed (Darlington 1965). Ecologically, the modern Lindisfarne dune system is currently characterized by a typical succession up to dune slack level (Ranwell 1972).

The development of the dune system would have been an environmental change with dramatic consequences: the dating of this change and the nature of the environment on Lindisfarne prior to this event is therefore of great local importance. Stratigraphic work and particle size analysis from in and around Green Shiel, the dunes and the buried ridge and furrow, has yielded groupings of the various sedimentary environments and allowed reconstruction. Samples from directly beneath building walls, beaches and dunes show that the sediments from the buildings are better sorted and more positively skewed than those from the modern environments. Although not indisputable, the geomorphological evidence from aerial photographs, documents and the buried soil strongly suggests that since the early medieval period the dunes have migrated from the north-east across Green Shiel and over the till cliff. A programme of thermoluminescence dating is planned.

There is no doubt that a substantial dune area had developed by the end of the sixteenth century, as Speed's map of 1610 clearly indicates the existence of such an area. The most likely period for the development of the Lindisfarne dune system is the fifteenth century; figure 6.4 shows that there is a peak of storminess in the 25 year period from 1475 to 1500 in north-west Europe,

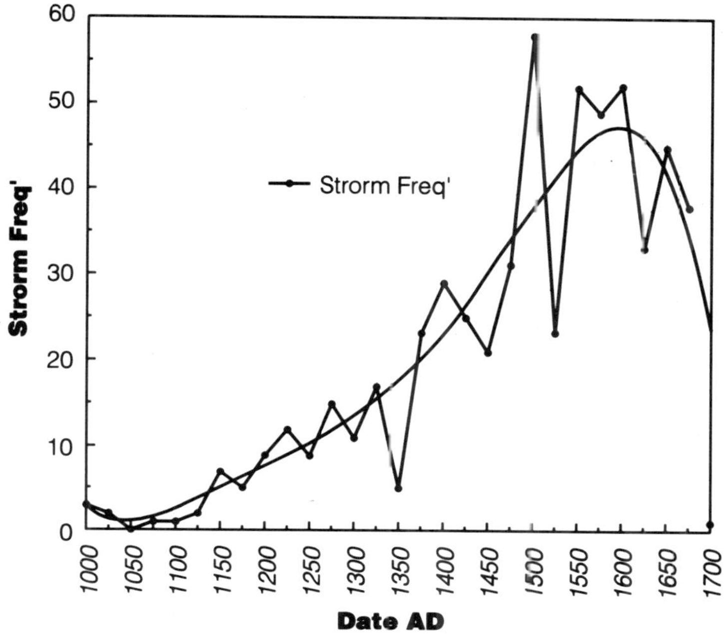


Figure 6.4 The frequency of storms in the Netherlands and northern Europe for the period AD 1000–1675 (mainly after Gottschalk 1971).

although the peak of the underlying trend is further into the sixteenth century (Gottschalk 1971, 1975, 1977).

*The digital terrain model.* Digital terrain models derived from an auger survey across Green Shiel and the adjacent ridge and furrow illustrate the topographical characteristics of the buried soil and the clay till now buried by sand (figures 6.5 and 6.6). It is therefore a visualization of the pre-sand dune environment. The above evidence suggests that Green Shiel was built adjacent to a stable, developed, soil in an environment that was relatively unaffected by aeolian processes. The lack of cereal remains from this soil and the evidence from seeds of weeds (especially *Rumex acetosa* L.), as well as phytolith data, may indicate pastoral land use. This ties in with the economic evidence from the settlement as it is presently understood. The broad width of the ridge and furrow is also an indication that it is probably no earlier than the post-conquest period (Baker and Butlin 1973: 34–5) and this alone is evidence that the Green Shiel environs were not subjected to the extreme sedimentation processes until the eleventh and twelfth centuries at the earliest.

Other evidence for an ameliorating climate during the early medieval period comes from a site nearby. Near Warkworth at Low Hauxley to the south of

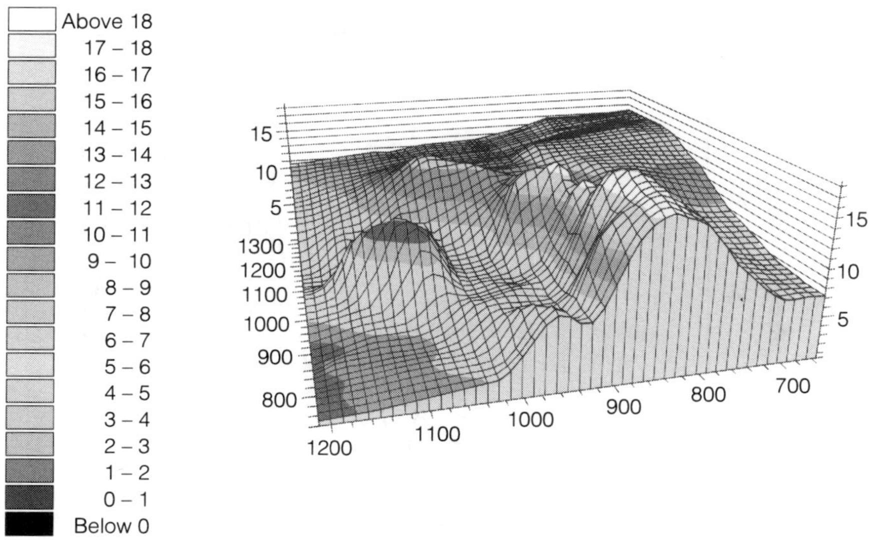


**Figure 6.5** A digital terrain model of the modern dune topography around Green Shiel.

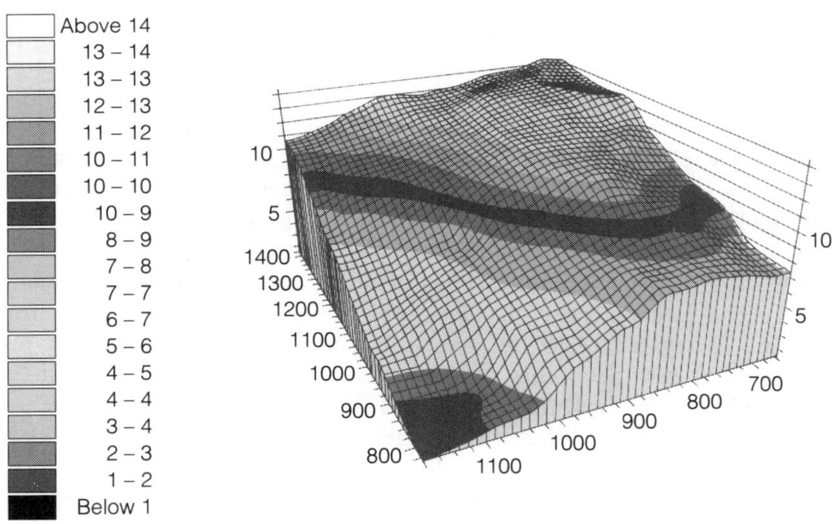


**Figure 6.6** A digital terrain model of the buried early medieval topography around Green Shiel.

Lindisfarne, the development of a dune slack is radiocarbon dated to the late tenth century (Frank 1982). This slack was subsequently covered by blown sand. On Lindisfarne itself the results of the biometric analyses of whelk assemblages also support this contention; specimens derived from archaeological contexts are significantly longer than modern examples, indicating that the shoreline was subjected to less intensive wave action during the early medieval period (Walsh 1988).

When all of this evidence is considered, there must be little doubt that the present dune system, which was clearly developing by the sixteenth century, probably did not exist during the early medieval period.

*The pollen diagram.* The final evidence pertinent to our discussion of the early medieval environment and economy on Lindisfarne is the pollen diagram. A core revealing 1.17 m of reed peat over sand and clay was taken from the middle of the Lough on the east side of the island. This is about 1400 m from Green Shiel and the edge of the ridge and furrow. The pollen diagram (figure 6.7) can be divided into three zones. The top of the diagram has very little pine and therefore pre-dates the reafforestation of the mainland, which began in the Kyloe hills at the turn of this century. Reafforestation is reflected in the sample from the moss pollster. Two bulk samples were radiocarbon dated at Belfast.

Zone 1 is dominated by Coryloid, in this case mostly hazel, and birch with low grasses and aquatics. All of the Coryloid pollen grains fall into the *Corylus* type or indeterminate categories and the local conditions, with a lack of wet-land around the site and calcareous soils both suggest that all the Coryloid pollen is derived from *Corylus* rather than *Myrica*. Willow is also present. Zone 2 sees a fall and then stabilization of Coryloid and birch and a gradual increase in grasses. The aquatics suggest the Lough had shallowed to a depth allowing colonization by emergent and submerged aquatics including water milfoil and pondweed. Dividing zones 2 and 3 is the most dramatic change in the diagram, a fall in Coryloid from 40 per cent TLP to under 10 per cent TLP and a fall in birch. There is a concurrent rise in grasses and sedges. The aquatics and sedges suggest a further shallowing of the Lough and encroachment by fringing vegetation, with the disappearance of milfoil probably because of seasonal drying out. Cereal-type grains increase, as do open-land species, such as the ribwort plantain and docks. It is interesting that alder remains approximately constant throughout the diagram.

Zone 1 is interpreted as reflecting a hazel/birch woodland probably to the north (i.e. windward) of the Lough. The only hazel on the island today is found in a few hedges and behind the dune field to the north-west. It therefore seems reasonable that the woodland was between the Lough and the Green Shiel site. The first quarter of zone 2 has been dated to AD 657–785 (UB-3585), and sees the shrinking and/or opening-up of the woodland and its partial replacement by grass and scrub. Fluctuations in the curves suggest this may have been episodic, as might be expected with human management. The change to zone 3 radiocarbon dated to just before AD 1270–1395 (UB-3646) and the open grass dominated landscape of today must represent almost complete clearance of the woodland. However, the rise in cereal-type pollen may well be just a

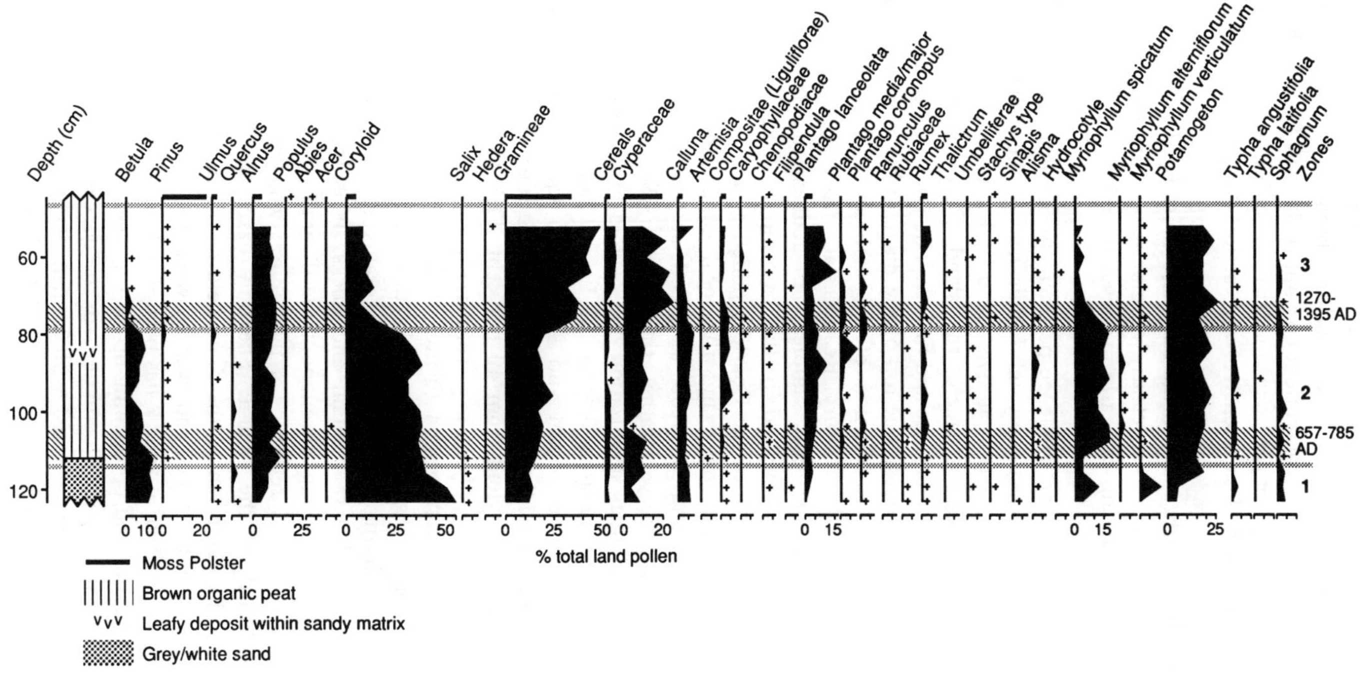


**Figure 6.7** The pollen diagram from the Lough.

reflection of the opening-up of the Lough and not necessarily any rise in arable cultivation. The background count of *Coryloid* and alder, both now rare on the island, clearly reflects a mainland input. Zone 3 can therefore be taken as reflecting the totally cleared landscape that exists on Lindisfarne today.

The near synchroneity of the earliest radiocarbon date with the arrival of the early monastic community in AD 635 and the Lough's rapid infilling is remarkable. There is no obvious physical explanation for the initiation of organic sedimentation in the Lough at this time, indeed sediments such as this are generally considerably older. In addition, this period is generally considered to have been marginally drier and warmer than the present (Lamb 1982). There is no obvious natural mechanism for the removal of older lacustrine or marsh sediments, and therefore the possibility that the Lough was artificially created must be considered. If so, the most likely reason would be the construction of a freshwater fish pond. The Lough may have been dug out of a spring depression and converted into a pond by controlling the outflow. The outflow has been controlled in recent times. The reasons why the monastic community might have done this are not hard to see, given the restrictions imposed upon the eating of meat by the church. A fishpond may have been a more secure and cheaper source of fish than sea-fishing in the early days of the monastery; alternatively such work might have guaranteed a constant water supply. Whatever the purpose, the clearance represents a significant labour investment. While this in no way alters the interpretation of the pollen diagram, it does reinforce the suggested intensive nature of agriculture and resource exploitation on the island in the eighth and nineth centuries.

The deforestation at the zone 2/zone 3 boundary is difficult to date at present, but the deposition rate in this type of environment is characteristically high (at 1.7 yr $mm^{-1}$). Since the pine rise is not present at the top of the diagram and there is no reflection of a plantation containing sycamore, alder, elder, birch, black poplar, mountain ash and willow which existed at the edge of the Lough by 1854 (Johnston 1875) it is suggested that the top of the diagram cannot be younger than the eighteenth century. There is also a sand layer at *c*.85 cm depth, which may reflect one or more major storms and progressive dune encroachment. This implies that zones 2 and 3 cover shorter rather than longer periods. Using the accumulation rate of 0.6 mm $yr^{-1}$, the deforestation and sand in-wash or in-blow would date to the thirteenth or fourteenth centuries. The reasons for removing such a valuable resource as the hazel woodland are not known, but may relate to the adoption of stone for building instead of hazel wattles. Hazelnuts were found amongst the refuse on the floor of building A at Green Shiel.

All of the palaeoenvironmental evidence combined indicates that during the early medieval period Lindisfarne was dominated by pasture with some arable, and areas of shrinking woodland. Pasture and possibly hazel woodland extended into the area now covered by dunes. There must be little doubt that the economically useful part of the island during the early medieval period extended almost as far as the north shore, as the shifting dune system would not have encroached into this area.

Although arable farming was being carried out on the island, there is no doubt that the rearing of cattle especially was an important economic activity as evidenced by faunal remains from Green Shiel (Virr 1989). The palaeo-environmental evidence, including the bone evidence, and fungal spores retrieved from the buried soil which may be associated with cattle, also highlight such practices in the northern area of the island.

*Other resources.* Lindisfarne and specifically the settlement at Green Shiel would not simply have been an important stock-rearing component of a larger estate. It would also have contributed other resources, most of which would be derived from the marine and littoral environments. There is much evidence from Green Shiel for the intensive exploitation of marine molluscs, although fish remains have not proved very abundant, possibly for taphonomic reasons.

### *Documentary sources*

Although early Northumbria in general and the Lindisfarne Community in particular are well documented by contemporary standards, the purpose of this documentation was essentially didactic and there is virtually no information in contemporary sources about land use, or even on a larger scale, about the organization of the monastic estates. The Community's title to lands and various other rights was established in later documents, especially that known as the *Historia de Sancto Cuthberto* (Hodgson Hinde 1868, Arnold 1882). However, even if these entitlements date from the late seventh century, the time of Cuthbert himself, as is frequently internally claimed, there is no information about the specific exploitation and management of resources, tillage and animal husbandry. The pre-conquest period in the north-east is badly served by the sort of documents more widely available for southern England: the charter evidence is minimal and there is of course no Domesday Book entry for Northumberland. However, it is wrong to assume that an island economy during the early medieval period would have approached some kind of self-sufficiency, or even attempted reliance, on the resources found within its own boundaries.

It is widely believed that early medieval settlement in the north of England was characterized by the multiple estate framework as identified by Jolliffe (1926) and then elaborated by Glanville Jones and others (Jones 1971, 1976; Barrow 1973; Winchester 1978; Newman 1984; Higham 1986: 288–94). Despite the debate over the origin of these estates (Gregson 1986) there is little doubt that they were an efficient way of maximizing a diverse range of economic resources from a wide geographical area (O'Sullivan 1984: 146–9). Therefore the economy on Lindisfarne itself should be considered as one unit within a wider network of economic relations which supported a much larger social body, in this case the Community of St Cuthbert and its dependants, whether centred at Lindisfarne, Chester le Street or Durham.

## THE TWELFTH TO SIXTEENTH CENTURIES

### *Non-documentary sources*

The clearance of the medieval priory unfortunately took place when the skills of archaeological excavation were in their infancy, and although the plan of the Benedictine monastery is understood, no work was undertaken on the pots or bones which might have contributed to our knowledge of the broader economic context. However, archaeological work in and around the village has also shed some light on the medieval economy (figure 6.2), although it offers a rather different perspective. Excavations by Brian Hope-Taylor and our own work at Jenny Bell's Well Midden indicate that the foreshore opposite St Cuthbert's Island was the scene of considerable activity in the past (Beavitt et al. 1990). A massive stone revetment, probably dating from the thirteenth century or earlier, was found underneath the midden, running parallel with the natural line of the cliff, and Hope-Taylor is believed to have excavated thirteenth- and fourteenth-century buildings immediately to the east of this. There are therefore some hints that the focus of the village may have changed.

Excavations have also taken place on the site of the present museum (O'Sullivan et al. 1985) and the fish bone from the late medieval levels supports the view that the fishing industry was well-developed: the bones of cod, haddock, eel, ling, roker, whiting and herring were found, and the size of fish indicates that both deep-water and inshore fishing were practised. The medieval pottery from these sites shows trading links with the Tees Valley and the port of Scarborough to the south, and inland with the Scottish borders (Bown 1985). From the late fifteenth century quantities of foreign imported pottery start to appear, although this may have at least as much to do with the decline of local industries as the success of those of the Rhineland and the Netherlands in capturing an overseas market. Holy Island's location would certainly have facilitated this trade; the impression of affluence which the presence of exotic goods often conveys is probably misleading in this instance.

### *Documentary sources: priory account rolls*

The following account of the historical evidence is derived primarily from study of the translations of medieval priory documents in the Durham archives published by James Raine (1857).

The re-foundation of the priory as a Benedictine cell of Durham towards the end of the eleventh century offers a new starting point from which to view changes in settlement and land use on Holy Island. The Durham archives contain many documents which relate to the history of the medieval priory. They are basically of two kinds: records of gifts and donations to the priory, and inventories and accounts of the possessions of the cell.

Although these records do offer some insight into economic activity on the island, they also have substantial limitations. First, they are incomplete. Although gifts of land and other rights would certainly have been carefully remembered we cannot assume that the written record, as it now survives, is intact, as not all such gifts would necessarily have been recorded in writing, particularly in the eleventh and twelfth centuries (Clanchy 1991). These centuries are not covered at all by the account rolls, which only survive from 1326. A further drawback is the fact that only a selection of the Holy Island accounts were included in Raine's study.

A second problem arises from the ecclesiastical structure of the Durham community, and also of Holy Island parish. Gifts to the Community of St Cuthbert in the later Middle Ages are in essence gifts to the Durham Chapter, which exercised a fairly tight control over its central finances. The specific benefices of the cell on Holy Island were few, and give a very restricted impression of resources. A further point of potential confusion and conflation stems from the fact that the parish of Holy Island contained four separate chapelries, those of Holy Island, Ancroft, Kyloe and Tweedmouth. The chapelry of Holy Island consisted of the constabularies of Holy Island itself, and of Goswick and Fenham on the mainland. It is only occasionally possible to be certain that references to tithes, pasturage and so on refer to the produce and husbandry of the island itself.

The priory was not a very popular billet for Durham monks, who seldom stayed more than a few years; they were also few in number – usually about a half-dozen – although the priory buildings would also have housed a number of monastic servants (Cambridge 1988). The origins of the present village are unclear and the size of the secular community on Holy Island is not known. At the end of the Middle Ages, a list of sixteenth century freehold properties in the village suggests that this was then of similar extent to the present, but the priory documents say virtually nothing about the lay population. The priory received the tithe income of the island, and had control of the church living. It also owned some property in the village.

Bearing these restrictions in mind, we can now look at some of the evidence for land use and economic activity on Holy Island provided by the documents. It must be emphasized that this is not based on systematic or comprehensive work, and relies solely on material published by Raine.

Shortage of fuel, and the necessity of purchasing this, is one fairly frequently reflected theme (table 6.1). One of the first donations specific to Holy Island Priory records the gift by one David de Houburne in the thirteenth century of 40 wain-loads of peat, to be cut from any part of the peat moss of Holburn. This gift was subsequently confirmed by his heirs and descendants on a number of occasions (AD 1273 and 1278: Raine 1857: 76). The later accounts record the regular purchase of substantial amounts of coal, as well as less costly fuel such as brushwood and heather, for both domestic consumption and some industrial use. The digging of peat at Holburn continues, as it features intermittently in subsequent account rolls.

Both of the industries which were developed on the island in the industrial revolution, fishing and lime production, are also attested on the island in medieval times (table 6.2, table 6.3). A small fishing and shipping industry was developed by the monks in the later fourteenth century, but even before this inventories of the priory stores indicate that salted fish played a substantial part in the priory diet: the very first inventory, of 1326, records the presence of 60 dried codfish. Slightly later inventories note materials for salting, and

**Table 6.1** Selected data about fuel use on Lindisfarne in the Middle Ages

I David de Houburne gave to God and St Cuthbert, of Halieland, forty four-wheeled wainloads of peat yearly, from his peat-moss of Houburne, to be taken from any part of it at their pleasure with pasturage for oxe and horse employed in their carriage, and right of road forever.

| *Date* | *Recorded details* |
|---|---|
| 1344–5 | 57½ chaldrons of coals for the brewhouse, lime kilns etc. ... Brush wood, fewel and bents bought, 43s 4d |
| 1346–7 | Fuel... 66 chaldrons of and a quarter of of coals, 104s 3d... 76 trusses of heather for the bakehouse and brewhouse, for lack of other fuel, 9s 10d... Brushwood bought at Ditchand 53s 4d... to men digging peats in Howburne Moss, 8s... |
| 1364–5 | 45 chalders of coals and 30 loads of faggots, £8 6s 8d |
| 1450–1 | Kitchen expenses... salt and fuel, £4 1s 5d |

substantial quantities of a wide range of fish, including lamprey, salmon, herring, haddock, cod and possibly ling, implying access to offshore fish. A fish tithe was collected from Tweedmouth in these early rolls (both salmon and white fish are mentioned); fish tithe was collected from Holy Island in 1361. However, it does not seem to be a constant feature, although it is regularly recorded for Tweedmouth.

The cell possibly became actively involved in the development of fishing and shipping as a response to its financial difficulties in the later fourteenth century. It became necessary to raise additional revenue to compensate for the decline of income from other sources, principally as a result of border troubles.

In 1376, the monks commissioned a boat. Two years later this boat was making a profit of 100s, as well as £10 received for corn transported south by this means. This inspired the Community to invest in another. The following year, the priory accounts note the expenses incurred by the two boats, including the wages of fishermen. Further vessels are acquired in the course of the next couple of years, and ten fishermen working during the winter months, were also paid. Tithe income remained low for some time, but the fishing and shipping venture continued at a fairly low key over the following decades; there are references to the purchase of half a ship, and some herring nets, in the 1390s, and some further investment in the early 1400s. The dubious profits from coastal wrecks are recorded at this time and the monks also tried to raise additional income from collecting boxes, possibly in an attempt to develop the pilgrimage trade. The boats owned by the priory in the fifteenth century seem to have been inexpensive and unlikely to be more than rowing dinghies, but the processing of fish within the priory enclosure is attested by the continued presence of salting materials, and a fish house is among the buildings listed just before the Reformation. In any case, it is clear that much of this activity represents participation in an already existing fishing industry. The village population must have provided the source for the fishermen and even some of the mariners required to man the boats.

**Table 6.2** Selected data about fishing on Lindisfarne in the Middle Ages

| *Date* | *Recorded details* |
|---|---|
| 1326 | In the larder... 60 dried codfish |
| 1339–40 | In the larder... 40 dried codfish and 80 haddock... |
| 1341–2 | In the larder... 50 codfish, 1,000 herrings, one quarter of salt of Poitou, a quarter of salt, and a kit containing 5 bolls of white salt... |
| 1346–7 | Kitchen expenses... lampreys, 2, 3s... codfish... herrings, 150... other white fish in abundance... salted salmon, 6, and as a necessary consequence, salt, 6½ quarters... |
| 1348 | 7 chaldrons, 3 bolls of salt, 7,800 red herrings, 198 kellinges and codlings |
| 1361 | Received for tithe fish on Holy Island, and above those taken in kind, 3s 9d |
| 1363 | Bought a boat, with oars, two anchors, one great and one small, with a cable, 61s 8d |
| 1376 | For building a ship and a boat, £17 11s |
| 1378–9 | Received for the profits of the ship this year, 100s<br>Received for corn carried to the south this year, £10<br>Expenses incurred by the ship & a new boat, £6 0s 9d |
| 1379–80 | Expenses incurred by the two boats, with the wages of fishermen, 19s 8d |
| 1380–1 | Received 106s 8d for the moiety of a ship sold profits of the ship and boat, 43s. sum of receipts, £128 5s 9d |
| 1381–2 | Bought the three parts of our ship, formerly sold by John Normamby, £6 13s 4d<br>A sloop bought of a certain scotchman with the oysters and other goods contained in it, 100s<br>A little boat or coble, bought and repaired, 16s 8d. A sail for the ship, 14s<br>To ten fishermen who fished in the boat during the winter, 24s |
| 1382–3 | Received for an old ship and sloop sold, £11 13s 4d |
| 1394–5 | Tithes of two fishing boats, 1 6s; old nets sold, 30s |
| 1396–7 | Goods obtained from a wreck by the prior's boat, sold for 30s |
| 1397–8 | Bought half a ship, with its tackle, £10 six herring nets £10 |
| 1398–9 | Sold salt fish and herrings 68s 6d... one boat purchased, and another lost upon the sea and redeemed, 29s |
| 1401 | Expended upon the boat, with nets, depyngs, and two sails, with their necessaries, £9 18s 10d |
| 1404–5 | Received for a herring boat, old, 40s |
| 1405–6 | Sold old nets, 30s |
| 1417–18 | Sold a boat, 8s |
| 1418–19 | A boat bought for the good of the house 13s 4d |
| 1425–6 | Repairs of a ship £13 6s 8d |
| 1533 | The fyshouse... 2 vats, 1 hogshead, 1 lyestonehammer, 30 large dishes... |

**Table 6.3** Selected data about quarrying and building on Lindisfarne in the Middle Ages

| *Date* | *Recorded details* |
|---|---|
| 1344–5 | Timber bought for the new chamber of the prior, 26s 2d |
| 1346–7 | Spars of oak, joists of oak, spars of fir and boards, each joist worth 2s 4d– £6 15s 6d<br>Iron Two bundles of steel containing 66 rods, 14d... 10 stone of iron of Spaingye at 7d per stone, 5s<br>10d... 18 stone of Weardale iron at 4d per stone, 5s 5d |
| 1348 | Instruments for the quarry... 2 gavelocks, 2 mells, 1 kevell, 5 wedges, 5 stoneaxes, 5 chisels, 5 small hammers for breaking limestone, an axe for cleaving wood, a hatchet... |
| 1367 | In the priory of Holy Island... 5 hatchets for cutting stones, 2 large hammers of iron, 2 hacks, 1 pulipike, 6 chisels, 2 trowels for the masons, 4 picks for scaters, 4 battering axes, 6 wooden hammers hooped with iron, ... 16 halters... one iron rake for the lime kiln... 8 limestone hammers... |

Extractive industry in the form of stone quarrying is also attested from the fourteenth century. The stone was generally used as building stone, but also burnt for lime. This extraction seems to have been, at least initially, in connection with fairly extensive building works on the priory itself, but some tools for lime-breaking are listed in the last inventory of the priory, in 1536. However, other building materials, including iron and the large timbers required for roofing posts, joists and so on had to be purchased.

Indirectly it is possible to infer that by the time of the first account rolls most of the island was already given over to pasture (table 6.4), although a systematic study of the original documents is really required here, as the extent to which the priory records reflect the pattern of the whole island is open to dispute.

First, there is virtually no sign of the paraphenalia of arable husbandry on the island itself: ploughs and grain were usually stored in the barn at Fenham on the mainland, where the monks had a mill, rather than in the priory. Such small stores of grain as were kept at the monastery seem to have been to supply the immediate needs of the bakehouse and brewhouse. Second, there is some evidence of fairly extensive pastures on the island. The recorded number of livestock in store varies considerably, and their precise whereabouts are rarely clear, but at least one entry, for 1376, refers to 20 oxen and cows for the larder, 54 pigs, young and old, and 100 muttons specifically in the island pastures. Third, the occasional purchase of straw for roofing and so on may also reflect a shortage of this commodity within the available resources, although it is perhaps more probable that straw of a particular type and quality was being sought. Most significantly, however, when tithe corn returns are available for the island itself, they suggest that only a fairly small percentage of the island acreage was under crop.

It is possible to calculate approximate acreages for crops on the basis of the value of the tithe as grain prices and crop yields are known, in general

**Table 6.4** Selected data about farming on Lindisfarne in the Middle Ages

| *Date* | *Recorded details* |
|---|---|
| 1326 | In the barn at Fenham... 30 quarters of wheat, worth per quarter, 4s... 40 quarters of barley at 3s... 70 quarters of oats, at 2s<br>At Fenham... two ploughs with all their requisites |
| 1362 | For covering the buildings and for straw bought of Thomas Bruke, 13s 8d |
| 1370–1 | At Fenham... 21 oxen, 3 coup-carts, 4 wains, 4 ploughs... 6 harrows, 2 carts with horses worth 100s, 6 chalders of wheat, 34 qualters of meslin and barley... |
| 1376 | At Holy Island, in the pastures, 20 oxen and cows for the larder, 54 pigs young and old, 100 muttons, each worth 2s |
| 1415–16 | At Holy Island in the granary, 10 bolls of wheat, 6 bolls of malt... |
| 1416 | In the granary 10 bolls of meal, 32 bolls of malt<br>In the manor of Fenham, 13 oxen... 2 ploughs... 3 collars... 3 coulters |
| 1450–1 | Corn tithes of the island recorded as 6s 8d |
| 1501–2 | Corn tithes of the island recorded as 20s |
| 1536 | Corn tithes of the island recorded as 20s |

terms, for this period. The proportion of acres given over to particular crops is the main uncertainty: barley, wheat and oats fetched different prices. Wheat was the most valuable crop, and a larger acreage was necessary to produce the equivalent tithe in oats or barley. Patterns of crop production vary substantially in different parts of the country, but there is some information available about the relative general popularity of different cereal crops in different areas. It is therefore possible to infer reasonable limits for the total acreage under crop by making a series of calculations. Work by Harold Fox (personal communication) on crop yields per bushel suggests that when the tithe was worth 20s, the maximum acreage under crop was probably about 94 acres, less than 10 per cent of the island's total area. It could have been more, if the proportion of oats was unusually high, but it could have been substantially less, if wheat formed a greater share of the crop. The earliest surviving detailed map of Lindisfarne (figure 6.8) shows an infield of *c.*48 acres in 1725, containing just the fields immediately to the north of the village; this may well reflect an arable acreage not very much less than that of three or four centuries before.

The documents provide no clue as to the date when extensive pasturage became the norm on the island. Figure 6.8 shows a simple infield/outfield system with a number of small arable fields clustered around the village. This system may have its origins in the fourteenth century. As we know nothing about agricultural practice on the island in the twelfth and thirteenth centuries, however, the possibility of continuity in land use from the early medieval period must also be considered. An assart from the waste is recorded in the early fourteenth century, probably somewhere near the Snook, but its purpose is not known. Also, at some point a population of rabbits was introduced; these

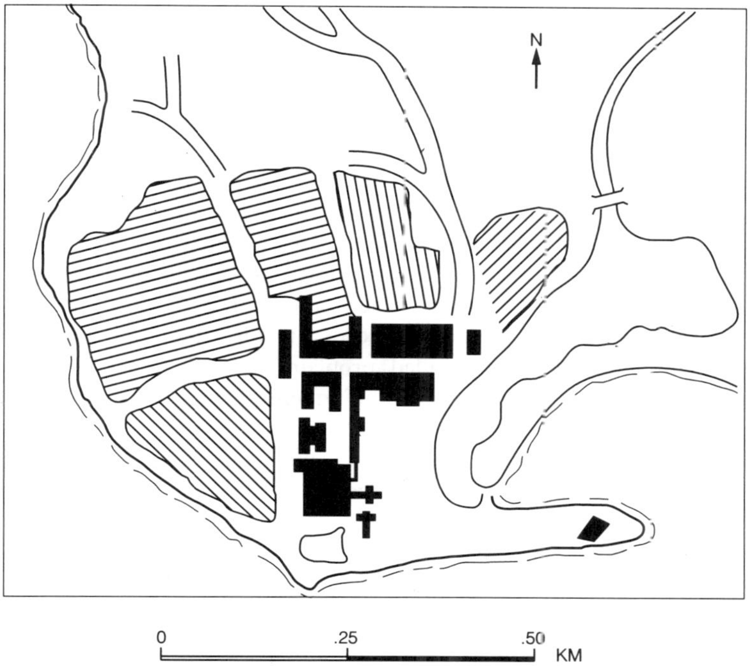


Figure 6.8 The infield and outfield system on Lindisfarne, *c*.1725.

are intermittently documented from the early fourteenth century, and they formed a source of income in late medieval and post-medieval times.

## CONCLUSIONS

Two problems lie at the core of this chapter. The first is the extent to which land-use practices on Lindisfarne were determined by local or external factors, whether these are environmental, economic, or social. The second, is the degree to which our appreciation of land use is limited, or even defined, by the kind of data which we use.

In the early middle ages we know nothing about the specifics of land use on Lindisfarne from the documents. We can only infer probable use on the basis of external analogy or make assumptions about practice based on a reading of monastic lifestyles. The archaeological and palaeoecological evidence for the early medieval period emphasizes the importance of stock-rearing as well as the exploitation of naturally occurring resources such as marine mollusca and the maintenance of restricted hazel woodland. There is no evidence for dramatic change within this period, although the date of the pollen digram from the Lough raises the possibility that, as a result of the establishment of the Anglo-Saxon monastery, the deliberate management of freshwater resources was introduced. External influences as a force for change are not apparent. However, all the general evidence for the organization of estates at this time indicates that the island would have been integrated within a much larger economic unit with which it was economically interdependent.

In the later Middle Ages it is possible to see economic practice on the island as a reflection of general trends in the later fourteenth century in England. The documents highlight the evidence for pastoralism, especially sheep farming and it is possible to see this as the local response to changes in farming practice in the wake of the Black Death and the rise in agricultural wages. The financial difficulties of the priory are clearly documented. There was a fall in rents of all kinds and a substantial drop in tithe income. These problems were common to many landlords at this time, but on Lindisfarne they were compounded by the added problem of border warfare. Short-lived responses such as the fishing and shipping enterprises, and the attempts to raise money on the reputations of patron saints, seem to be clear reflections of a fairly urgent attempt to raise cash revenue and can most easily be seen as responses to external economic, political and social factors. However, it is perhaps in the nature of documents that they highlight external, immediate and highly visible factors at the expense of the longer term processes of economic change and continuity. In the fourteenth and/or fifteenth centuries the island saw dramatic local environmental change, losing a third of its farmland to an encroaching dune system. This transformation is undocumented, unless one counts the growth of revenue from the coneygarth.

Our examination of two periods from the middle ages using primarily non-documentary sources for the earlier and documentary sources for the latter has implied a certain amount of continuity in economic practice, arable farming never having been an important element on the island itself. Without knowledge of earlier land use, the temptation to assume that external factors brought about change in the fourteenth century on Lindisfarne would prove overwhelming. However, with some insight into early medieval farming practice on the island it is clear that the picture may not be that simple. In both periods Lindisfarne was clearly a part of a wider social and economic network, not a dependance of the mainland elements of this network, but a crucial social centre and important associate within the economic system.

In any region there may be both internal and external causes of land-use change. Indeed in integrated economies they are often one and the same; a land-use change may be caused by a change in prices in response to shifts in supply and demand. Documents may detail these economic changes in terms of the value of stock, remuneration, taxes and so on (figure 6.9; O'Riordan 1981). It is not until the rise of environmental concern in the industrial age that land use is commonly recorded (or appreciated) for non-monetary purposes. Early maps only show it as part of the demarcation of ownership or stock-taking. Change is generally only recorded through its economic or social impact or cause. Because many documents are not simply ledgers of the environment but ledgers of economic interest they can frequently only be related to non-documentary evidence of land-use change through an understanding of the economics of the social system in question. In general, documentary and

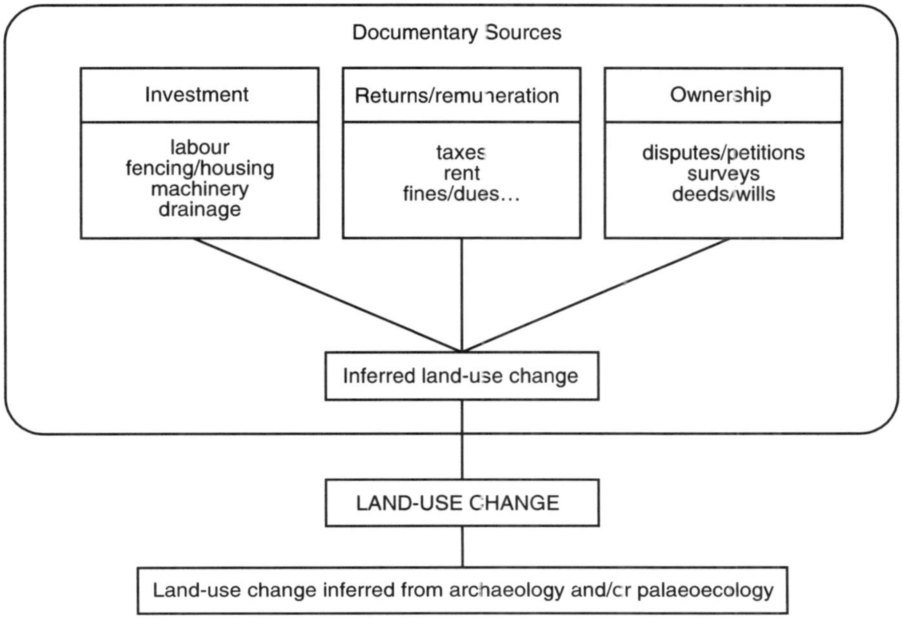


**Figure 6.9** A model of the various factors influencing the survival of different types of evidence of medieval land use on Lindisfarne.

non-documentary sources are not comparable; they can only ever be at best complementary, and may well be contradictory. The search for the documentary trail that calibrates or even validates a particular palaeonvironmental method such as a pollen diagram is a search for the Holy Grail. Documentary sources can provide possible interpretations through analogy, but this is limited and predominantly concerns only human interference. Through documenting labour requirements and expenditure we can use documents to create a 'taskscape' (*sensu* Ingold 1991) which through other means adds another dimension to the reconstruction of landscape. These limitations on the environmental precision or meaning of documents exist in addition to the myriad of social biases and deliberate inaccuracies contained within them (Dyer 1988). By integrating the study of documents from Lindisfarne with archaeological and palaeo-environmental evidence, we hope to gain a fuller understanding of the economy and social system which created the taskscape of the island.

## REFERENCES

Arnold, T. (ed.) (1882), *Symeonis Monachi Opera Omnia* (Rolls series, London, vol. 1)

Baker, A. R. H. and Butlin, R. A. (1973) *Studies of Field Systems in the British Isles* (Cambridge University Press, Cambridge)

Barrow, G. W. S. (1973) *The Kingdom of the Scots* (Edward Arnold, London)

Beavitt, P., O'Sullivan, D. and Young, R. (1990) 'Fieldwork on Lindisfarne, Northumberland 1980–1988', *Northern Archaeology* 8: 1–23

Bown, L. (1985) 'The pottery', in O'Sullivan et al. (1985), op. cit.

Cambridge, E. (1988) *Lindisfarne Priory and Holy Island* (English Heritage, London)

Clanchy, M. (1991) *From Memory to Written Record* 2nd edn (Edward Arnold, London)

Darlington, J. (1965) 'Lindisfarne, Northumberland: physiography and vegetation', Unpublished postgraduate dissertation, University College London

Dyer, C. C. (1988) 'Documentary evidence: problems and enquiries' in Astill, G. and Grant, A. (eds) *The Countryside in Medieval England* (Blackwell, Oxford) pp. 12–35

Frank, R. 1982 'A Holocene peat and dune-sand sequence on the coast of north-east England – a preliminary report', *Quaternary Newsletter* 36: 24–32

Gottschalk, M. K. E. (1971, 1975, 1977) *Stormvloeden enrivieroverstromingen in Nederland*, 3 vols (van Gorcum, Assen)

Gregson, N. (1986) 'The multiple estate: some critical questions', *Journal of Historical Geography* 11(4): 339–351

Higham, N. (1986) *The Northern Counties to AD 1000* (Longman, London)

Hodgson Hinde, J (ed.) (1868), *Symeonis Dunelmensis Opera et Collectanea* (Surtees Society, Durham, vol. 51)

Ingold, T. 1991 'Temporality of the Landscape', unpublished paper given at the Theoretical Archaeology Group Conference at Leicester, December 1991

Johnston, G. (1875) 'Our visit to Holy Island in May, 1854', *History of the Berwickshire Naturalist's Club* 7: 27–52

Jolliffe, E. A. (1926) 'Northumbrian institutions', *English Historical Review* 41: 1–42

Jones, G. R. J. (1971) 'The multiple estate as model framework for tracing early stages in the evolution of rural settlement' in Dussart, F. (ed.) *L'Habitat et les Paysages Rureaux d'Europe*, Les Congrès et Colloques de l'Université de Liège, LVIII (Liège, 1971) 251–67

Jones, G. R. J. (1976) 'Historical geography and our landed heritage', *University of Leeds Review* 19: 53–78

Lamb, H. H. (1982) *Climate, History and the Modern World* (Methuen, London)

Newman, R. (1984) 'The problems of rural settlement in Northern Cumbria in the Pre-Conquest period', in Faull, M. (ed.) *Studies in Late Saxon Settlement* (Oxford University Department for External Studies) 155–76

O'Riordan, T. (1981) *Environmentalism* (Pion, London)

O'Sullivan, D. (1984) 'Pre-Conquest settlement patterns in Cumbria', in Faull, M. (ed.) *Studies in Late Saxon Settlement* (Oxford University Department for External Studies) pp. 143–54

O'Sullivan, D. (1990) 'The plan of the early Christian monastery on Lindisfarne: a fresh look at the evidence' in Bonner, G., Rollason, D. and Stanncliffe, C. (eds) *St Cuthbert, his Cult and his Community to AD 1200* (Boydell and Brewer, Woodbridge) pp. 125–42

O'Sullivan, D. and Young, R. (1991a) 'The early medieval settlement at Green Shiel, Northumberland', *Archaeologia Aeliana*, 5th Series 19: 55–70

O'Sullivan, D. and Young, R. (1991b) 'The early medieval settlement at Green Shiel, Lindisfarne: an interim report on the excavations, 1984–91', *Archaeology North* 2: 17–21

O'Sullivan, D. et al. (1985) 'An excavation in Holy Island village, 1977' with contributions by E. Allison, M. Archibald, L. Barn, A. Donaldson, D. Huggins, A Locker and J. Rackham, *Archaeologia Aeliana* 5th Series 13: 27–116

Ranwell, D. (1972) *Ecology of Salt Marshes and Sand Dunes* (Chapman and Hall, London)

Raine, J. (1857) *The History and Antiquities of North Durham* (G. Andrews, Durham)
Virr, M. (1989) 'An investigation into some selected cattle remains from Green Shiel, Lindisfarne', unpublished B.A. dissertation, University of Leicester
Walsh, K. (1988) 'An analysis of the marine mollusca from the Green Shiel site on Lindisfarne (Holy Island)', unpublished postgraduate dissertation, Department of Archaeology, University of Leicester
Winchester, A. J. L. (1978) 'Territorial structure and agrarian organisation in medieval and sixteenth century Copeland, Cumberland', unpublished doctoral dissertation, University of Durham

7

# A Rough and Cold Country: Farming and Landscape Changes in Northern Snowdonia Since *c*.1600

Mark Brayshay and Andrew Williams

ABSTRACT

Employing manuscript records and field investigations, this chapter reconstructs the historical development of an area of northern Snowdonia in Wales, which was part of the vast Vaynol Estate for almost three centuries. The changing pattern of farms and farming is reconstructed and the profound influence of the physical environment on the processes of change is considered. Farm holdings characteristically stretched in linear fashion from the base of the slope to the mountain summits, thereby occupying a share of both the better and the poorer quality land. During the eighteenth and nineteenth centuries, population growth and industrial expansion caused the number of farms to increase; many were exceptionally marginal. The dual 'quarrying–farming' economy had collapsed by 1900 and the pattern of farms has since been rationalized, but the imprint of the past remains. Case studies are made to show the intimate relationship between the physical environment and farming, past and present.

## INTRODUCTION

The appearance of the present-day landscape of northern Snowdonia owes much to the processes of change which have operated during the past three or four centuries. Despite its apparent immutability, a series of influences have combined to bring about reorganizations in the pattern of settlements, farms and fields in this mountainous and remote area of Wales. Indeed, while vestiges of the earlier medieval landscape remain, the human geography of the area is largely a product of the modern era and reflects decisions made by owners and occupiers, past and present, in response not only to the physical capability of the land for agriculture and primary extractive industry, but also to profound

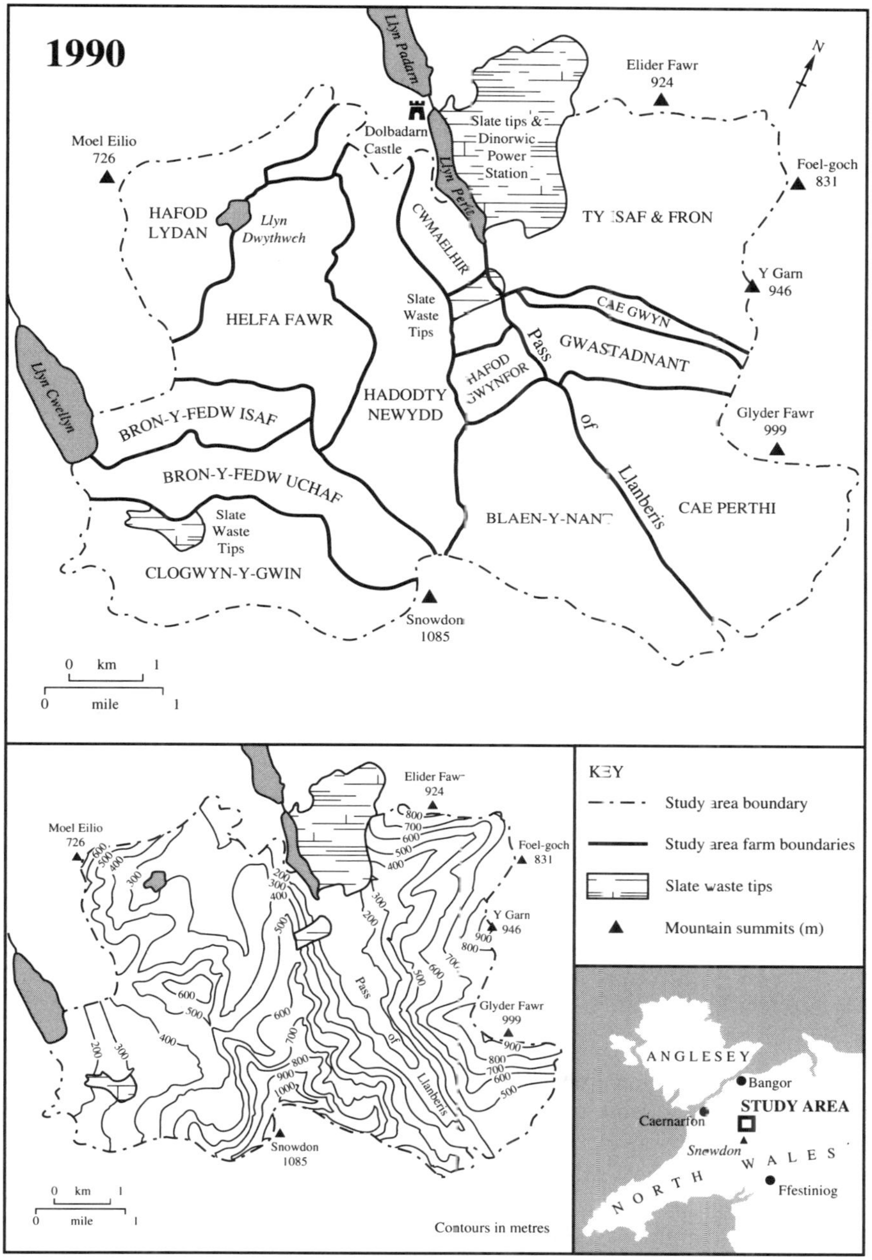


Figure 7.1 The study area, northern Snowdonia, Wales. The modern pattern of farms is shown in relation to relief.

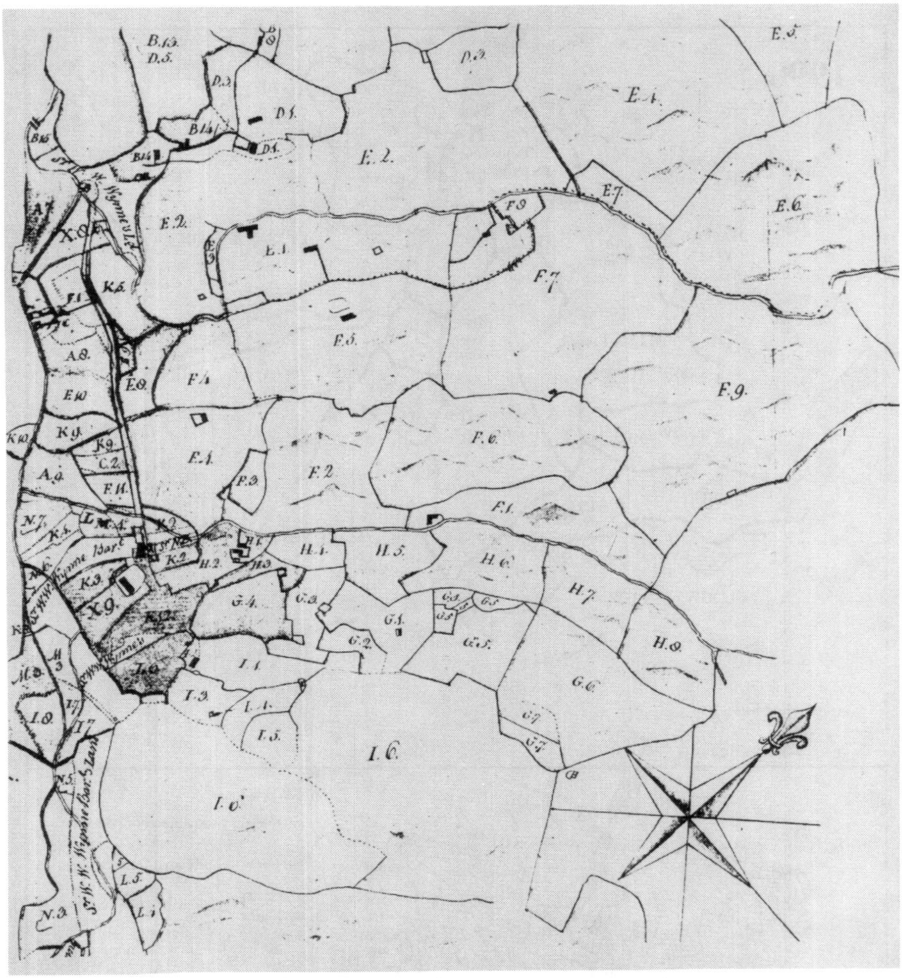

**Plate 7.1** An extract from Vaynol MS 4055: *A Survey of the Vaenol Estate*, Vol. I, 1777 (reproduced by permission of the Gwynedd County Record Office).

shifts in local economic, demographic and social circumstances. The aim of this chapter is to examine the variety of factors which together have influenced the decisions taken about landholdings and which have therefore helped to shape the appearance of this part of the Snowdonia landscape. A key objective is to explore the links between the process of human occupation and development of this marginal, upland area and the detailed characteristics of the physical landscape. Attention will be focused on an area comprising the great natural 'fortress' of land encircled by the five mountain peaks of Yr Wyddfa (Snowdon), Glyder Fawr, Y Garn, Elider Fawr and Moel Eilio. It includes the dramatic Pass of Llanberis and, indeed, lies mostly within the parish of Llanberis, although a small portion extends into the neighbouring parish of Bettws Garmon (Figure 7.1). Until 1967 this area was part of the vast Vaynol

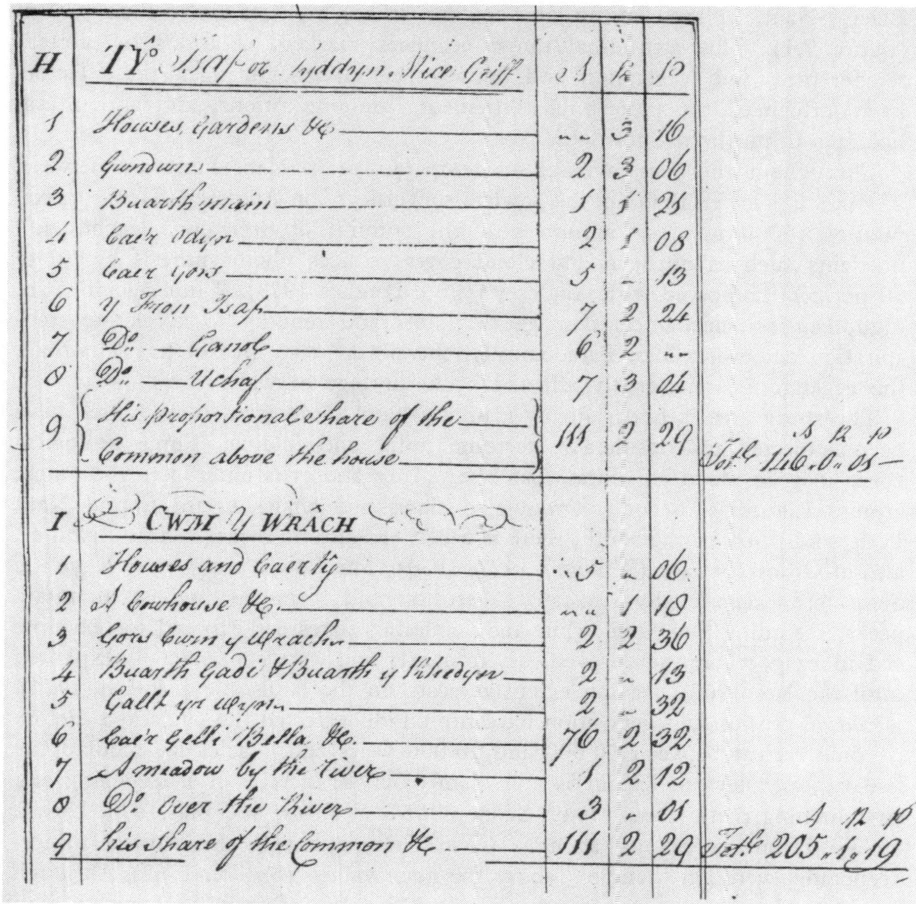

| | | A | R | P | |
|---|---|---|---|---|---|
| H | TY Asaf or Lyddyn Nico Griff. | | | | |
| 1 | Houses, Gardens &c. | .. | 3 | 16 | |
| 2 | Gwndwn | 2 | 3 | 06 | |
| 3 | Buarth main | 1 | 1 | 25 | |
| 4 | Cae'r Odyn | 2 | 1 | 08 | |
| 5 | Cae'r Gors | 5 | – | 13 | |
| 6 | y Fron Isaf | 7 | 2 | 24 | |
| 7 | Do. — Ganol | 6 | 2 | .. | |
| 8 | Do. — Uchaf | 7 | 3 | 04 | |
| 9 | His proportional share of the Common above the house | 111 | 2 | 29 | Totl 146.0.05 |
| I | CWM Y WRACH | | | | |
| 1 | Houses and Gaerty | .. | 5 | .. 06 | |
| 2 | A Cowhouse &c. | .. | 1 | 18 | |
| 3 | Gors Cwm y Wrach | 2 | 2 | 36 | |
| 4 | Buarth Gadi & Buarth y Kiledyn | 2 | – | 13 | |
| 5 | Gallt yr Wern | 2 | – | 32 | |
| 6 | Cae'r Gelli Bella &c. | 76 | 2 | 32 | |
| 7 | A meadow by the River | 1 | 2 | 12 | |
| 8 | Do. Over the River | 3 | – | 01 | |
| 9 | his Share of the Common &c. | 111 | 2 | 29 | Totl 205.1.19 |

**Plate 7.2** A further extract from Vaynol MS 4055: *A Survey of the Vaenol Estate*, Vol. I, 1777 (reproduced by permission of the Gwynedd County Record Office).

estate, and its recent landscape history may therefore be recovered in considerable detail by studying *inter alia* the surviving surveys and other records of the Vaynol properties (Parry 1970, Roberts 1973; and see plates 7.1 and 7.2). Evidence drawn from these remarkably comprehensive documentary records will be linked in this study to findings derived from a detailed field survey of the soils, vegetation, topography and present-day farms of the area.

## THE PHYSICAL LANDSCAPE

In this area, relief exerts a fundamental control on all aspects of the physical environment. Great variations in altitude occur within very short distances and the landscape is dominated by steep slopes and cliffs. Thus, for example, over a distance of less than 2.5 km from the floor of the pass of Llanberis at

Blaen-y-Nant, to the summit of Snowdon, the land rises more than 930 m (figure 7.1). This attitudinal range produces marked variations in rainfall, temperature, soil, vegetation and drainage conditions. In turn, these factors have influenced the pattern of settlement, communications, and agricultural holdings in northern Snowdonia.

Throughout the year, average air temperatures in Nant Peris are between 2° and 5° lower than in the lowlands of Lleyn or Anglesey. Heavy precipitation (5000 mm per annum near the summit of Snowdon) leaches soil nutrients such as nitrogen and cloud cover reduces photosynthesis by up to 30 per cent compared with Anglesey (Elfyn Hughes 1973). Wind velocities are high, causing mechanical damage to plants and tending to accelerate transpiration and water loss. Paradoxically, therefore, in an area of heavy rainfall, the vegetation is frequently affected by a shortage of water.

The study area is underlain by Cambrian and Ordovician rocks which have been subjected to considerable tectonic uplift and folding. Three geological zones may be identified. First, Cambrian slates and grits outcrop to the north, around Llanberis. Second, Ordovician slates and shales occur around Nant Peris, and third, to the south, there is a belt of volcanic rocks, mainly rhyolites, also of Ordovician age (Williams 1927, 1930). The Lower and Upper Rhyolitic Series are exceptionally resistant to erosion and form the highest mountain peaks, including Snowdon, while the Cambrian slates are exposed for example at Dinorwig, where they proved easy to quarry and were commercially exploited until the late 1960s, leaving extensive scars on the landscape (Lindsay 1974).

Glacial erosion and deposition have profoundly affected the landscape. Before its final retreat between 50 000 and 10 000 years ago, glacier ice scoured the two major valleys of Llanberis and Nant-y-Betws, thereby over-deepening and straightening them (Lewis 1949, Seddon 1957, Whittow and Ball 1970, Unwin 1973). Tributary valleys, such as Cwm Gafr, although also subject to ice-deepening, were left 'hanging' above the main valleys (Seddon 1962). Although the ice action was predominantly erosional, patches of drift do occur at high levels. There are late-stage moraines across the mouth of many of the corries and some sediment was brought down the valleys by later fluvial action (Unwin 1975). Periglacial processes were also active, producing extensive scree slopes in some sections of the Llanberis Pass (Ball and Goodier 1970).

The great variation in soil types in the study area reflects the marked changes in local climate and relief (Ball 1968, Ball et al. 1969). Brown earths, the most fertile soils, occupy the valley bottoms around Nant Peris and Nant-y-Betws. Large spreads of alluvium cover the fluvioglacial boulders in the valley floors, while drift material is encountered at the base of the valley sides. These silty loam soils are well-structured and well-drained, but generally shallow at only 30–40 cm deep. Small pockets of brown earth soils are occasionally encountered at high altitudes in the study area, probably associated with base-rich volcanic ash deposits, and picked out by farming communities for exploitation and settlement in otherwise unattractive marginal locations.

The relatively high base status of the valley floor soil supports *Agrostis–Festuca* grassland, but the fields have mostly been somewhat improved and are these days sown with Italian ryegrass (Perkins 1978). Gleyed soils, adjacent to the major rivers, tend to be rush-infested, especially by *Juncus effusus* and *Juncus squarrosus*. Acid brown earths, although extensive elsewhere in Britain's uplands, are relatively rare in Snowdonia. They occur, however, in the study area not only on steep slopes of 20–25° above Nant Peris, at altitudes between 225 and 300 m, where they have formed in slaty drift, but also to the north-east of Gwastadnant, at an altitude of 600 m, where they are associated with pockets of base-rich drift. These acid brown earths are characterized by a humose topsoil and an ochreous coloured Bs horizon, both with a silty-loam texture. Although bracken infestation is a considerable problem in such areas today, these soils are also capable of supporting *Agrostis–Festuca* grassland. Peaty podzols occur at higher altitudes on the slate geology, where they are associated with a relatively thick blanket of glacial drift. They support a vegetation community dominated by purple moor grass, or *Molinea caerula*. At altitudes between 325 and 600 m, on moderate slopes of 15–20°, the surface layer of peat is characteristically 10–15 cm in depth. Here the natural sward is dominated by mat grasses (*Nardus stricta*) with some *Molinea caerula*. The grazing is poor and lacks nutrients and thereby offers a very low livestock carrying capacity. Writing in 1846, Thomas Rowlandson recognized the constraints that the physical environment imposed on farmers in the area which obliged them to focus exclusively on rearing livestock. Snowdonia, he observed,

> is essentially a pastoral district... there are many farmers... who do not grow one single ounce of grain: nay, I was assured that some did not even cultivate a potato-patch (Rowlandson 1846: 566).

The Snowdon mountain sheep farms were small and impoverished. Rowlandson advised that liming would have been 'beneficial', but the farmers were 'too poor to pay for the lime' (Rowlandson 1846: 569).

## LANDLORDS AND TENANTS: THE EVOLUTION OF OWNERSHIP AND OCCUPANCY OF THE VAYNOL ESTATE

Before the conquest of Wales by Edward I in the thirteenth century, the Pass of Llanberis was part of the ancient township of Dinorwig, belonging to the native princes of Gwynedd. Approximately 30 miles in circumference, Dinorwig appears to have been co-extensive with the parishes of Llanberis and Llandeiniolen and was probably the largest unit of its kind in north Wales (Jones Pierce, 1940). In the period before 1282, the area was occupied by bond or 'boon' communities subject to the prince and, even after his property escheated to the English crown, the bondmen remained on two-thirds of the Dinorwig lands, tied to the soil in their reckoned land *tir cyfrif* settlements (Roberts 1949, Jones 1964, Roberts 1977). The remaining third of the property was described in 1352 as 'freehold in the crown' and was granted separately to a lessee who, in turn, let individual holdings to tenants. However, after the Edwardian conquest, the bondmen of Gwynedd frequently petitioned the king for freedoms and rights comparable with those prevailing elsewhere under

English law, as well as a general release from their service obligations (Beverley Smith 1966). Henry VII finally granted a Charter of Liberties in March 1507, emancipating the bondmen and granting them the right to buy and sell land. Gavelkind (partible inheritance) was replaced by primogeniture by letters patent in October 1504. However, when English burgesses residing in the walled towns of north Wales objected to the enfranchisement of the bondmen, the king's council, meeting in February 1509, rescinded all the liberties granted to the Welsh since 1504. The English burgesses and the Welshmen were required to appear before the king to resolve their dispute but, unfortunately, Henry VII died before this crucial meeting took place. The status of the Charter therefore remained uncertain and gavelkind persisted until well into the sixteenth century (Beverley Smith 1966).

Although as early as the reign of Henry IV the freehold portion of Dinorwig had been leased from time to time to 'fermiers', in 1535 the crown finally decided to grant a lease on the entire property for a term of 40 years to Henry Webb. The resolution of legal disputes regarding the rights of the Welsh tenants was thereafter a matter for the crown lessee and not for Henry VIII himself. In 1564, some years before Webb's lease expired, Elizabeth I re-granted the property for 21 years to Thomas Wynn ap William (Thomas Williams), a younger son of the Cochwillan family who, in 1558, had already inherited former episcopal lands, including Maenol Bangor (Vaynol), from his uncle. Indeed, it is probable that before 1564 he had also acquired an interest in part of Dinorwig as one of Webb's sub-tenants (Jones Pierce 1940, Barnes 1970). In any case, in 1571, the queen's grant was changed again and converted to a lease for three lives in succession, to Thomas, William and Simon Williams.

During the next 50 years the Williams family began a ruthless remodelling of the Vaynol lands, including those in Dinorwig. A number of tenants with theoretical ancient rights of succession were peremptorily evicted and replaced by others who were obliged to pay substantial entry fines and rentals. The fragmentation of holdings produced by gavelkind was thus gradually eradicated and the last vestiges of the pre-conquest bondlands were removed. While legal suits brought by displaced tenants in the 1590s delayed the process, by 1627 the modernization was complete; a new network of farms was in place, and a major agrarian reorganization had been effected in northern Snowdonia.

In 1696, when the last surviving lessee, Sir William Williams, died childless, the Vaynol Estate was entailed and bequeathed as a life interest to Sir Bourchier Wrey and his two sons. On the expiry of these three named lives, the property was to revert to the crown. Between 1696 and 1756, when Sir Bourchier Wrey's second son, the Reverend Chichester Wrey, died, Vaynol was subject to over-exploitation and stagnation. Holdings were put out at rack rents, timber was indiscriminately felled, and neither repairs nor improvements were carried out (Aris 1977). However, in anticipation of its reversion, the crown granted the property to John Smith of Tedworth in Hampshire, in recognition of his service as Chancellor of the Exchequer in the reign of William III and as Speaker of the House of Commons for the first two Parliaments of Queen Anne. Although Smith did not himself survive to enjoy the fruits of the crown's generosity,

Table 7.1 The Vaynol Estate and farm units in the study area, 1696–1967

| *Date* | *Total area of Vaynol Estate (ha)* | *Annual rent income yielded by the Estate* | | | *Number of farm units in the Llanberis and Bettws Garmon[a] study area* |
|---|---|---|---|---|---|
| | | £ | s | d | |
| 1696 | –[b] | 2 410 | 4 | 11 | 12 |
| 1777 | 12 116.8 | 12 132 | 4 | 8 | 27 |
| 1832 | 15 238.1 | 13 527 | 15 | 8 | 34 |
| 1869 | 15 454.0 | 19 901 | 1 | 5 | 51 |
| 1967 | 8 731.8 | | –[c] | | 15 |

[a] Land in Bettws Garmon was added to the Estate by 1777.
[b] The exact size of the Estate in 1696 is unknown.
[c] The Estate was valued in excess of £1 million.

*Sources*: Gwynedd Record Office: Vaynol MSS 3643, 3751, 4055, 4063, 4067, 4202, 4195, and 4427. Vaynol MS XSC/807 Sale Catalogue. *The Times*, 14 July 1967 and 21 July 1967.

the Estate ultimately passed to his son and heir, William Smith, who controlled it until 1764. Vaynol was then bequeathed to William's nephew, Thomas Assheton of Ashley in Cheshire, who subsequently took the surname Assheton-Smith and inaugurated a period of positive and innovative management. An Estate survey was carried out in 1777, new 21 year leases were introduced incorporating improvement clauses, considerable extra land was gradually acquired, and the slate-quarrying industry was substantially developed (Vaynol MSS 4055, 4057, 4063 and 4066). Thomas Assheton-Smith died in 1828 and was succeeded by his son of the same name. A new Estate survey was then made in 1832 which indicates that the property was 26 per cent larger than it had been in 1777 ((Vaynol MSS 4067 and 4376; see table 7.1). In addition to slate quarrying, output from copper mines in Nant Peris added to the Vaynol income in the 1830s (Bassett and Davies 1977). Indeed by 1843, Vaynol was wealthy enough to lend money to the Corporation of Caernarfon and to become mortgagee of extensive urban property (Roberts 1973, Aris 1977). The Estate built a railway to link the slate quarries with Port Dinorwig and, when the second Thomas Assheton-Smith died in 1858, Vaynol was widely known as 'slate country' where 2400 men were employed in the quarries. Output from Dinorwig had risen from 20 000 tons per annum in 1826, to more than 100 000 tons by the early 1860s (Lindsay 1974, Bassett and Davies 1977).

When a third great survey was made in 1869 on the orders of the then owner, George William Duff Assheton-Smith, Vaynol had reached its zenith in size and prosperity owning 15 454 h (38 158 acres) of land spread across 27 parishes from the Vale of Conway to the Lleyn peninsula and extending into Anglesey (Vaynol MSS 4108 and 4195). However, the tenant farmers were severely affected by the agricultural depression of the 1870s and rents quickly fell into arrears. A slump in the slate industry in the 1880s and 1890s made matters worse, and out-migration led to a decline in population. Thus, by the beginning of the twentieth century the Estate was no longer profitable. In 1914 land near Caernarfon, as well as in Pwllheli and Anglesey, was sold. Some 3078 h (7600 acres) had been shed by 1919. During the interwar period, Vaynol was maintained in reasonably good order but only at great expense, and by the 1950s the problems of dereliction of buildings, neglected drainage, vacant smallholdings and impoverished soils required drastic action (Richards 1961). A new agent was appointed in 1956 and, together with the owner, Sir Michael Duff, a radical modernization and rationalization plan was devised (Stuart 1961). In common with practices occurring elsewhere in Wales, Vaynol farms were amalgamated to form economic units, smallholdings were extinguished, rents were doubled, and new tenancy agreements were devised (Parry 1985). However, notwithstanding the optimism expressed in the early 1960s, the new arrangements proved inadequate and, in 1967, the remaining 8732 h (21 560 acres) of the Estate were sold privately to Broadland Properties of Scarborough. The Ministry of Agriculture acquired Snowdonia in 1969 and since then the farms have been re-sold individually, mostly to their tenants (Parry 1970). Those within the borders of the Snowdonia National Park have also entered into agreements with the government regarding management and access (Parry 1970).

## THE ESTABLISHMENT OF PERMANENT FARMS

Evidence of prehistoric settlement survives in the area as hut remains at Cwmdwythwch, but little is known for certain regarding the character of any associated farming (figure 7.2). Farms and settlements permanently occupied throughout the year do not appear to have existed in the mountains of Snowdonia until the medieval period. The 1352 *Record of Caernarvon* mentions four 'havotries', or summer-season farms, located near to the prince's stronghold of Dolbadarn Castle. Two of these can be related to the remains of medieval long-houses at Cwmbrwynog and Cwmdwythwch, located on moderate slopes in the largest single block of land below 300 m in the area. While these sites seem well-suited to exploit the short growing season at high altitude, by the thirteenth century climatic deterioration probably meant that even these more carefully chosen high-altitude locations were too poorly drained, cold and exposed to permit year-round occupation (Parry 1978, 1981; Wigley et al. 1981). Increasing summer wetness and declining temperatures were significant factors in reducing the potential for successful cropping in upland areas (Bell and Ogilvie 1978, Ingram 1978, Lamb 1982, Campbell 1991). However, by the mid-fourteenth century, there is evidence of farmers from the lowlands customarily moving their sheep, goats and cattle into the mountain pastures of northern Snowdonia on the kalends (1st) of May, thereby allowing the meadow grass to grow in the fields of the *hendref* (lowland farm) and a hay crop to be taken to supply winter fodder when the livestock returned there in September (Roberts 1959, Elfyn Hughes 1973). The practice paralleled the shieling system operated in upland areas of northern England and in the Scottish highlands (Miller, 1967, Whittington 1973, Roberts 1977). Transhumance of this kind

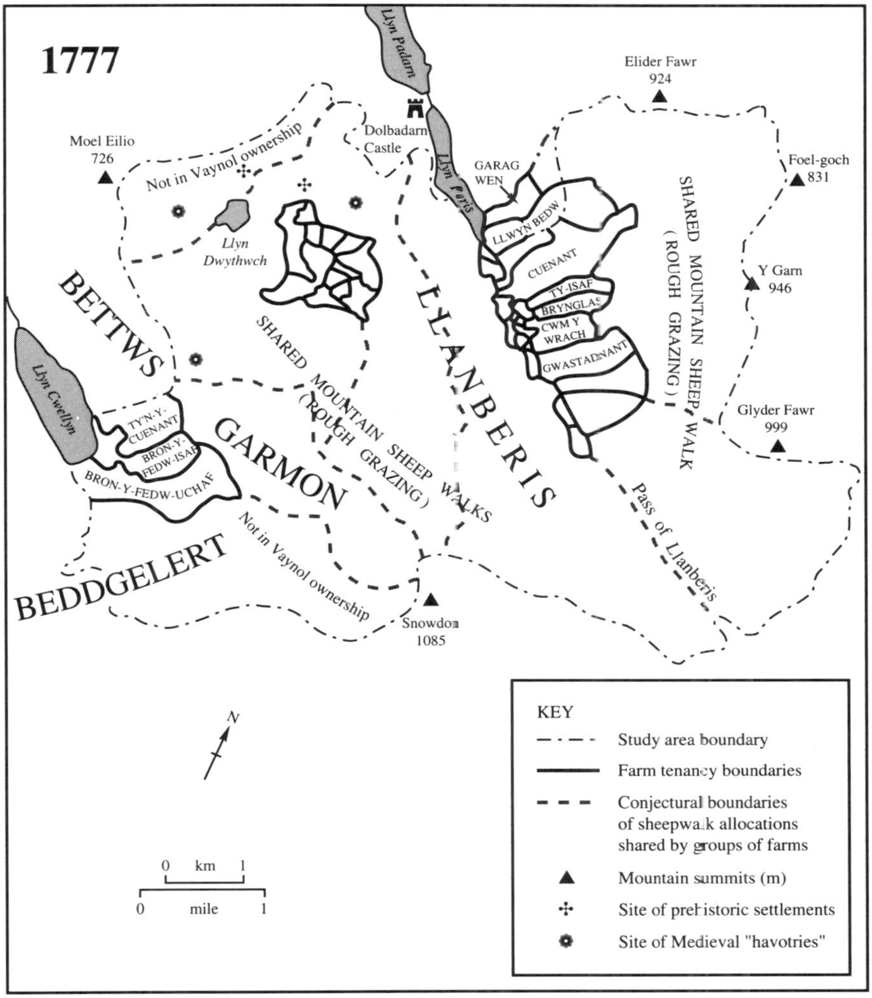


**Figure 7.2** The pattern of farms in the Pass of Llanberis, Cwmdwythwch and Bettws Garmon, 1777 (based on Vaynol MS 4055, Gwynedd County Record Office). In-bye land, *ffridd* and mountain sheepwalk are shown.

depended on plentiful land resources in both the lowlands and the uplands. However, in medieval Wales the partible inheritance system – gavelkind – led to the subdivision of lowland holdings until the units were too small to support a family. At that point, independent farms were often created in the uplands, severed from their former lowland counterparts. Although the precise date when such permanent farms were established in northern Snowdonia is not known, 12 holdings listed in that part of the Vaynol rent roll of 1696 which refers to the study area can be identified as permanently inhabited holdings arranged along the lower slopes of the Pass of Llanberis. In view of the litigation already noted, instituted by the 'ancient' tenement holders of Dinorwig following their eviction and replacement on the orders of the Vaynol Estate in the period after 1564, it is reasonable to assume that the farms listed in the 1696 rent roll were established at least by the mid-sixteenth century and probably some years earlier.

The survey of the Vaynol Estate made for Thomas Assheton-Smith in 1777 includes maps of every farm then located in the study area. All 12 of the 1696 permanent holdings can be identified, together with a further five occupying similar adjacent sites in the Pass of Llanberis (figure 7.2). In contrast to the medieval 'havotries' which avoided the badly drained bottom land of Llyn Dwythwch, each farm in the Pass was allocated a portion of 'in-bye' land on the floor of the valley where a hay crop, or potatoes, could be cultivated. In those few cases where a holding did not extend as far as the valley bottom, a separate area of *tir coch* (brown soil) was assigned to it elsewhere. Above their in-bye fields, all farms had a portion of *ffridd*, or holding land, on the lower slopes of the mountains where livestock could graze in early spring, allowing the in-bye meadows to recover and begin the production of hay grasses well before the commencement of the brief growing season of the pastures on the unenclosed 'sheepwalks' at higher altitudes. This mountain grazing was held in common by groups of farms, each assigned a share of sheepwalk in proportion to their enclosed land below (Rowlandson 1846, Dodd 1926–7, Roberts 1973). In a pattern similar to that observed in other upland areas, the farms of northern Snowdonia therefore characteristically stretched in linear fashion from the valley bottom to the mountain summits (Eyre 1957, Hodgson 1969, Roberts 1977). The relationship between the three '*management units*', local topography and edaphic conditions was extremely close. Thus the *in-bye fields* occupied the brown earth soils associated with the base-rich valley floor alluvium and the glacial drift material on the lowest slopes of the mountain. Where poor drainage produced gleying, a network of ditches could yield notable improvements and the brown soils responded well to applications of manure (Hyde-Hall 1809–11, Rowlandson 1846). The *ffridd* was located primarily on brown podzolic soils overlying glacial drift deposited on moderate to steep slopes immediately above the valley floor and extending to 325 m (about 1000 feet). In hollows, and on flatter parts of the slope, poorly drained peaty gleys occur. Both are relatively deep soils which respond to drainage and fertiliser application. There is evidence that the band of brown podzolic soils still supported a significant tree cover until the land was leased to the Wrey family in the period after 1696. Thereafter, the remaining useful timber was felled, sold, and not replaced by tenants obliged by their landlord to meet high rents. Above the *ffridd*, the *sheepwalk* occupied areas of mainly peaty podzols, peaty gleys and peat rankers. The mountain soils support only a poor and nutrient-deficient sward with an extremely brief growing season and low carrying capacity. In the area as a whole, the growing season lasts less than 200 days from May until October. Above 325 m it is even shorter (Elfyn Hughes 1973; Perkins 1978).

Studies elsewhere indicate that the enclosure process was generally under way in North Wales by the mid-eighteenth century (Edwards 1963). Vaynol documents show that all the lower land of northern Snowdonia had been enclosed by 1777 and the entire sheepwalk was allocated. In addition to those in the Pass of Llanberis, another seven permanent farms had also been established near Llyn Dwythwch, above the marshy floor of the *cwm*. Although used as a *hafodydd* in medieval times, this area was extremely marginal as a location for permanent settlement. Indeed, a Vaynol Estate rental of 1771 shows that segments of meadow located 1.5 km away adjacent to Llyn Padarn had been assigned to the Dwythwch farms to enable a supply of winter livestock fodder to be cultivated. Transhumance from the lowlands to this part of Snowdonia must therefore have been entirely squeezed out by this time and the process had probably commenced much earlier during the sixteenth-century reorganizations effected by Thomas and William Williams. Thus, by the later eighteenth century, the agricultural viability of tenancies in the study area depended critically on the balance between livestock numbers and the winter-fodder production of the in-bye land (Elfyn Hughes 1973). This was noted by Edmund Hyde Hall during his visit to Llanberis in 1810:

> Even for hay there appears but small room, and the quantity of stock which can be kept through the winter must therefore be proportionately limited (Hyde-Hall 1809–11: 185).

Although Thomas Assheton-Smith and his successors encouraged improvements, the scope to challenge the fundamental geographical controls was small. Assheton-Smith described his property as 'a rough and cold country; a tract of bogs and rocks': and these physical constraints directly expressed themselves in the structure and appearance of the human landscape. Severe limitations on farming continued in this cold and remote location (Williams 1941). Despite its ruggedness, however, northern Snowdonia has always been a fragile landscape, easily damaged by over-exploitation. During the course of the nineteenth century, largely in response to the pressure of population growth and the development of smallholdings for slate quarrymen, the number of farm units established in the study area almost doubled; many possessed little or no in-bye land and were therefore obliged to cultivate hay crops in the fields of the *ffridd* in a vain effort to sustain livestock during the winter.

## POPULATION GROWTH AND ITS IMPACT ON THE LANDSCAPE

A five-fold increase in population occurred in Llanberis parish between 1801 and 1881 (figure 7.3). While many slate workers came from outside the area and lived in barracks at the quarry, considerable numbers were local men. Extractive industries thus provided for a 'dual economy' whereby farming could be combined with mining and quarrying. The manuscript census enumerators books, dated 1841–81, show that many households contained both farmers and quarrymen (Census 1841, PRO HO 107/1393; Census 1851, PRO HO 107/2515; Census 1861, PRO RG 9/4330; Census 1871, PRO RG 10/5716; Census 1881, PRO RG 11/5566). Moreover, in order to provide homes and added income as an encouragement to the industrial workforce, Vaynol created many

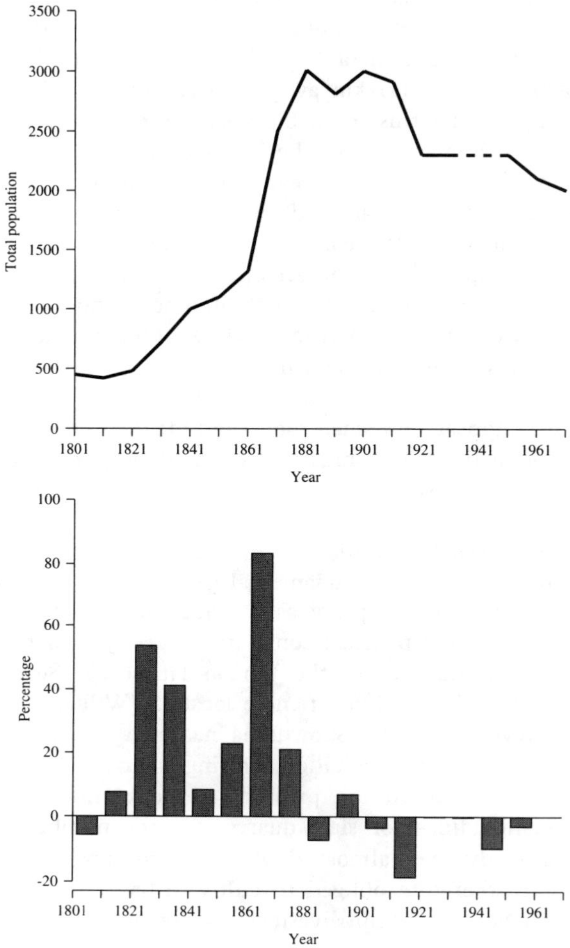


**Figure 7.3** Population growth in Llanberis Parish, 1801–1971. The upper graph shows absolute change; the lower graph shows decennial percentage change.

new crofts and permitted the building of industrial cottages on the land of existing agricultural tenancies in the Pass of Llanberis. The number of farm units grew from 34 in 1832, to 51 by 1869, incidentally yielding a 47 per cent increase in rent income for the Estate (table 7.1, figure 7.4). Expansion was particularly concentrated in Nant Peris and the Llanberis Pass. There can be little doubt that the landscape was seriously overcrowded. However, the process of morcellation also affected the farms near Llyn Cwellyn in Bettws Garmon. Spoil heaps from the quarries encroached on the available farmland

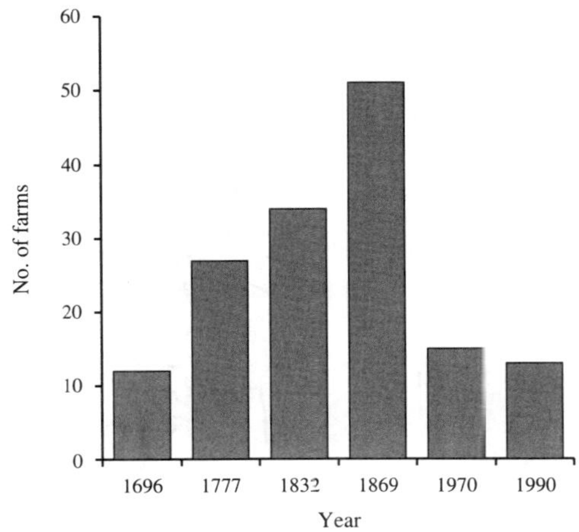


Figure 7.4 The changing number of farms in the study area, 1696–1990 (based on field surveys and Vaynol MSS 3643, 4055, 4202, 4195 and XSC 807, Gwynedd County Record Office).

and, in Nant Peris, much crucial in-bye land was annexed for crofts and for cottage building (Vaynol MSS 2903). Several old-established farms were fragmented and had inadequate in-bye land and *ffridd* in the key morphological units (figure 7.5). Their ability to yield an income sufficient to meet Vaynol rents and to feed a family was thereby disastrously impaired. In the 1890s foreign slate began to compete successfully for a share of the market and the quarry workforce in northern Snowdonia contracted sharply (Bassett and Davies 1977). The area's dual economy disintegrated and, together with the impact of the prevailing depression in agriculture, tenant farmers could not pay their rent. Although the Estate reduced some tenant debts, and waived others, decline was unavoidable. Out-migration ensued and, by the beginning of the twentieth century, the total population had begun to fall.

## TY ISAF, CEUNANT AND FRON: A CASE STUDY OF THREE FARMS

The processes of change leading to the establishment of permanent farms, their subdivision and fragmentation, and finally their reorganization and rationalization, may be reconstructed in detail by means of a case study of three neighbouring farms which are typical of the area as a whole. A detailed examination of Ty Isaf, Ceunant and Fron makes possible the further examination of the close relationship between the structure of farm units and the physical environment. Neither Ty Isaf, nor Ceunant and Fron were mentioned in the 1696 Vaynol rentroll. However, the neighbouring holdings of Llwyn Bedw, Garag Wen, Bryn Glas, Cwm-y-Wrach and Gwastadnant were by then in existence, and it is probable that the farm referred to as 'Cwm Gafr' in 1696

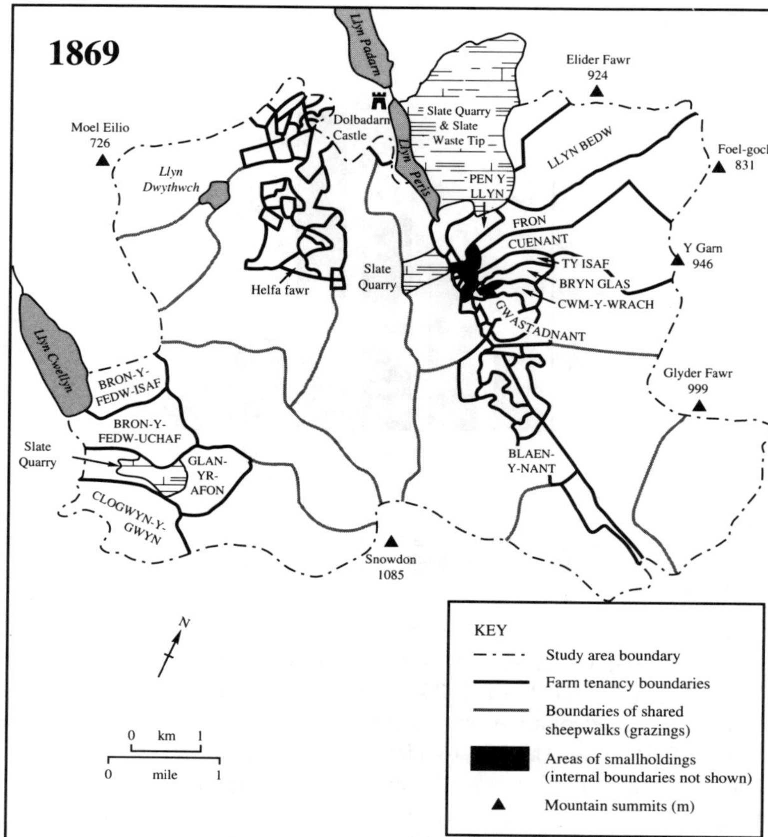


**Figure 7.5** The pattern of farms in the Pass of Llanberis, Cwmdwythwch and Bettws Garmon, 1869 (based on Vaynol MS 4195, Gwynedd County Record Office).

was the direct antecedent of the three case study holdings. The Afon Gafr drains from the *cwm* to the river in the valley below (figure 7.6). The names Ty Isaf, Ceunant and Fron appear first in the Vaynol Survey of 1777. At that time, Ceunant and Fron were regarded as a single, large holding of 843 acres, including almost 645 acres of open mountain sheepwalk; a position prevailing until at least 1823 according to the valuations and notebooks made by the Vaynol Estate Steward (table 7.2). Ty Isaf was much smaller (146 acres) although its in-bye land was clearly judged to be better because some was valued at £1 per acre (table 7.3 and figure 7.7). The different values attached to the three main farm management units minutely reflected the basic morphology of the landscape and thus its agricultural potential. Even the individual field names recorded in the 1777 Vaynol survey indicate a detailed appreciation of topography and edaphic conditions (plates 7.1 and 7.2; Barnes 1970). Whereas in-bye land, capable of yielding hay, might be worth upwards of 15 shillings per acre, *ffridd* was rarely valued at more than 10 shillings per acre (usually considerably less). Reflecting its high value, farmhouses and

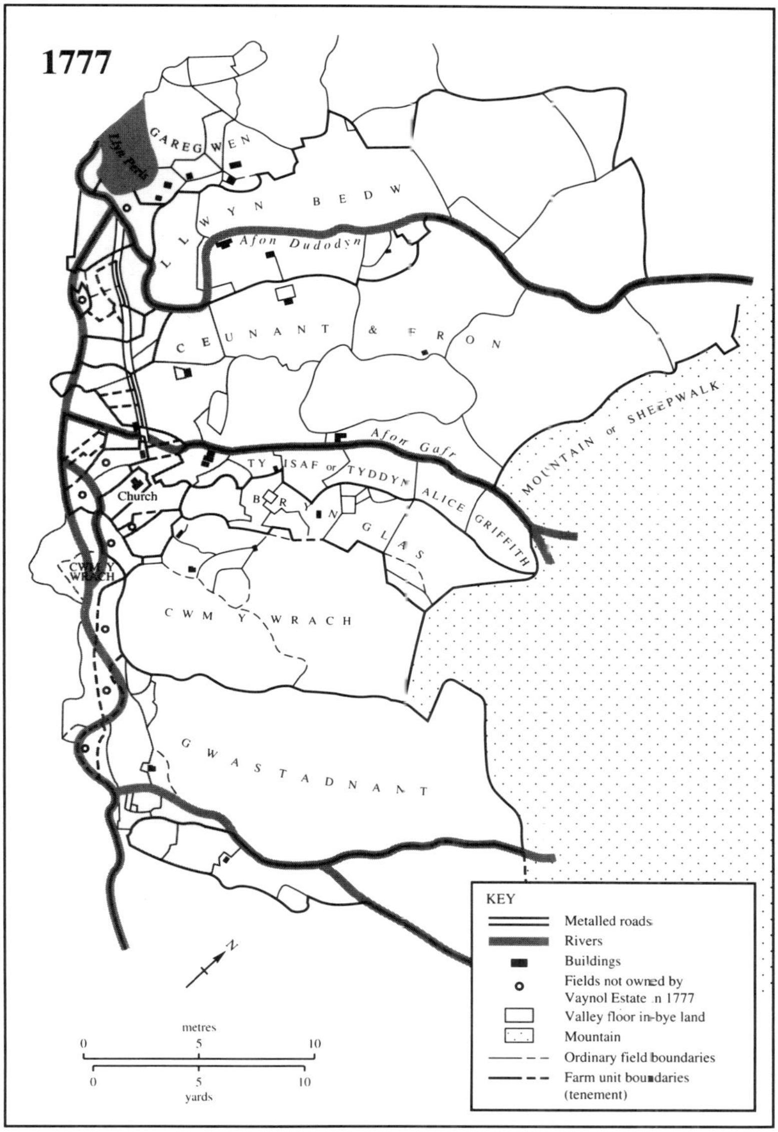


**Figure 7.6** The pattern of farms in Nant Peris, 1777 (based on Vaynol MS 4055, Gwynedd County Record Office).

out-buildings were traditionally not built on in-bye land, but instead were erected in the *ffridd* above. *Molinea caerula* and *Nardus stricta* mountain pastures, located above the 325 m contour, were valued at less than sixpence an acre. Transect field surveys carried out annually between 1982 and 1986 indicate how precisely local land capability had been judged in the past (figure 7.7). Soil type, soil acidity and the predominant vegetation community were

**Table 7.2** Ceunant and Fron, Nant Peris, in 1823

| | *Ceunant and Fron (David Ellis and others)* | *A* | *R* | *P*[a] | *s* | *d* | *£* | *s* | *d* | *Crop* |
|---|---|---|---|---|---|---|---|---|---|---|
| 1 | House and Caer Drain | 12 | 2 | 0 | 3 | 6 | 2 | 2 | 0 | Pasture |
| 2 | | 16 | 0 | 19 | 5 | 0 | 4 | 8 | 4 | Pasture |
| 3 | | 1 | 2 | 10 | | | | | | Pasture |
| 4 | Cowhouse and garden | 18 | 2 | 14 | 15 | 0 | 13 | 18 | 9 | Hay |
| 5 | Cae Ty'n Llwyn | 23 | 0 | 11 | 12 | 0 | 13 | 16 | 9 | Hay |
| 6 | Caer Gwyddel | 21 | 3 | 37 | 5 | 0 | 5 | 9 | 10 | Pasture |
| 7 | Cae Rhos | 36 | 0 | 30 | 5 | 0 | 9 | 0 | 11 | Pasture |
| 8 | Buarth | 1 | 1 | 19 | 8 | 0 | – | 10 | 11 | Pasture |
| 9 | Maes Gwyn | 50 | 0 | 29 | 2 | 6 | 6 | 5 | 5 | Pasture |
| 10 | Cae Esgair | 13 | 0 | 10 | 2 | 0 | 1 | 6 | 1 | Pasture |
| 11 | | 1 | 2 | 21 | 16 | 0 | 2 | 13 | 0 | Hay |
| 12 | | 1 | 2 | 29 | | | | | | Hay |
| 13 | Share of sheepwalk | 644 | 3 | 16 | – | 5 | 14 | 0 | 0 | Pasture |
| Totals | | 843 | 0 | 5 | | | 73 | 12 | 0 | |

*Sources*: Gwynedd Record Office: Vaynol MS 4066, Mr Evans' Valuation, 1823; Vaynol MS 4055. Estate Survey 1777; Vaynol MS 4063, Mr Jones' Survey and Valuation, 1810–11.

[a] Measurements in acres, rods and perches.

**Table 7.3** Ty Isaf in 1823

| | *Ty Isaf (Ellis James)* | *A* | *R* | *P*[a] | *s* | *d* | *£* | *s* | *d* | *Crop* |
|---|---|---|---|---|---|---|---|---|---|---|
| 1 | House, and yard and croft | – | 3 | 16 | – | | 1 | 0 | 0 | |
| 2 | Gwndwn | 2 | 3 | 6 | 20 | 0 | 2 | 15 | 9 | Hay |
| 3 | Buarth Main | 1 | 1 | 21 | 16 | 0 | 1 | 2 | 1 | Hay |
| 4 | Caer Odyn | 2 | 1 | 8 | 14 | 0 | 1 | 12 | 4 | Hay |
| 5 | Caer Gors | 5 | 0 | 13 | 10 | 0 | 2 | 10 | 7 | Hay |
| 6 | Y Fron Isaf | 7 | 2 | 24 | 8 | 0 | 3 | 1 | 2 | Pasture |
| 7 | Y Fron Ganol | 6 | 2 | 0 | 6 | 0 | 1 | 19 | 0 | Pasture |
| 8 | Y Fron Uchaf | 7 | 3 | 4 | 5 | 6 | 2 | 2 | 10 | Pasture |
| 9 | Share of sheepwalk | 111 | 2 | 29 | 0 | 5 | 2 | 11 | 5 | Pasture |
| Totals | | 146 | 0 | 1 | | | 18 | 15 | 1 | |

*Sources*: Gwynedd Record Office: Vaynol MS 4066, Mr Evans' Valuation, 1823; Vaynol MS 4055, Estate Survey 1777; Vaynol MS 4063, Mr Jones' Survey and Valuation, 1810–11.

[a] Measurements in acres, rods and perches.

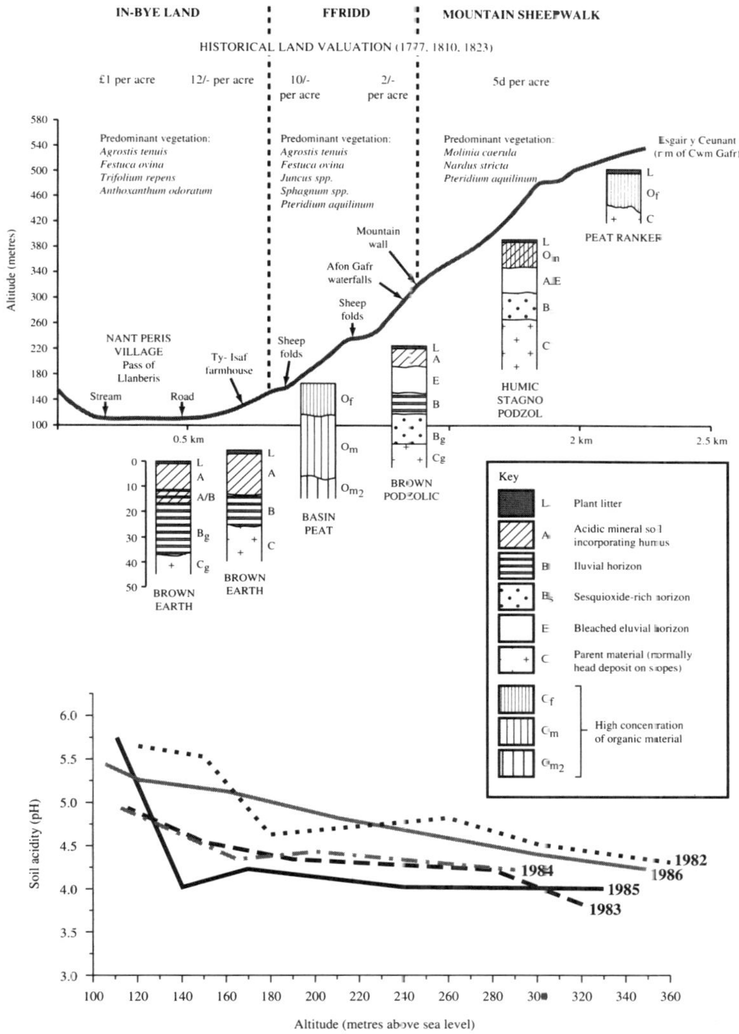


Figure 7.7 A generalized model of the vegetation, soil type and soil acidity in the in-bye fields, *ffridd* and mountain grazing of Ty Isaf farm, Nant Peris. (The related range of rents per acre payable in the period 1777–1823 by Vaynol tenants for in-bye land, *ffridd* and mountain sheepwalk are shown.)

all found to change markedly from the in-bye fields in the valley, through the *ffridd*, to the mountain grazing above. Moreover, variations in the values historically placed on each individual field still accurately reflect different soil, vegetation and drainage conditions. Thus, although modern improvements have

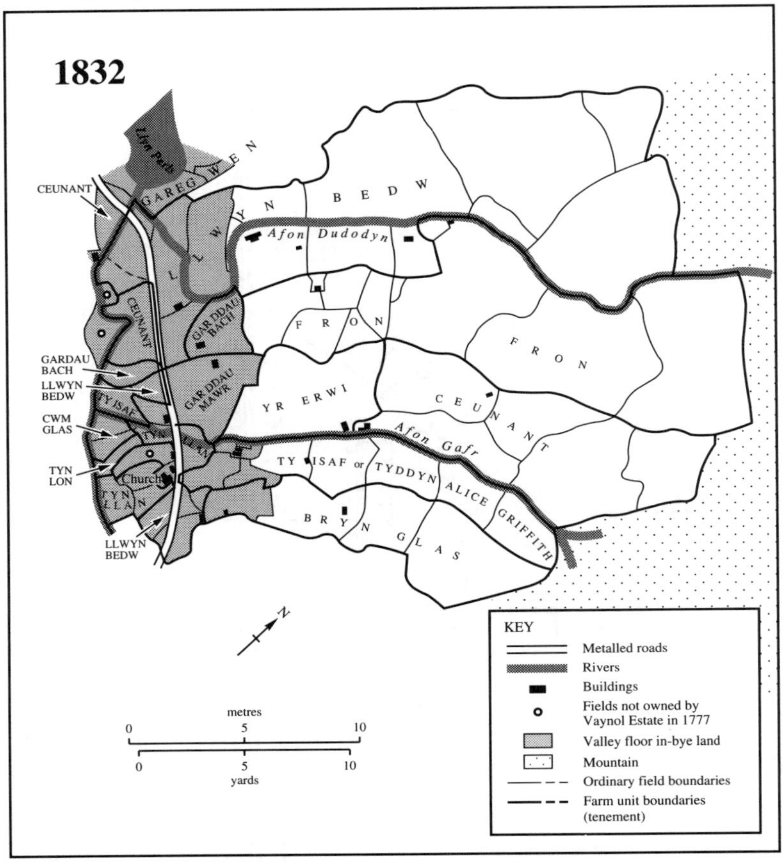


**Figure 7.8** The pattern of farms in Nant Peris, 1832 (based on Vaynol MS 4202, Gwynedd County Record Office). In-bye land, *ffridd* and mountain sheepwalk are shown.

been attempted, such as artificial drainage and heavy applications of lime and fertiliser, the underlying contrasts remain strong and the physical controls on the environment are as powerful as ever.

By the time of the 1832 Estate survey, minor reorganizations of the fields of Ty Isaf had been made, and more significant changes had affected Ceunant and Fron (figure 7.8). The crofts of Garddau Mawr and Garddau Bach were established on two of the former valley floor fields, while Yr-Erwi occupied the lowest field of Ceunant *ffridd*. Moreover, additional cottages had been built on the land of all three farms, and the census of 1841 indicates that these provided accommodation primarily for slate quarry workers and their families. In all, there were seven dwellings on the three farms, inhabited by a total of 50 people. Ceunant and Fron were separately tenanted in 1832, but both were agriculturally rather marginal, and all three holdings had been noticeably affected by the expansion of the slate industry. Hay was being cut as high as 425 m. The quality was undoubtedly extremely poor, although probably

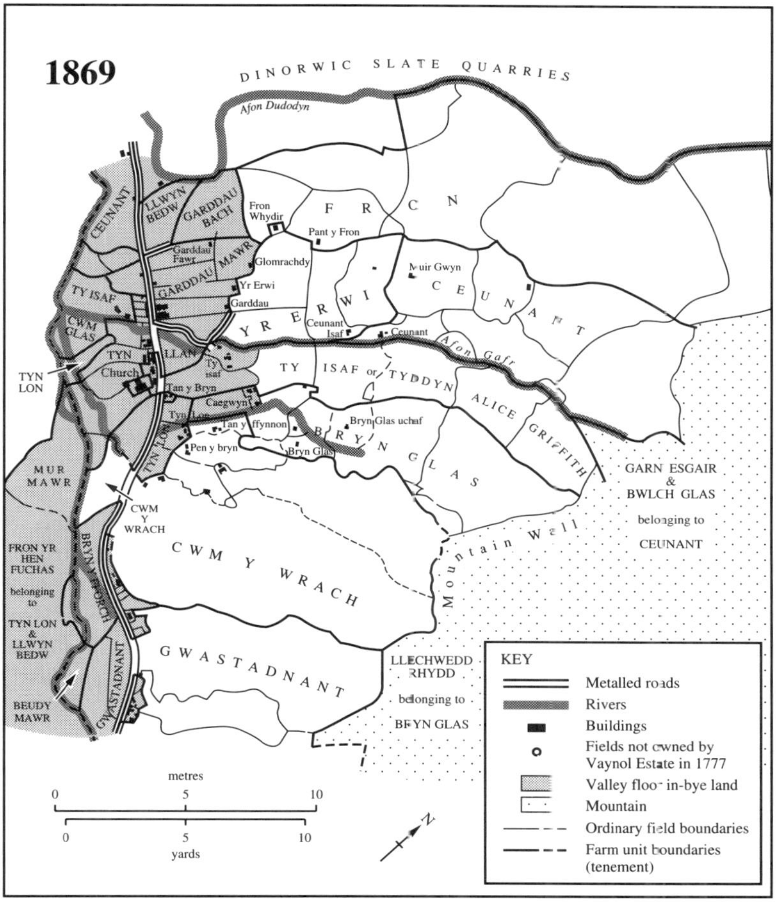


Figure 7.9 The pattern of farms in Nant Peris, 1869 (based on Vaynol MS 4195, Gwynedd County Record Office). In-bye land, *ffridd* and mountain sheepwalk are shown.

better than might be obtained today. After all, in the mid-nineteenth century, the alpine pastures were grazed much more heavily throughout the year, thereby reducing the dominance of mat grasses and encouraging the better species to colonize at altitude. The recent practice of removing the majority of ewes to overwinter in the lowlands has undoubtedly brought about a deterioration in the quality of the mountain sward.

The full impact of the expansion of the Dinorwig quarries is revealed in the 1869 Estate survey by the further extensive encroachment by cottages and crofts on the in-bye land and *ffridd*, not only of Ty-Isaf, Ceunant and Fron, but also neighbouring farms (figure 7.9). Moreover, slate spoil had by then engulfed the farm of Lwyn Bedw, which survived only as a tiny croft to the south-east of the Afon Dudodyn.

Deprived of good hay-producing land, and therefore operating an extremely marginal farm, the household heads, William Hughes and Ellis Davies, who resided with their families in the two Ceunant dwellings, both described themselves as 'slate quarriers' in the census of 1861. Ten years later, Davies' widow, Jane, is recorded as a 'farmer', but her five adult sons were quarrymen. The cottage of Fron was occupied by Robert Griffiths, who described himself in three censuses (1851–71) as a farmer, but his neighbour, Owen Owens, whose household of 12 lived in the principal farmhouse of Fron, was recorded as a 'quarryman'. Four of Owens' sons and his son-in-law were similarly employed. Clearly, their flock of sheep was of secondary importance and only alluded to in the 1871 census after the death of Owen Owens, when his widow Laura headed the household and described herself as a 'farmer'.

In a landscape where quarrying and farming had become inextricably linked, the decline of the former irrevocably damaged the latter. Job losses in Dinorwig undermined the viability of the much subdivided holdings of Nant Peris. Agriculture stagnated in the early years of the twentieth century until attempts were made to rationalize farm units. The process has continued ever since. In 1985 a study of mid-Wales showed that the number of farms in that region had declined by 43 per cent since 1951, and that on average each farmer was looking after 2.5 times more land and livestock. Broadly similar trends also affected northern Snowdonia. Thus, following the sale of the Vaynol Estate in 1967, Ty Isaf, Ceunant and Fron were acquired by their occupiers. However, by the late 1980s the three holdings had been amalgamated as a single farm and, in a sense, the pattern of the seventeenth-century landscape had been restored. Although the modern farm is not known as 'Cwm Gafr', as it was in 1696, the land occupied is the same. Taking advantage of modern hill farm sheep subsidies, very large flocks now graze the mountain pastures of Cwm Gafr in summer. However, in winter, the majority are transferred to rented keep on the lowlands of Lleyn or Anglesey. Thus, even transhumance, or a rather peculiar modern version of it, has been reintroduced.

Three centuries of profound change have left an indelible imprint. Indeed, northern Snowdonia is an archetypal landscape palimpsest. The striking physical and ecological evidence of past farms and settlement which survive today can be understood only by studying it in conjunction with surviving documentary material. Together these complementary approaches shed light on the making of a remarkable landscape.

## REFERENCES


Aris, M. (1977) 'The Vaynol Estate', in Bassett, T. M. and Davies, B. L. (eds) *Atlas of Caernarvonshire* (Gwynedd Rural Council, Caernarfon) pp. 135–9

Ball, D. F. (1968) 'Soils of Wales: the distribution of major soil groups', *Transactions of the VIIIth Congress on Soil Science*, Bucharest 5: 69–79

Ball, D. F. and Goodier, R. (1970) 'Morphology and distribution of features resulting from frost-action in Snowdonia', *Field Studies* 3: 193–218

Ball, D. F., Mew, G. and Macphee, W. S. G. (1969) 'Soils of Snowdon', *Field Studies* 3: 69–107

Barnes, F. A. (1970) 'Settlement and landscape changes in a Caernarvonshire slate-quarrying parish', in Osborne, R. H., Barnes, F. A. and Doornkamp, J. C. (eds) *Geographical Essays in Honour of K. C. Edwards* (University of Nottingham, Nottingham) pp. 119–30

Bassett, T. M. and Davies, B. L. (eds) (1977) *Atlas of Caernarvonshire* (Gwynedd Rural Council, Caernarfon)

Bell, W. T. and Ogilvie, A. E. J. (1978) 'Weather compilations as a source of data for the reconstruction of European climate during the medieval period' *Climatic Change* 1: 331–48

Beverley Smith, J. (1966) 'Crown and community in the principality of North Wales in the reign of Henry Tudor', *Welsh History Review* 3: 145–71

Campbell, B. M. S. (ed.) (1991) *Before the Black Death: Studies in the 'Crisis' of the Early Fourteenth Century* (Manchester University Press, Manchester)

Census Enumerators Books, PRO HO 107/1393 1841 (Public Record Office)

Census Enumerators Books, PRO HO 107/2515 1851 (Public Record Office)

Census Enumerators Books, PRO RG 9/4330 1861 (Public Record Office)

Census Enumerators Books, PRO RG 10/5716 1871 (Public Record Office)

Census Enumerators Books, PRO RG 11/5566 1881 (Public Record Office)

Dodd, A. H. (1926–7) 'The enclosure movement in North Wales', *Bulletin Celtic Studies* 3: 210–38

Edwards, J. W. (1963) 'Enclosure and agricultural improvements in the vale of Clwyd 1750–1875', unpublished M.A. thesis, University of London

Elfyn Hughes, R. (1973) 'Studies in sheep population and environment in the mountains of North West Wales', *Journal of Applied Ecology* 10: 107–12

Eyre, S. R. (1957) 'The upward limit of enclosure on the east moor of north Derbyshire' *Transactions of the Institute of British Geographers* 23: 61–74

Hodgson, R. I. (1969) 'Medieval colonisation in northern Ryedale, Yorkshire' *Geographical Journal* 135: 44–54

Hyde-Hall, E. (1809–11) *A Description of Caernarvonshire*, ed. from MS in the Library of University of Bangor by E. G. Jones (Gwenlyn Evans, Caernarvon 1952)

Ingram, M. J. (1978) 'The historical study of climate: a report on the work of the climate research unit at the University of East Anglia', *Local Population Studies* 21: 24–30

Jones, G. R. J. (1964) 'The distribution of bond settlements in North-West Wales', *Welsh History Review* 2: 19–36

Jones Pierce, T. (1940) 'Notes on the history of rural Caernarvonshire in the reign of Elizabeth', *Transactions of the Caernarvonshire Historical Society* 2: 35–57

Lamb, H. H. (1982) *Climate, History and the Modern World* (Methuen, London) pp. 178–230

Lewis, W. V. (1949) 'Glacial movement by rotational slipping', *Geografiska Annaler* 31: 146–58

Lindsay, J. (1974) *A History of the North Wales Slate Industry* (David & Charles, Newton Abbot)

Miller, R. (1967) 'Land use by summer shielings' *Scottish Studies* IX: 193–219

Parry, Bryn R. (1970) 'The Vaynol Estate: end of 400 years of history', *Royal Welsh Agricultural Society Journal* 16: 15–8

Parry, M. L. (1978) *Climatic Change, Agriculture and Settlement* (Dawson, Folkestone)

Parry, M. L. (1981) 'Evaluating the impact of climatic change' in Delano Smith, C. and Parry, M. (eds) *Consequences of Climatic Change* (Department of Geography, University of Nottingham) pp. 3–16

Parry, M. L. (1985) *Mid-Wales Upland Study* (Countryside Commission, Cheltenham), 56–8

Perkins, D. F. (1978) 'Snowdonia grassland: Introduction, vegetation and climate' in Heal, O. W. and Perkins, D. F. (eds) *Production Ecology of British Moors and Mountain Grasslands III: Snowdonia Grasslands* (Springer-Verlag, New York) pp. 289–96

Richards, A. R. (1961) 'Facing up to the sixties: Vaynol, part I – the problem', *Journal of the Chartered Land Agents Society* 60: 126–32

Roberts, B. K. (1977) *Rural Settlement in Britain* (Dawson, Folkestone)

Roberts, R. A. (1949) 'The historical development of hill farming in Wales', *Agricultural Progress* 24: 25–9

Roberts, R. O. (1973) *Farming in Caernarvonshire around 1800* (Caernarvonshire Record Office, Caernarvon)

Roberts, R. O. (1959) 'Ecology of human occupation and land use in Snowdonia', *Journal of Ecology* 47: 317–23

Rowlandson, T. (1846) 'Agriculture of North Wales', *Journal of the Royal Agricultural Society* 7: 553–89

Seddon, B. (1957) 'Late glacial cwm glaciers in Wales', *Journal of Glaciology* 3: 94–9

Seddon, B. (1962) 'Late glacial deposits at Llyn Dwythwch and Nant Ffrancon, Caernarvonshire', *Proceedings of the Royal Society* B 244: 459–81

Stuart, D. R. M. (1961) 'Facing up to the sixties: Vaynol, part II – the solution', *Journal of the Chartered Land Agents Society* 60: 154–62

*The Times* (1967a) 'Snowdon is sold privately' 14 July 1967, p. 3

*The Times* (1967b) 'The Commons: new owners on Snowdon', 21 July 1967, p. 8

Unwin, D. J. (1973) 'The distribution and orientation of corries in northern Snowdonia', *Transactions of the Institute of British Geographers* 58: 85–97

Unwin, D. J. (1975) 'The nature and origin of the corrie moraines of Snowdonia', *Cambria* 2: 20–33

Vaynol, MS 2903. Notebook of Vaynol Estate Agent: complaints and requests, 1871–4 (Gwynedd County Record Office)

Vaynol, MS 3643. A Rentroll of the estate of Sir William Williams of Vaynol in the county of Caernarvon – Baronet deceased, 1696 (Gwynedd County Record Office)

Vaynol, MS 3751. Michaelmas Rental of the Estates of Thomas Assheton Smith in Llanberis, 1799 (Gwynedd County Record Office)

Vaynol, MS 4202. Valuation of the Vaynol Estate *c.*1832 (Gwynedd County Record Office)

Vaynol, MS 4055. A Survey of the Vaynol Estate. Vol. I. 1777 (Gwynedd County Record Office)

Vaynol, MS 4057. Survey and Abstract of the Vaynol Leases in the Parish of Llanberis, 1778 and *c.*1800 (Gwynedd County Record Office)

Vaynol, MS 4063. Survey and Valuation by Mr Walter Jones, 1810–11 (Gwynedd County Record Office)

Vaynol, MS 4066. Valuation by Mr Evans, 1823 (Gwynedd County Record Office)

Vaynol, MS 4067. A Survey of the Vaynol Estate. Vol. I. 1832 (Gwynedd County Record Office)

Vaynol, MS 4108. Particulars for Valuation of Lands in the Parish of Llanberis; the property of George William Duff Assheton Smith, 1868 (Gwynedd County Record Office)

Vaynol, MS 4195. A Survey of the Vaynol Estate. Vol. III. 1869 (Gwynedd County Record Office)

Vaynol, MS 4376. John Millington's Account with Thos. A. Smith, esq. for the Rents of his Estates in Caernarvonshire and Anglesey for one year due Lady Day, 1832 (Gwynedd County Record Office)

Vaynol, MS 4427. John Millington's Account Book, 12th November 1869 (Gwynedd County Record Office)

Vaynol, MS XSC 807. Sale Catalogue, 1967 (Gwynedd County Record Office)

Whittington, G. (1973) 'Field systems of Scotland', in Baker, A. R. H. and Butlin, R. A. (eds) *Studies of Field Systems in the British Isles* (Cambridge University Press, Cambridge) pp. 530–79

Whittow, J. and Ball, D. F. (1970) 'North West Wales', in Lewis, C. A. (ed.) *Glaciation of Wales* (Longman, London)

Wigley, T. M. L., Ingram, M. J. and Farmer, G. (eds) (1981) *Climate and History: Studies in Past Climates and their Impact on Man* (Cambridge University Press, Cambridge)

Williams, D. (1927) 'The geology of Snowdon', *Quarterly Journal of the Geological Society of London* 83: 346–431

Williams, D. (1930) 'The geology of the country between Nant Peris and Nant Ffrancon', *Quarterly Journal of the Geological Society of London* 86: 191–233

Williams, D. (1941) 'An eye-witness account of agrarian conditions in Caernarvonshire during the Napoleonic era', *Transactions of the Caernarvonshire Historical Society* 72–84

# 8

# Social, Economic and Environmental Change in the Fen and Fen-edge Region of East Cambridgeshire in the Nineteenth century

Robin A. Butlin

ABSTRACT

This chapter examines the human–physical environment linkages in four parishes in East Cambridgeshire in the nineteenth century, with special reference to the social, settlement, land-use and land-drainage systems which characterized these fen and fen-edge environments. The forces for change are examined in detail, including attempts at improvement of land drainage systems through new technologies and management systems, and agricultural change through the enclosure movement, fluctuating market demands for agricultural produce, and a declining labour force. Traditional land-use systems and poverty were, however, remarkably persistent, particularly among the smallholders and agricultural labourers. Data and information sources used in the study include records of land drainage, early field surveys and regional syntheses of the British Geological Survey, together with evidence from enclosure, tithe and estate records.

## INTRODUCTION

The dynamics of agrarian landscape and socio-economic change in the fen and fen-edge regions of Cambridgeshire in the nineteenth century are complex. Their experiences were not untypical of fen, marsh and moorland settlements in Western Europe at the time, and similar to those of the Netherlands and northern Germany, for example, in relation to drainage technology, poverty, styles of living and emigration. Their specific characteristics derive, however, in large measure from a complex interaction between local environmental and social conditions (the processes of which were increasingly well understood as time progressed, but the technical means for mastering which were very slow to develop and be implemented). The processes of social and economic change reflected both local and wider national and international experiences of economic growth and decline. Of particular importance in the fen and fen-edge regions was the presence of large numbers of labourers and smallholders, living at the margins of existence, who were undoubtedly affected by land drainage improvements, enclosures and changes in marketing and employment opportunities generated by the increases in population of the large industrial cities. During the nineteenth century major changes in population density, land-use, and (partly through enclosure) the scales of landholding were experienced that transformed the face of the region. The coming of the railway both opened up distant urban markets for horticultural products and also facilitated the movement of people, some out of the region in search of employment, and others as visitors to the region to follow their naturalist interests in the study of the wetland habitats, flora and fauna of the region.

The aims of this chapter are to trace, using a wide range of documentary, published government survey, and cartographic and environmental evidence, the principal features of socio-economic and environmental change in the course of the nineteenth century in four fen-edge parishes of Cambridgeshire. Particular attention is paid to the open-field arable and pasture systems of the higher land, the fenland economy, and the problems of land drainage in the face of natural hazards, changing technology, and the constant problem of drainage maintenance and administration.

The region of this study lies in the southeastern part of the Fenland basin, in the eastern part of the county of Cambridgeshire and on the border with the county of Suffolk (figure 8.1). The Fens form a 1300 square mile (3367 km$^2$) basin, within which are a number of 'islands' of upland, including the Isle of Ely. The surface of the fen country is flat, and often lies below sea level, its pumped drainage system, the major protection fron flooding, taking the water from field drains to the embanked rivers which flow to the Wash along courses much altered and straightened by human engineering. The geological bases of the region are sedimentary rocks of the Jurassic and Cretaceous periods, notably Gault Clay, Cambridge Greensand, Lower, Middle and Upper Chalk, and a range of superficial deposits from the Pleistocene and Recent periods, including till, head and river-terrace deposits, fen peat, fen clay, mere deposits and river alluvium (Hodge and Seale 1966).

The surface soil of the fenland is peat-derived. The most extensive type in the region here considered is the Adventurers' Series, a basin peat soil comprising a peaty loam topsoil of dark brown to black colour of about 1 m in depth. This soil is much subject to wind erosion and is a poor support for houses and roads, the preferred basis being the silt banks of the rodhams or former watercourses (Seale 1975: 8). The peat soils are most extensive in the parishes of Soham, Isleham and Wicken. Important physiographical and pedological features of the region are the remains of former meres – shallow lakes in the fens which accumulated shell marl deposits through the bicarbonate-rich waters of the fen rivers flowing from the upland calcareous rocks. A good example is to be found to the west of Soham at the site of Soham Mere, a former fenland lake of *c.*1170 acres (474 h) in area, contained on three sides by a low Gault Clay ridge, and drained in the seventeenth and eighteenth

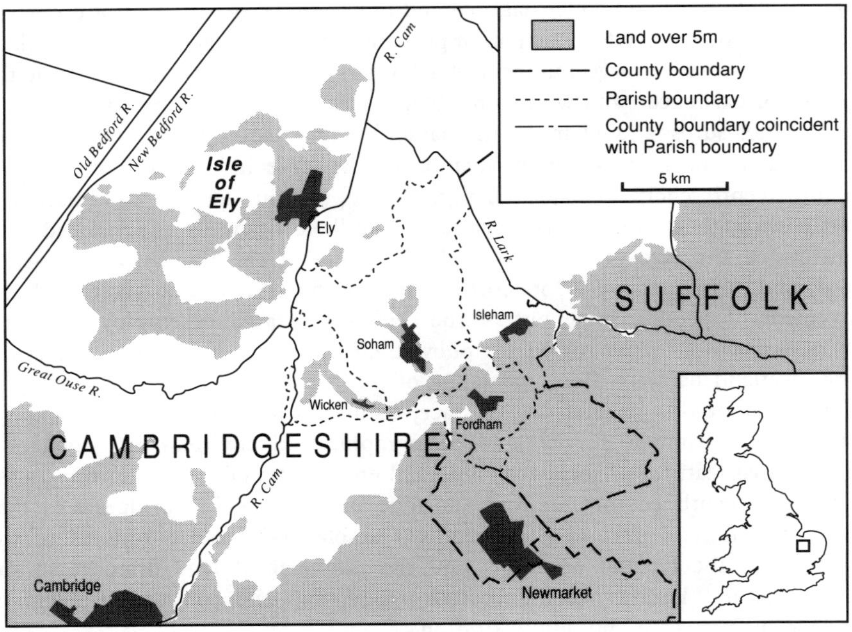


**Figure 8.1** A location map.

centuries. Between the fen soils and the upland soils are transitional soils known in the region as skirt or skirtland soils, mainly peaty and humose gleys, including the Wicken Series, a grey–brown clay loam or clay (Hodge and Seale 1966). The upland soils are extremely varied in Wicken, Fordham, Isleham and Soham, and for the most part are brown calcareous soils, brown earths, and rendzinas, and include the Soham Series – brown calcareous soils. These were the areas that contained the former open arable fields up to the period of enclosure.

The area is within the catchment of the River Great Ouse, the main southern drainage channel of the Fens, which has been frequently and extensively modified by the creation of cut channels since the seventeenth century (figure 8.1). It rises in Northamptonshire and thence flows (formerly by means of a very tortuous course) to the North Sea at King's Lynn. Its main Fenland tributaries are the Cam, Lark, Little Ouse, Wissey and Nar. The Cam and the Lark (or Mildenhall river) are of central importance in this particular study. The other main Fenland rivers are the Nene, Glen, Welland and Witham. They have a total catchment area of about 6000 square miles (156 000 km$^2$), and discharge into a lowland area which is only one-fifth of the size of the catchment. Very large amounts of water are thus regularly discharged from the uplands, particularly in winter, into the lowland fenland basin, the coastal end of which is barred by silts at heights of about 10–15 feet (3.0–4.5 m) above sea level, the differential between peat and silt partly reflecting the shrinkage of the inland peats after drainage. The peats themselves are generally no more than about 5 feet (1.6 m) maximum above sea level, and in that part of the study region between the Lark and the Ouse they are below sea level. Small differences in height have in the past frequently been a precarious margin between safety and disaster.

The principal settlements of the study area are the villages of Wicken, Isleham, and Fordham and the small town of Soham. They are all located on the fen edge, abutting the chalk upland in the south-east of the Fens. All the parishes but Fordham include large areas of low-lying peat fenland. Soham lies on a low narrow peninsular ridge and is surrounded on three sides by fen. The settlements themselves are located on marginally higher ground, but the fenland sections are at or below below sea level, and thus have a long history of inundation by water from inland catchments. The combination of higher and lower land in each of the parishes has encouraged a mixed rural economy throughout historical time.

The villages on their slightly elevated dry sites were, until the early nineteenth century (and longer in the case of Soham), surrounded by open arable fields, pasture moor and meadow, while the fenland areas provided summer pasture, peat, reeds, fish and wild fowl.

## ECONOMIC AND SOCIAL CONDITIONS IN THE NINETEENTH CENTURY

From the early nineteenth century to 1851 there was a conspicuous increase in population, reflecting improved drainage of the fens and the intensification of cultivation, and the broadening of employment opportunities in services related to agriculture and in small towns such as Soham (Butlin 1990b). With few exceptions the peak census year for the fenland villages and towns was 1851, with the populations reducing thereafter through migration to the industrial cities of the North and through emigration overseas (Table 8.1).

In Cambridgeshire in general by this time, most of the open fields had been enclosed, drainage was improving, and farming, at least on the uplands, had the appearance of being of good quality. Industrial demand from the growing industrial areas of England increased, and movement of farm produce was facilitated by the rapid growth of railways in the region from mid-century onwards.

The larger farmers of Cambridgeshire benefitted from the period of High Farming more than the many agricultural labourers and smallholders, who still lived a very marginal existence. S. Jonas wrote in in 1846: ‘Comparing the present system [of agriculture] with the former it is astonishing to mark the increased wealth our present improved system brings to the state’ (Jonas 1846: 35), contrasting with Caird’s discouraging analysis of agriculture in the county in 1850–1 (Caird 1852: 477). Jonas (1846, 35) reports on an increase in the population of the county of 75 113 (from 89 364 to 164 459) between 1801 and 1841, of whom many had ‘found employment in agricultural pursuits, or as tradesmen, mechanics, and artisans, in supplying the wants of those engaged in agriculture’. He says that there had been much improvement in the agriculture of the county as a whole:

**Table 8.1** Population in four Cambridgeshire parishes, 1801–1891

| | *1801* | *1851* | *1891* |
|---|---|---|---|
| Soham | 2,004 | 4,706 | 4,138 |
| Fordham | 700 | 1,584 | 1,284 |
| Isleham | 1,212 | 2,236 | 1,691 |
| Wicken | 614 | 995 | 716 |

*Source*: Censuses of Population.

> Few counties, if any, have improved more in cultivation than Cambridgeshire has lately done. All the open common-fields have been enclosed (with the exception of five or six parishes), and instead of a system of cropping so exhausting to the land as a fallow and two white-straw crops in succession, with other men's flocks of sheep eating up your food and preventing improvement, we now see the land farmed on the four-course system – the best that can be adopted, unless on very fine land (Jonas 1846: 38).

He stresses, among other points, the importance of the London market, especially for the sale of sheep.

Darby (1938) has suggested that the agrarian experience of the county of Cambridgeshire in the nineteenth century was greatly similar to that of the rest of England, with two periods of improvement and prosperity (the early part of the century up to 1815 and 1838–74), and two periods of depression, (1815–38, and from 1874 to the end of the century). Despite bad weather and harvests in the early 1840s, and the advent of the potato blight in 1845, there were signs of improvement in agrarian conditions by the late 1840s. The last quarter of the nineteenth century again brought agricultural depression, and Cambridgeshire was badly affected, as the accounts of the Poor Law Commissioners and various Royal Commissions on agriculture and agricultural conditions testify. In 1880 a visitor to the county described the causes of depression as being a combination of four or five years' poor harvests and importation of wheat from North America, the regional consequences being the insolvency or near-insolvency of most of the Fenland farmers, and a significant increase in the amount of unlet land on the uplands (Darby 1938: 127). Data from the parliamentary enclosure awards indicate very complex tenurial and proprietorial arrangements in this micro-region which undoubtedly require further study. At Isleham, for example, there were large numbers of freeholders with very small holdings in the open fields and modest use-rights in commons and fens (evidenced earlier in the fenland anti-drainage and anti-commons appropriation riots of the mid-seventeenth century, and dating in all probability from the eleventh and twelfth centuries). There were few large estates with aristocratic or upper gentry landowners in this region. The continuation well into the nineteenth century of traditional systems of cultivation and tenure, especially the preservation of the extensive commons, probably owes much to the defence of their use by the many cottagers and commoners. The manorial structure was quite complex: the territory of Soham, the largest settlement in the area, comprised at least three manors in the mid-nineteenth century, and there had been more in the past. There was also evidence of pressure on land in the form of squatting – illegal settlement – by the poor on the commons, including those at Soham. Another measure of the degree of independence in this area was the extent of non-conformist religion, non-conformity being associated with, as demonstrated by Mills in his study of the numbers of smallholders in the Cambridgeshire village of Melbourn in the nineteenth century (Mills 1984: 489), 'a widespread social distribution of landownership'.

## AGRICULTURE AND FIELD SYSTEMS

An account in May 1842 of an estate in Isleham in the ownership of King's College, Cambridge gives an idea of some of the important issues of agricultural management in this region at the time:

> The above estate is principally open-field land, and land of first rate quality for turnips, barley, wheat and clover, it would be much improved by an inclosure... It is in a most excellent state of cultivation and does the tenant great credit... The fen land pays a drainage tax of 3s 6d per acre; it is fen land of the worst description, and no clay near to improve it. The common for eighteen score [360] sheep extends over the fallow field and the land being good sheep land makes it valuable, for that sort of keeping is not subject to rot sheep. I would by all means recommend an inclosure... . (KCC MSS, ISL/24).

In some respects, this epitomizes the dilemma of the tension between the would-be reformers and those who opposed agricultural reform. There was progress, but it was slow, and awaited the structural changes effected by enclosure (where it occurred) and further improvement in drainage. The four parishes of the area here studied all possessed open-field arable systems in the early nineteenth century. For Soham, a survey of 1656 shows four open fields, NoDitch, North or Metlam, Downe and Clipsall Fields, the largest being the North Field. Elsewhere, to the north and north-east, were several grazing commons, and to the south and south-west were intercommons with the adjoining parish of Fordham (Barringer 1991: 7, Butlin 1993: 174). Charles Vancouver in 1794, in his *General View of the Agriculture of the County of Cambridge*, identifies 900 acres (364 ha) of open-field arable south of Soham and 300 acres (121 ha) in Soham's large North Field. In May 1837, when the parish was viewed in connection with the commutation of tithes by J. Mocus, an assistant Tithe Commissioner, the open arable land was measured at 1022 acres (413 ha) (PRO IR 18 13632). Soham, a thriving market town, had been the site of settlement since Neolithic times, and was a classic fen-edge parish, with cereal-growing and sheep-rearing on the higher land combined with a fenland economy on the lower ground. Its North Field survives unenclosed to the present day (plate 8.1), although the management system has changed to that of individual

**Plate 8.1** Aratral curve, North Field, Soham, Cambridgeshire in 1987 (photograph R. A. Butlin).

cultivation and cropping). The assistant tithe commissioner in 1837 wrote of the system of management of the arable as:

> ... one of almost constant cropping, a fallow being scarcely known, and a very small proportion of turnips grown, indeed there are scarcely any sheep in the Parish though I should think it is well adapted for them. Some of the land is cropped invariably with alternate crops of Wheat and Beans the principal part of the open Arable land is a fine working deep loam – part of it good Turnip land which if inclosed would be particularly valuable, though not prized as it deserves (PRO IR 18 13632).

In the neighbouring village of Fordham there were at the time of enclosure in 1820 six open fields, one of which (Clipsall Field) was intercommoned with Soham. Isleham had four open fields – Dunstall Field, Great South Field, Little South Field and Inhams Field – described by Vancouver (1794: 33) as 'four distinct, open, common fields... proper for the cultivation wheat of barley, oats, rye, peas, clover and turnips...', and the open fields and strips survived to the the time of their enclosure by authority of the General Enclosure Act of 1845 (the enclosure award was agreed in 1854), and are shown as they appear in the pre-enclosure plan drawn up in 1848 (CRO Q/RD c 74), in figure 8.2.

Wicken had three open fields comprising a very small area of arable. The lands enclosed are described in the enclosure award of 1849 as 'divers common fields, commonable, Lammas and waste lands', over which there were claims

Figure 8.2 The open fields at Isleham, Cambridgeshire, in 1848, immediately prior to enclosure (from CRO Q/RD c 74).

to rights of sheepwalk and shackage. The open arable fields were High Fen Field, Afterway and Padney Field (CRO Q/RD c 69; R.a 10).

Crop rotations in the region were described in George Cooke's *Topographical and Statistical Description of the County of Cambridge* (1820: 27): 'The general mode of management of arable land is to take two crops and a fallow. In some parishes the lighter lands, after lying two years under rye-grass, trefoil and Dutch clover, are winter fallowed for Tartarian oats, or summer tares. In others the first year fallow, dung, sheep folding, and lighter manures; for the second year wheat, winter fallow, wheat stubble, a light land dressing during the spring; for the third year crop of barley; barley stubble sown the fourth year, with peas, oats and lentils'. The folding of sheep on fallow and stubble was a common practice in these chalk 'upland' parishes. The Tithe File for Wicken includes a report of a meeting on 20 April 1838, which records, with reference to a farm formerly belonging to Spinney Abbey:

> To this farm is attached a most extraordinary right of Sheep-Walk or Field Course over all the Open Fields and Commons which in *Practice*

> at least is rigidly exclusive – no other Persons being allowed to turn any Sheep upon them. This right is exercised – from 9 days after the Fields are cleared from the Corn – to the 6th of April – No occupier being permitted to plough his land until after that time (PRO IR 18. 13665).

This economic system and its associated landscapes were transformed in three of the four parishes between 1790 and 1854 by Parliamentary enclosure (see below).

The fenland pastures continued to be of great importance for the grazing of horses, sheep and cattle, especially in the summer, and were of particular significance for the smallholders and the poor, despite the attendant hazards of foot-rot and easily spread diseases among the animals (Rowell 1986). The natural products of the fens, including turf, sedge, fish and fowl, were of great importance to the lives of the poor, and significant elements of the dual upland/lowland economy of the fen-edge parishes. The intensive exploitation of fen products, however, sometimes led to resource depletion. The intensive cutting of peat by the poor in Wicken Poor's Fen for use as fuel, for example, notwithstanding the attempted restrictions and limitations on this practice by the Court Leet, meant that the Fen was extensively dug out by the 1830s. The use, production and value of a particular type of sedge (*Cladium mariscus*) has also been carefully documented by Rowell (Rowell 1982: 1986, Rowell and Harvey 1988). He shows that its main uses were for thatching and kindling (the sedge at Wicken was used for thatching), although it was also used as fuel and as a surfacing material for trackways.

The poorer quality peat fen at Soham was estimated by the Assistant Tithe Commissioner in May 1837 as 4334 acres (1754 ha), and the better quality 'fen and bordering land arable' at 2166 acres (876 ha), the latter being increasingly used for arable rotations. He described the poorer fenland in these terms:

> The Fen Land also varies much, a large proportion being very poor peat fen, of little or no value, a great distance from the Village – occasionally growing corn of poor quality and very uncertain as to its quantity. Of this a large proportion has been, and is, in the course of being cut up as Turf for fuel, which is carried on to an extent I have never before witnessed. Part of the land so treated has been levelled, and does not appear much, if any, less productive than the part not disturbed but there are many acres, probably between 200 and 300 – now entirely unprofitable... . These [poor peat] lands... might be vastly improved by good farming, as they are now effectually drained by a steam engine – The roads are I conceive quite impassable in Wet seasons being nothing more than the black peat of the neighbourhood (PRO IR 18 13632).

## PARLIAMENTARY ENCLOSURES

As Turner (1980: 85) has shown, Parliamentary enclosure in Cambridgeshire as a whole was undoubtedly late when compared with other counties in eastern and midland England: 'Of the mainstream counties Cambridgeshire was clearly the peculiar one with such late enclosures: 19.6 per cent of it was enclosed after 1830, a statistic which easily distinguishes it from all other counties... . [of which] as much as 14.6 per cent was open field arable, a remarkably late and extensive survival of ancient open fields'. The reasons for this are complex. There was opposition to enclosure from the freeholders, especially on the chalk uplands, who were reluctant to surrender their rights of sheep-folding on the open-field arable and pastures, and also from larger landowners and clergy because of the anticipated effects on the future economic condition of the poor, especially the cottagers and agricultural labourers. Bendall (1992: 23) has also, following the Postgate thesis of 1964, suggested that the maintenance of a high percentage of arable land through to the mid-nineteenth century and the flexibility of the open-field practices, which accommodated mixed corn crops, pulses and the extension of ley farming, all facilitated late enclosure. Other factors included the amount of common pasture, the proximity to Cambridge, and the large number of small owner–occupiers.

Of the four parishes here studied, three were partially enclosed by private and general acts of Parliament, and two (Wicken and Isleham) were enclosed late. The earliest enclosure was for Fordham, the private Parliamentary Act dating from 1809 and the enclosure award from 1820, affecting 4050 acres (1639 ha), including old enclosures. The Wicken Act was passed in 1840, and the award made in 1849, affecting 3908 acres (1581 ha), of which 834 (337 ha) were arable, meadow and pasture and 2962 (1, 200 ha) were fenland and ancient enclosed land. The enclosure for Isleham under the provisional order of the General Enclosure Act of 1845, affected 1439 acres (582 ha) of mainly open-field arable, the very substantial area of fenland (about 2600 acres/1052 ha) being omitted. The allotments in the Isleham enclosure award finally agreed in 1854 (CRO Q/RD c 74) comprised 75 of less than 10 acres (4 ha), of which 62 were less than 5 acres (2 ha), ten from 11 to 20 (4.5–8.0 ha), four of 21–30 (8.5–12.0 ha), two of 41–50 (16.5–20.0 ha), four of 51–75 (20.5–30.0 ha), two of 76–100 (30.7–40.5 ha), and two of 101–150 acres (41.0–60.5 ha). Two allocations of land were made in the enclosure award to King's College, Cambridge, and Pembroke College, Cambridge, respectively. The tenure of land, as revealed in the enclosure awards, was a mixture of freehold and copyhold. There were relatively few sizeable allotments in the Fordham award for 1820. Only 13 allotments were in excess of 100 acres (40 ha), including those of the vicar of Fordham, Trinity Hall, Cambridge, Christ's College, Cambridge, the Bishop of Ely (420 acres/170 ha), the lord of the manor – the largest allocation of 520 acres (210 ha) – and the rest to individual farmers. The majority of allotments were very small: there were 90 freeholds of less than 5 acres (2 ha), for example.

Further evidence on land tenure and size of holdings in this region is provided by Parliamentary enquiries of the late nineteenth century. The transcript of the minutes of evidence contained in the *Reports from the Select Committee and Others on Allotments, Small Holdings and Peasant Proprietors* (BPP 1888–

94) for Friday 31 May 1889 contains the verbal and written evidence from Mr Charles Bidwell, Land Agent and Surveyor:

> ...the parish of Soham is a parish very well known to me... It contains 12,526 acres [5076 ha]; and there are 195 occupiers of under one acre [0.40 ha] (these are chiefly small gardens, with which I need not trouble you); there are 77 occupiers of plots under five acres [2 ha]; there are 34 occupiers of plots under 10 acres [4 ha]; there are 43 occupiers of plots under 20 acres [8 ha]; and there are 57 occupiers of plots under 50 acres [20 ha]. There is a considerable quantity of this land occupied by little men who own the fee... . These little men where they own the land have not done so well, nor have they stood the bad seasons as well as the men who rent the land... . It is very good land, accessible to good roads, and convenient to the railway; the parish is, curiously enough, an uninclosed parish. A great many of these small strips of land under an acre in size, lie in different parts of the parish. It is very extraordinary how these little men have struggled and worked through the bad times we have experienced lately (p. 275).

He describes how the produce of the increasing horticulture of the parish (green crops, carrots and roots, including potatoes) went to Newmarket, and refers also to supplementary activities such as hackney carting and coal merchanting. He mentions 238 acres (96 ha) of unenclosed common land in Soham where they are allowed to turn out horses and cattle: a 'substantial help to them in keeping their stock' (p. 275).

The enclosures effected dramatic changes in the landscape. The old open fields and common meadows and pastures disappeared, and were replaced by square fields held in severalty. New roads were constructed, often metalled with stone dug locally, and old driftways and footpaths were stopped up. Enclosure did not, however, affect the whole of the parishes studied. Soham did not undergo Parliamentary enclosure, the poor's commons were unaffected, and Isleham Fen, a very large area, was not included in the enclosure award, although the tithe award map of 1848 (CRO 311/P1) clearly shows that it remained subdivided into many holdings, some of them extremely small.

The continuation of a smallholding structure into the twentieth century has been demonstrated by Hill in her study of the history of Isleham Fen (on the west bank of the River Lark north of Isleham village) in the 1930s (Hill 1990). At that time the village was characterized by general hardship, subsistence activity, isolation, lack of family continuity from the previous century (indicating a high degree of population mobility), lack of social distinction between smallholders and farmworkers, high rates of marriage mobility and adherence to the Baptist Chapel. In another paper, Hill has also queried some of the traditional views about the habits of the fenland inhabitants, including incest, and the consumption of opium, and therein has perhaps provided a basis for a revisionist social history of that region (Hill 1992).

## FENLAND DRAINAGE AND ITS EFFECTS

The main environmental hazards of the Fens included the inadequate drainage systems and the problem of flooding of the lower land. These were continuations of problems already familiar from previous centuries (Butlin 1990a). The combination of sea floods of the coastal areas of the Fens through periodic North Sea storm surges (Lamb 1991) and the frequency of flooding of the inland Fens by the large volumes of water, especially in winter, draining from the surrounding uplands, sustained the necessity of finding more effective means of draining the excess water from the land. Improved drainage required not only a better pumping system to lift the water to the raised main drainage channels, however, but also solution of problems such as the management of the outfalls of rivers to the sea, the maintenance of drainage channels, the complexity of administrative authorities and the crucial question of financing maintenance and improvements. The technological questions of the capacity or incapacity of the drainage 'engines' (windmills) to remove the water remained, the latter magnified by the continued problem of peat shrinkage after drainage, (Darby 1983, Butlin 1990a). There was no general agreement on these issues, and indeed it is worth noting that some areas of the Fens actually benefitted from poor drainage, notably the areas where the cutting of turf, reeds and sedge was an important part of the subsistence and commercial economy.

An account of the drainage problems of the Fenlands in the early nineteenth century was given in a review by John Rennie, a civil engineer:

> The rivers above described [the Ouse, Nene, and Cam], rising at a great height in the interior country, and their courses, until they enter the Fens, being also high, the floods come down with great rapidity, and when they enter the low lands, where the fall is gentle, and the channels shallow and narrow, the water cannot get off; it therefore accumulates in the washes, and against the banks, frequently making breaches, and over flowing large districts of land. This took place in the Spring of 1808, that damage was done to the extent of One Million [Pounds]' (Rennie 1810: 3).

He indicates that the channels of the Rivers Stoke, Brand or Little Ouse, and the Lark or Mildenhall, were in a bad state, and these rivers had very weak banks. A major reason for the poor state of drainage of the Fenland in the late eighteenth and early nineteenth centuries was the poor outfall of the Ouse at King's Lynn, comprising a great meander bend and a channel full of sandbanks (Darby 1938: 117–18). The solution adopted was a straightening of the bend by a new cut: Eau Brink Cut was authorized in 1795, and completed in 1821. It was extended further out to sea by the Marsh cut of 1852, and later reinforced by training walls.

The main progress in drainage was via the various attempts to improve the river outfalls into the sea and by introduction of steam-pumping, from about 1820 onwards. Although critical of past attempts at drainage improvement, and providing an engaging and somewhat impractical and idealistic alternative

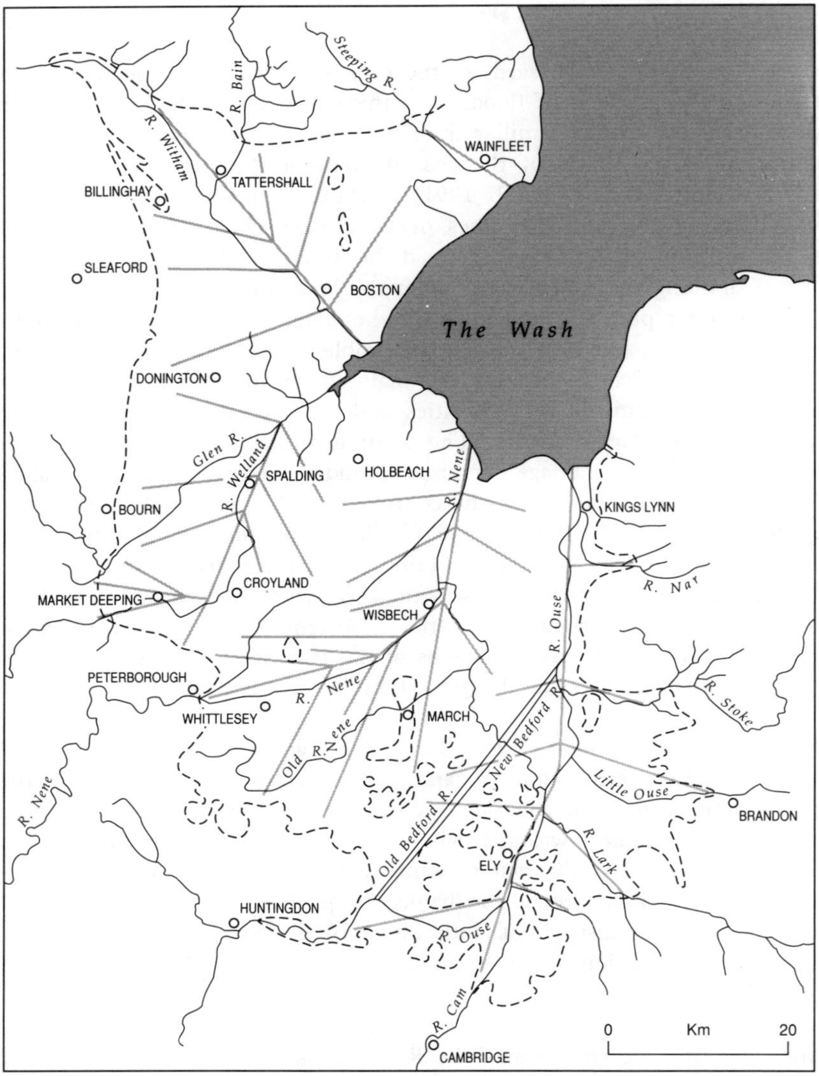


**Figure 8.3** An 'ideal' fenland drainage scheme (from Skertchley 1877: 177).

to the straightening of river-courses (shown in figure 8.3), S. B. J. Skertchley, in his Geological Survey Memoir, *The Geology of the Fenland* (1877), recognized the need for further drainage improvements and the potential of the steam pump. He states (regarding the South Level of the Fens) that:

> Water-mills became very general, and indeed were so numerous as to impart a peculiarity of the landscape. But in the early part of this century the intermittent and uncertain action of the wind was found too precarious a motive force to rely on, especially as the evil had further

**Plate 8.2** An old fen drainage mill at Wicken Fen, *c.* 1870 (from Skertchley 1877: 50).

> increased, and steam mills began to be erected. They were simply large lifting wheels worked by steam, but the more modern ones, introduced about the year 1850, are centrifugal pumps. Windmills have now almost disappeared from the Lincolnshire fens, but they still linger in the Middle and South Levels (Skertchley 1877: 51).

He provides, in the form of an etching of Wicken Fen (50), an illustration both of a small drainage mill and the state of the undrained fen (plate 8.2)

In addition, Gardiner records, in his *A History, Gazetteer and Directory of Cambridge* (1851: 391), that:

> Before the drainage of the fens, there was a large lake or mere at Soham, through which was in ancient times a dangerous passage by water to Ely. This mere, which contains 1,369 acres [554 ha], is now drained by a steam engine of 40 horse power, and has long been converted into tillage, and is very fertile. Middle Fen, also near Soham, and comprising about 7000 acres [2833 ha], is now drained by a steam engine of 60 horse power.

The effect of the steam-engines on the landscape was direct and indirect: direct in the sense that these beam engines within their tall brick-built houses and with adjacent boiler- and scoop-wheel houses became important features of the landscape, and indirect in that they effected the disappearance of the windmill and, of course, the gradual removal of excess water. By 1852 only

170 windmills remained in the Bedford Level and its environs, most of the pumping being carried out by about 43 steam engines (Darby 1983: 176–7).

Darby (1938) suggests that the late eighteenth century had been a time of scepticism about the economic prospects for the Fens, on account of the ineffectiveness of its main drainage machine – the windmill. In contrast, improvement of the agriculture in the Fens through drainage and better management was anticipated by agricultural writers of the early nineteenth century, including Gooch in his *General View of the Agriculture of the County of Cambridge* (1811). His account stressed the actual and potential profitability of the Fens, particularly because of the high price of peat turf, sedge (partly because straw had become very expensive) and of reeds.

More efficient drainage was, however, a mixed blessing. The draining of the fens and meres of the region caused rapid shrinkage of the drained peat, and the lowering of the land surface by about 3 m, seen most dramatically by the Holme Post, an iron column from the Crystal Palace, 5.5 m in length, which was driven into the peat just south of Whittlesey Mere in 1851, until its top was level with the surface, to assess the extent of shrinkage after the drainage of the mere by the installation of an Appold centrifugal pump in 1851. By 1871 the peat had shrunk 2.1 m, and after the installation of a more powerful pump in 1877, the shrinkage by 1892 was 3.5 m (Grove 1962: 111, Hutchinson 1980, Sheail and Wells 1983). Although in some areas the peat had shrunk to a point where claying – the mixing of underlying clay with the peat to give the soil more body – became a practicable and rapidly adopted facility, enabling the more efficient cultivation of wheat, coleseed, oats and turnips (Darby 1938: 120–1), in others where there was no convenient clay or sand to mix with the peat the land remained unimproved.

The effects of this new drainage technique on the landscape in the longer term facilitated more extensive cultivation of wheat oats and potatoes, and the development of market gardening, including the growing of fruit. Although smallholdings predominated, especially in the Fens, the time was not far off, especially after the First World War, when massive consolidation of holdings into the hands of wealthy agricultural speculators would take place, and a new era of capital- and labour-intensive, large-scale farming would occur.

One aspect of the historical geography of Fenland drainage in the later nineteenth century that seems to have received very little attention is that the problem of lack of water, or drought, especially in summer. Skertchley (1877: 52) observes that:

> As to the effect of the present system there cannot be a doubt respecting its efficacy for draining, indeed the Fens now suffer more frequently from drought than flood. It seems never to have struck people that the Fens could be too much drained, yet in December 1874, I walked dryshod along the bottom of drains that ought to have had six feet of water in them, and saw pits sunk four feet below the bottom to obtain water... . Ely was in the autumn for three days waterless, in consequence of the lowness of the river [Ouse], and it is not unusual to see water pumped into the Fens to moisten the pasture in the vicinity of that city... .

He affirmed the need to regulate the discharge of water in order to control aridity, complaining that none of the drainage Acts relating to the Fenland gave authority to store water against times of shortage.

Drainage also effected major changes in the vegetation and ecology of the fens. Woodland was formerly the natural vegetation of the upland areas and the climax vegetation of the fens but, as Hodge and Seale (1966: 18–20) have indicated, 'Fluctuations in water-level... have interrupted or reversed the Fenland succession so that the history of the vegetation is complicated, and open water, reed and sedge swamp, and fen woodland or carr have existed together in varying proportions, especially around the margins'.

Sheail and Wells (1983) have demonstrated in a detailed historical study the effects of drainage on the ecology of the fenlands of nearby Huntingdonshire: a progressive reduction in the number and diversity of wildlife species (in the case of waterfowl partly through the increase in wildfowl shooting, after improvements to the gun had been made), and of butterflies by overcollecting. There was a general impoverishment of habitats, including those for fish.

The arrival from the 1840s, many by railway, of increasing numbers of naturalists in the fens, especially Wicken Fen, subjected the flora and fauna to reduction of species by collecting. Conybeare, in his *A History of Cambridgeshire*, describes some of these effects on Wicken Fen:

> And even in Wicken Fen, though the sedge and the rushes and the reeds still rise, as of old, from the water-logged soil, the swarming animal life which was once so conspicuous a feature of the district can be seen no longer. Now and then a Swallow-tail butterfly is still to be beheld flitting over the water-courses and reedbeds; but its companion, the Great Copper, is utterly extinct, along with the whole population of ruffs and reeves and water-fowl of every tribe, whose cries and motions made a never-ceasing stir throughout what is now the most silent expanse in England (Conybeare 1897: 253).

New crops and rotations were introduced even to the more economically marginal of the newly drained peatlands, and the drainage of the peat fens also produced even more extensive changes in peatland management. The gradual abandonment of much older traditional uses such as peat and sedge-cutting resulted, unless deliberately countered, in new ecological successions. Wicken Fen, the oldest nature reserve in Britain, is an interesting example. The system of land use in Wicken up to the time of enclosure in 1849 involved the cultivation of a small area of open field and enclosed arable, together with the use of various areas of fen for peat-cutting, the cutting of reeds and sedges, and common grazing for cattle. These management systems created a variety of fen habitats, including open water habitats in the drainage ditches or dykes, the lodes (ancient waterways), reed-beds, sedge-fields, and fen carr and woodland:

> In the 17th century the area was cut extensively for sedge and only a small amount of peat was dug. With increasing land drainage, peat

> cutting became the main activity on the fen and this continued until 1847 when the local Enclosure Acts were passed and the practice began to decline. By 1870 very little peat was dug or sedge cut and carr gradually began to encroach across the peat waste (Barringer et al. 1991: 3).

The increase in the spread of carr following the decline in peat- and sedge-cutting, necessitated that the modern management of Wicken Fen, in order to effect as great a restoration as possible to its former conditions, has included the revival of reed-, sedge- and peat-cutting in some areas, and the raising of the water level, while elsewhere encouraging the natural succession from sedge fen to carr and fen woodland.

## DRAINAGE ADMINISTRATION

One of the ubiquitous features of drainage management in the Fens, and indeed elsewhere in England, was the anachronistic nature of the administrative systems. This is a point which Darby stresses in his major works on the Fens:

> Looming above the technical questions of outfall difficulties and their associated internal problems was the legacy of antiquated administrative arrangements. Commissioners of Sewers [medieval in origin] still continued to look after the drainage of some areas, and they had power to levy rates to sustain their operations. Their ancient rights were codified and their working clarified by the Sewers Act of 1833, but this did nothing to affect the fundamental basis of drainage administration (Darby 1983: 170).

The Bedford Level Corporation's powers, established in the seventeenth century, were modified in the eighteenth and nineteenth centuries. Of the Level's three parts (North, Middle and South), North Level had been virtually autonomous since 1754, and was totally independent in 1857, as was the Middle Level in 1810, leaving the Corporation with responsibility only for the South Level, which ceased to exist in 1920, with the Ouse Drainage Act which created the Ouse Drainage Board. In addition, there were internal drainage boards such as the Eau Brink Commissioners and the South Level Commissioners (1827). The Land Drainage Act of 1861 changed the powers of the Commissioners of Sewers and created new authorities in the form of elected Drainage Boards:

> These were to have all the powers of Commissions of Sewers but, instead of being appointed by the Crown, they were to be elected by people who paid the drainage rates. The Act, however, did nothing to remove the outstanding difficulty that bedevilled all attempts at effective draining. This was the lack of a single unified control along the whole length of a river (Darby 1983: 172).

Drainage taxes were levied, although the poorer fen-users paid at lower rates, and the drainage boards continually attempted to improve the drainage maintenance systems and occasionally to effect more substantial capital investments, but the major solution of the problems of fen drainage had to await the greater statutory powers, greater financial investments, and the new drainage authorities of the twentieth century.

## POVERTY AND LIVING CONDITIONS IN THE FENS IN THE NINETEENTH CENTURY

Poverty was widespread in the Fens and in much of East Anglia. Conybeare, writing in 1897 states that 'Cambridgeshire has become for the time one of the poorest counties in England' (Conybeare 1897: 269) and there are many accounts of the extent and consequences of poverty in and for the fen-edge settlements. Each of these had its private charities and public poor relief systems, involving the Old Poor Law until 1834 and the New Poor Law thereafter. Soham, the largest of the four settlements, figured prominently in the history of the workhouse movement in Cambridgeshire (Hampson 1934).

At the beginning of the nineteenth century, the percentage of the population of each settlement receiving poor relief was as follows: Isleham 5.7 per cent, Soham 21 per cent, Fordham 7.2 per cent, and Wicken 11.1 per cent (in Soham over half the paupers were children) (Munby 1967: 41). A short entry on Soham in the evidence from Alfred Power's report to the Poor Law Commissioners, published in 1834, reads:

> I had the advantage of meeting with the parish officers of Soham here, one of the largest parishes in the county. The rates have been as high as £3000 but the extent of the parish is 12,000 acres, and the population in 1831 was 3,667. At the present time there are about 150 labourers unemployed in this parish; they are sent to the road and the gravel-pits. At one time last year there were as many as 200. This continues about two months; during the rest of the year there is full employment (Power 1834: Appendix A, 249).

On 10 January 1852, the Soham entry in the *Cambridge Chronicle* stated: 'The poor are in a most deplorable state, owing to lack of employment, they wander the streets, objects of regret and commiseration'. In November of the previous year, under the heading of 'Free Trade', the reporter stated that 'Emigration is still going on, some fathers going alone, and leaving their children to be supported by the parish. Four and five families receiving parochial relief'. (Turner 1985: 18). On 28 October 1853, the reporter described the outbreak of 'Asiatic Cholera' in Soham, which also affected Isleham very badly, although the outbreak was over by the end of November that year. Each of the settlements had its own charities, going back to Tudor times, and in addition Soham had a very large area of common land which provided a certain amount of grazing for the animals of the poor, together with the areas of fen where turf could be won.

The intensity of use of the poor's common may give some indication of the pressure of population on land by the mid-nineteenth century. East Fen at Soham seems to have been particularly prone to squatting (illegal settlement), evidenced by an entry in the *Cambridge Chronicle* for 26 March 1864: 'Common Land. Not many parishes have such extensive commons nor so much waste land for the poor as Soham. On proportions of these persons have from time to time erected sheds, the sheds converted into cottages and later into substantial houses, and bit by bit land has been taken in for gardens' (Turner 1985: 28). The use of the 'poor's commons' had been a subject of careful scrutiny by the manorial courts of Soham and Fordham since the seventeenth century, although undoubtedly the allocation of grazing on these lands was too great a temptation, resulting in overgrazing, deterioration of quality, and the easy transmission of diseases between animals.

Acts of Parliament of 1758, 1789 and 1800, involving the drainage of the Soham or Middle Fen District, made provision for the reduction in drainage tax for the poor of Soham and Isleham (CRO R 64/5 Archers). The 1758 Act refers to:

> fifty acres of fen land belonging to the working poor of Isleham, and seventeen acres of land belonging to the hospital, and other poor of Isleham, and as the charging the said lands with the same tax as is hereby imposed on other lands within this district, would be taxing them more than they are able to pay; it is hereby further enacted, That the said lands shall never be taxed or charged with more than half the tax wherewith other lands within the same district are or shall be charged by virtue of this act (15).

The act continues with this theme of the poor (15–16) and states that because of the extensive use of the common and lands in Soham and Isleham by the poor, it would be very difficult to collect these drainage taxes, and the Commissioners were therefore empowered 'to sever so much of the said common and lands from the rest thereof, as shall be sufficient to raise the taxes charged thereupon, and to let the same from year to year'. The Acts of 1789 and 1800 continue this provision.

The environment of the Fens, the densities of population. and the low standards of living meant that many of the inhabitants had poor health, being subject to malnutrition, rheumatic fevers or agues, malaria, cholera and smallpox. Malaria was a common problem in most marshland areas of England. A report to the Privy Council, by Dr George Whitley in 1863 (*A Report as to the Quantity of Ague and other Malarious Diseases now Prevailing in the Principal Marsh Districts of England*), shows the prevalence of this type of disease in the lower areas of Kent, Essex, the Cambridgeshire and Lincolnshire Fens, Somerset, Lancashire and the Scottish Borders (Bruce-Chwatt and M. de Zulueta 1980: 133–4). M. Dobson's (1980) study of the geography of malaria in England indicates that the indicative contemporary term 'ague' (which literally means 'acute'), may have covered many fevers, including malaria, typhus fever, enteric fevers and influenza, and that there is some uncertainty as to whether English mosquitoes could transmit strains of malaria to humans. Attacks of 'ague' were, however, not usually fatal but very debilitating, at least as the evidence available for the counties of Essex and Kent suggests. Dobson contends that it was true *Plasmodium* malaria, which continued to the end of the nineteenth century, in the English marshes, its decline resulting from better drainage and the extensive use of quinine. At Soham and Isleham in 1853 there was a major epidemic of cholera, partly due to large populations and poor living and drainage and sanitation conditions. Smallpox was also common in the nineteenth century.

One remedy against the evils of life and poverty in the fens was the use of opium. Initially drunk in the form of 'poppy-head tea', it was also eaten. It was generally available in chemists shops in the fen areas, with very large quantities sold and taken, so that there was extensive resistance to the 1868 Pharmacy Act from pharmacists in the fen districts. Opium poppies had long been grown locally, although imported commercial opiates later became important. In their book on *Opium and the People*, Berridge and Edwards (1981) have a separate chapter on the Fens. Hence:

> For much of the first half of the [nineteenth] century, the Fens remained an unhealthy, marshy area, where medical assistance was limited, especially for the poor, and where large numbers of the working population of the countryside was prone to 'ague', 'painful rheumatisms' and neuralgia..., [and they cite a contemporary account that asserts that] 'There was not a labourer's house... without its penny stick or pill of opium, and not a child that did not have it in some form' (Berridge and Edwards 1981: 38–9).

## CONCLUSIONS

In the nineteenth century, the relations between the human populations of the Fenland region of eastern England and their varied and hazardous natural environment were influenced by a complex set of factors. Technological and increasingly market-driven economic pressures, together with the need to drain and reclaim land for more efficient and profitable production, further accelerated processes of land-use transformation dating back to Roman times (but effectively to the seventeenth-century capitalized reclamations). This intensification of land use – of arable farming in particular – had profound consequences for the natural ecology of the fens, including wildlife habitats, although periods of economic depression, for example in the late nineteenth century, did allow some reversion back towards less intensive forms of land management. Increased access to growing urban markets of the industrial North, the Midlands and London, and the increasing commodification of land through the processes of rural capitalism, brought pressures to reform land tenure and production systems of the islands and higher land of the Fens, which were transformed from late medieval open-field and communal management systems to those based on enclosure, individual tenure, and modern crop rotations. The Fenland economy and society did not, however, yield easily to these pressures for transformation, and complex systems of landholding persisted, together with the large numbers of smallholders, cottagers and squatters, who maintained their marginal existence in social and economic conditions characterized by poverty and exposure to disease and epidemics. All this occurred within a hazardous, frequently flooded and waterlogged environment, which was gradually mastered as drainage efficacy increased, but was always subject to the setbacks of extreme events and also to problems resulting from more effective drainage, such as drought, especially during the summer, and the shrinkage and wind erosion of the surface peat. This was also a period that witnessed early attempts to conserve the fenland environment for scientific investigation and public education.

An understanding of these processes is of interest in its own right for historical geographers and environmental historians, but it is also vital as a platform from which to develop ideas and methods for the conservation and restoration of precious wetland environments.

## ACKNOWLEDGEMENTS

I am most grateful to Dr Sarah Bendall, Dr Neil Roberts and Professor John Sheail for their helpful comments on an earlier draft of this chapter. Permission to cite from documentary records is acknowledged from The Provost and Fellows of King's College, Cambridge and the Cambridgeshire County Record Office, Cambridge. Crown copyright material from the Public Record Office, Kew, is reproduced with the permission of the Controller of Her Majesty's Stationery Office.

## DOCUMENTARY SOURCES

*Cambridgeshire County Record Office, Cambridge (CRO)*

Parish of Wicken, Enclosure act, map and award: 1840/1849. Q/RD c 69; R.a 10
Parish of Fordham, Enclosure map and award, 1820. Q/RD z 9
Parish of Isleham, Map of open fields, pre-enclosure map, 1844. P/515
Parish of Isleham, Tithe award map, 1848. 311/P1
Parish of Isleham, Enclosure award, 1848 plan Q/RD c 74
Drainage records, Soham or Middle Fen District. R 64/5 Archers

*Manuscript, King's College, Cambridge (KCC)*

Survey of an estate in Isleham belonging to King's College, Cambridge, May 1842. KCC MSS ISL/24

*Public Record Office, Kew (PRO)*

Tithe Files: Soham, PRO IR 18 13632; Wicken, PRO IR 18 13665

REFERENCES

Barringer, C. et al. (1991) *East Anglian Studies II: History of the Environment* (The Open University, Milton Keynes)

Barringer, C. (1991) 'Field Guide 7. Soham: a relict medieval landscape', in Barringer et al. (1991), op. cit.

Bendall, A. S. (1992) *Maps, Land and Society: a History with a Carto-bibliography of Cambridgeshire Estate Maps*, c.*1600–1836* (Cambridge University Press, Cambridge)

Berridge, V. and Edwards, G. (1981) *Opium and the People* (Allen Lane, London)

BPP (British Parliamentary Papers) (1888–94) *Reports from the Select Committee and Others on Allotments, Small Holdings and Peasant Proprietors*

Bruce-Chwatt, L. J. and Zulueta, J. de (eds) (1980) *The Rise and Fall of Malaria in Europe* (Oxford University Press, Oxford)

Butlin, R. A. (1990a) 'Drainage and land use in the Fenlands and Fen-edge of northeast Cambridgeshire in the seventeenth and eighteenth centuries', in Cosgrove, D. and Petts, G. (eds) *Water, Engineering and Landscape* (Bellhaven Press, London) pp. 54–76

Butlin, R. A. (1990b) 'Small-scale urban and industrial development in north-east Cambridgeshire in the nineteenth and early twentieth centuries', in Sporrong, U. (ed.) *The Transformation of Rural Society, Economy and Landscape* (Stockholm University, Stockholm) pp. 217–26

Butlin, R. A. (1993) *Historical Geography: Through the Gates of Space and Time* (Edward Arnold, London)

Caird, J. (1852) *English Agriculture in 1850–51* (Longman, Brown, Green and Longmans, London)

Cooke, G. (1820) *Topographical and Statistical Description of the County of Cambridge* (Cambridge)

Conybeare, E. (1897) *A History of Cambridgeshire* (Elliot Stock, London)

Darby, H. C. (1938) 'The nineteenth century', in Darby, H. C. (ed.) *The Cambridge Region* (Cambridge University Press, Cambridge) pp. 116–34

Darby, H. C. (1983) *The Changing Fenland* (Cambridge University Press, Cambridge)

Dobson, M. (1980) ' "Marsh fever" – the geography of malaria in England', *Journal of Historical Geography* 6(4): 357–90

Gardiner, R. (1851) *A History, Gazetteer and Directory of Cambridgeshire ... Including the Isle of Ely* (Cambridge)

Gooch, W. (1811) *General View of the Agriculture of the County of Cambridge* (Board of Agriculture, London)

Grove, A. T. (1962) 'Fenland', in Mitchell, J. B. (ed.) *Great Britain: Geographical Essays* (Cambridge University Press, Cambridge) pp. 104–22

Hampson, E. M. (1934) *The Treatment of Poverty in Cambridgeshire* (Cambridge University Press, Cambridge)

Hill, P. (1990) *The History of Isleham Fen in the 1930s* (privately published, printed by Piggott Printers, Cambridge)

Hill, P. (1992) 'Who were the Fen people?', *Proceedings, Cambridge Antiquarian Society* LXXXI: 97–114

Hodge, C. A. H. and Seale, R. S. (1966) *The Soils of the District Around Cambridge* (Agricultural Research Council, Memoirs of the Soil Survey of Great Britain, England and Wales, Harpenden)

Hutchinson, J. N. (1980) 'The record of peat wastage in the East Anglian fenlands at Holme Post, 1848–1978 A.D.', *Journal of Ecology* LVIII: 229–40

Jonas, S. (1846) 'On the farming of Cambridgeshire', *Journal of the Royal Agricultural Society VII*: 35–72

Lamb, H. H. (1991) *Historic storms of the North Sea, British Isles, and Northwest Europe* (Cambridge University Press, Cambridge)

Mills, D. R. (1984) 'The nineteenth-century peasantry of Melbourn, Cambridgeshire', in Smith, R. M. (ed.) *Land, Kinship, and Life-cycle* (Cambridge University Press, Cambridge) pp. 481–518

Munby, L. M. (1967) *Village History in Staploe Hundred* (Soham Village College: University of Cambridge Board of Extra-Mural Studies, Cambridge)

Postgate, M. R. (1964) 'The open fields of Cambridgeshire', unpublished Ph.D. thesis, University of Cambridge

Power, A. (1834) 'Soham Parish', in *Report from Alfred Power Esq. to the Commissioners Appointed by his Majesty to Inquire into the Administration of the Poor Laws. First Report from the Commissioners on the Poor Laws. British Parliamentary Papers, XVIII, Appendix A* (London) pp. 249–50

Rennie, J. (1810) *Report and Estimates on the Improvement and Drainage and Navigation of the South and Middle Levels of the Great Level of the Fens* (London)

Rowell, T. A. (1982) 'Wicken Poor's Fen: origins and land use history' (Bulletin No. 37, Cambridgeshire Local History Council, Cambridge) pp. 7–9

Rowell, T. A. (1986) 'Sedge (*Cladium mariscus*) in Cambridgeshire: its use and production since the 17th century', *Agricultural History Review* 34: 111–42

Rowell, T. A. and Harvey, H. J. (1988) 'The recent history of Wicken Fen, Cambridgeshire, England: a guide to ecological development', *Journal of Ecology* 76: 73–90

Seale, R. S. (1975) *Soils of the Ely District* (Memoirs of the Soil Survey of Great Britain, England and Wales, Harpenden)

Sheail, J. and Wells, T. C. E. (1983) 'The Fenlands of Huntingdonshire, England: a case study of catastrophic change', in Gore, A. J. P. (ed.) *Mires: Swamp, Bog, Fen and Moor. B. Regional Studies* (Elsevier, Amsterdam) pp. 375–93

Skertchley, S. B. J. (1877) *The Geology of the Fenland* (Memoir of the Geological Survey, HMSO, London)

Turner, M. (1980) *English Parliamentary Enclosure* (Dawson, Folkestone)

Turner, M. (1985) *Soham Chronicle 1787–1899* (Hobbs, King and Parr, Soham)

Vancouver, C. (1794) *General View of the Agriculture in the County of Cambridge: with Observations on the Means of its Improvement* (W. Smith, London)

9

# Woodland History and Plant Indicator Species in North-east Yorkshire, England

Richard Gulliver

ABSTRACT

The occurrence of 35 herbaceous woodland plant species has been recorded in 160 woods in north-east Yorkshire, England, together with a range of other site attributes. Printed and map sources have also been utilized. By combining these various lines of evidence it has been possible to divide the woodlands into five main age categories, plus conifer woods, coniferized woods and wood pasture. Some woodland features, such as woodbanks and genuine, long-established, mixed-species coppice, which in many parts of England are valuable indicators of ancient woodland status, occurred at too low a frequency to be useful for the recognition of most or all of the ancient woodland. By contrast, both ancient and primary woodland could be rapidly identified and differentiated by their complement of non-arboreal indicator species. The application of the methodology utilized in this study will allow the amounts of woodland in the different age categories to be quantified, and human impact on woodland resources and on the landscape to be assessed. The importance of utilizing a multi-disciplinary approach is emphasized: the relative value of the component techniques used in determining woodland age is, however, likely to vary from one geographical region to the next.

## INTRODUCTION

The last three decades have seen great development in the study of Britain's woodland history, which has included investigations of their prehistory (e.g. Simmons and Innes 1981, 1988), soil profiles (e.g. Ball and Stevens 1981), woodland invertebrates (e.g. Steele and Welch 1973, Kerney and Stubbs 1978, Stubbs 1982, Garland, 1984), lichens, (e.g. Rose 1976), case studies of woodland herbs (e.g. Peterken and Game 1984), case studies of trees strongly associated with ancient woodland (e.g. Pigott 1969), traditional uses of timber products (e.g. Linnard 1982), medieval cycles of coppicing (e.g. Rackham 1971), and medieval woodland distribution patterns (e.g. Wightman 1968). The subject has also been reviewed by Marren (1990), Peterken (1981), Rackham (1986, 1990) and Watkins (1990).

These developments have run alongside an awareness that ancient and primary woodland (defined subsequently) is being lost to agriculture, conifer afforestation, mineral extraction and urban development (Spencer and Kirby 1992). In response to this trend in England and Wales, the former Nature Conservancy Council (now English Nature and the Countryside Council for Wales) established the Ancient Woodland Inventory to document the extent and nature of surviving ancient woodland over 2 ha in extent (Spencer and Kirby 1992). The scale of this undertaking has meant that only a proportion of ancient woods could be surveyed in the field. For the rest, ancient woodland status was deduced from early Ordnance Survey (OS) maps, plus published and manuscript sources where available (e.g. Carter 1987). The Ancient Woodland Inventory has demonstrated the great regional variation in the extent of ancient woodland; and the characteristics of ancient woodland have also been shown to vary greatly from one region to the next (Peterken 1981, Rackham 1990).

### *Age categories of woodland*

Woodland in Britain is divided into four main age and origination categories, based on the work of Peterken (1981) and Rackham (1980, 1986, 1990).

*Primary woodland* has had tree cover present on the site since before 4000 BC when humans first started to make a serious impact on the British landscape. However, the nature of the tree cover and the composition of the shrub and herb communities are likely to have changed radically over time.

*Secondary woodland* is woodland on a site which had previously been non-wooded (e.g. grassland, arable or moorland) usually for a very long period of time. Tree colonization may have been spontaneous, or by planting or sowing. Some secondary woodland has been in existence for many centuries, for example, some sites date back to the end of the Romano-British period. Conversely some may only be a few decades old.

*Ancient woodland* is woodland which, it is assumed, has been present in the medieval period, either (a) because documentation about the wood survives from the medieval period, or (b) because maps and/or documentation exists in the Tudor or Stuart period, and this is taken to indicate that the wood was also present in the medieval period. In a few cases it may be possible to date a woodland directly from ring counts of surviving trees. Peterken (1981: 12) requires the documentary evidence to have existed by AD 1600 for a wood to be classed as ancient, whereas Rackham (1980, 1986, 1990) uses AD 1700 as the cut-off date for designation of ancient woodland status, and stresses the importance of the wood as a component of a complete ancient landscape. Where it can be clearly demonstrated (usually from archaeological evidence) that the site was once non-wooded, the term 'ancient secondary woodland' is applied. Spencer and Kirby (1992) subdivide ancient woodland into two categories: semi-natural ancient woodland, in which the canopy has arisen by natural regeneration from native trees, and other ancient woodlands, for which the current canopy has been planted, in part or in whole – albeit there is a continuity of use as woodland at the site.

*Recent secondary woodland* is secondary woodland which is post-ancient in origin (Peterken 1981: ch. 5, Peterken and Game 1984). The use of the words 'wood' or 'woodland' in this chapter denote broad-leaved woodland; conifer woods are explicitly described as such.

### *Plant species characteristics of ancient and primary woods*

Plant species specifically associated with primary woods (identified using cartographic, documentary, archaeological and topographical evidence) in central Lincolnshire were studied by Peterken (1974), and six categories of species fidelity were recognized:

1 those confined to primary woodlands
2 those confined to primary woodlands and sites believed to have once been primary woodlands
3 those almost confined to primary woodlands, very rarely colonizing secondary woodlands
4 those found at most localities in primary woodland, rarely colonizing other sites
5 those found at most localities in primary woods or other long established habitats
6 native trees and shrubs confined to primary woods and ancient hedges

Peterken and Game (1984) surveyed the total vascular plant species complement of 89 ancient and 273 secondary woods, also in central Lincolnshire. Each species could then be graded from 100 per cent faithful to ancient woods (i.e. an absolute indicator) through neutrality to being found preferentially in secondary woods. Hence, in central Lincolnshire, a near-perfect indicator system exists alongside an excellent set of documentary sources for woodland history. However, the indicator species technique is especially useful when only the minority of woods can be dated from documentary sources (as in north-east Yorkshire), indicator species can then be identified at these marker woods and the results extrapolated to other woods. The technique founders on circular arguments when virtually no historical maps or documents exist. Supplementary evidence, such as the existence of woodbanks (discussed subsequently) and long-established coppice stools, can, however, also be deployed (Rackham 1980, Peterken 1981).

## STUDY AREA AND METHODS

The area investigated (Figure 1) consisted of the eastern half of the old (pre-1974) North Riding of Yorkshire (east of easting SE/NZ 40) plus the northern half of the old East Riding of Yorkshire (north of northing SE/TA 40) which together constitute the region of north-east Yorkshire and which includes the

**Table 9.1** The category and number of occurrences of the 35 study species

| *Plant category (see text)* | *Scientific name* | *English name* | *Number of occurrences in 160 woods* | *Number of occurrences in 40 post-1850 rectilinear woods* |
|---|---|---|---|---|
| MC | *Mercurialis perennis*[a] | Dog's mercury | 115 | 15 |
| MC | *Hyacinthoides non-scripta*[a] | Bluebell | 110 | 12 |
| MC | *Oxalis acetosella*[a] | Wood sorrel | 105 | 13 |
| F | *Primula vulgaris*[a] | Primrose | 95 | 3 |
| F | *Veronica montana*[a] | Wood speedwell | 82 | 5 |
| F | *Allium ursinum*[a] | Ramsons | 80 | 4 |
| F | *Anemone nemorosa*[a] | Wood anemone | 79 | 2 |
| F | *Sanicula europaea*[a] | Sanicle | 70 | 5 |
| A | *Carex sylvatica*[a] | Wood sedge | 66 | 2 |
| A | *Chrysoplenium oppositifolium* | Opposite-leaved golden saxifrage | 56 | – |
| A | *Adoxa moschatellina* | Moschatel | 47 | 5 |
| A | *Luzula sylvatica*[a] | Great hairy woodrush | 46 | – |
| A | *Galium odoratum*[a] | Woodruff | 42 | 4 |
| A | *Orchis mascula*[a] | Early purple orchid | 40 | 4 |
| A | *Luzula pilosa*[a] | Hairy woodrush | 37 | 1 |
| A | *Campanula latifolia*[a] | Giant bellflower | 35 | – |
| A | *Melica uniflora*[a] | Wood melick | 29 | – |
| A | *Carex pendula*[a] | Pendulous sedge | 28 | – |
| A | *Geum × intermedium*[b] | Hybrid geum | 27 | – |
| A | *Viola reichenbachiana*[a] | Early dog-violet | 25 | – |
| A | *Carex remota*[a] | Remote sedge | 22 | – |
| I | *Lathraea squamaria* | Toothwort | 12 | – |
| A[c] | *Chrysoplenium alternifolium* | Alternate-leaved golden saxifrage | 9 | – |
| I | *Hordelymus europaeus* | Wood barley | 9 | – |
| A | *Geum rivale*[a] | Water avens | 8 | – |
| I | *Convallaria majalis*[a] | Lily of the valley | 7 | – |
| A[c] | *Hypericum pulchrum*[a] | Beautiful Saint John's wort | 7 | – |

**Table 9.1** *(continued)*

| *Plant category (see text)* | *Scientific name* | *English name* | *Number of occurrences in 160 woods* | *Number of occurrences in 40 post-1850 rectilinear woods* |
|---|---|---|---|---|
| I | *Paris quadrifolia*[a] | Herb Paris | 6 | – |
| A | *Stellaria nemorum* | Wood stitchwort | 6 | – |
| A[c] | *Vicia sylvaticum* | Wood vetch | 6 | – |
| I | *Aguilegia vulgaris* | Columbine | 5 | – |
| A[c] | *Melampyrum pratense*[a] | Cow-wheat | 5 | – |
| A[c] | *Actaea spicata* | Baneberry | 4 | – |
| I | *Lamiastrum galeobdolon*[a,d] | Yellow archangel | 4 | – |
| A | *Gagea lutea* | Yellow star of Bethlehem | 3 | 1[e] |

[a] Species which in central Lincolnshire have more than 50 per cent of their woodland stations in ancient woods (Peterken and Game 1984).
[b] *Geum* × *intermedium* was not included in the calculations when *Geum rivale* was also present.
[c] These species only occurred in ancient or primary woods.
[d] *Lamiastrum galeobdolon* can also occur in woods as a result of dumping unwanted plant material from gardens (Packham 1983).
[e] In a riverside rectilinear post-1850 wood.

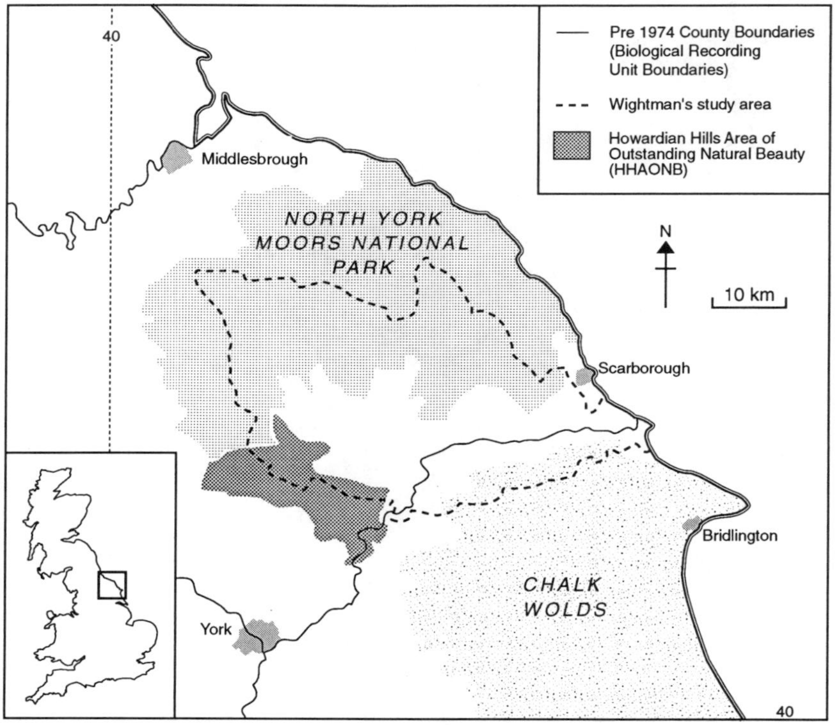


**Figure 9.1** The study area of north-east Yorkshire.

North York Moors National Park and the Howardian Hills Area of Outstanding Natural Beauty. Initially, a sample of 50 woods was surveyed and notes made of the plant species present. From these data it was possible to identify 35 herbaceous plants (all spermatophytes) which were typically present in woodland and ware only rarely to be found in other habitats, and which tended to be absent or very rare in recent secondary woods. These 35 species, known as the 'study species', are listed in Table 9.1. The pilot study further indicated that plant data gathered on the whole site basis (rather than from uniform sized plots) would probably yield meaningful results. A further 110 woods (none of which occurred on markedly acid soils) were then investigated and the study species recorded, together with the presence or absence of eight species of shrub (the study shrubs). These were field maple (*Acer campestre*), hazel (*Corylus avellana*), dogwood (*Cornus sanguinea*), spurge laurel (*Daphne laureola*), spindle (*Euonymus europaeus*), wild privet (*Ligustrum vulgare*), bird cherry (*Prunus padus*) and guelder rose (*Viburnum opulus*). In addition, the presence of two types of tree was noted, genuine wild crab apple (*Malus sylvestris*) and native hybrid lime (*Tilia cordata* × *platyphyllos*), together with details of the main canopy species. Nomenclature follows Stace (1991). Notes were also made of field archaeological features, and of indications of previous periods of woodmanship, such as well-established coppice stools and boundary pollards.

Evidence of natural slumping on slopes was recorded, and a range of features extracted from the 1 : 25 000 OS maps; for example, proximity to a present-day civil parish boundary, the shape of the external boundary of the wood, and the presence of watercourses.

Entries in the Domesday Book (Faull and Stinson 1986) were used to establish the presence on estates of woods associated with particular townships in 1086. Printed sources (e.g. North Riding Record Series (NRRS) 1894, and Wightman 1968) were examined to determine which woods were known to be present in the medieval and subsequent periods. Wood names and woodland sites appear to show a high degree of historical continuity: for instance, of the ten woods shown by Wightman (1968) to have been present in the Pickering–Newton–Lockton–Thornton area *c.*AD 1300, seven are named in the Duchy of Lancaster surveys in the early seventeenth century (NRRS 1894, Counsell 1990), while the remaining three had almost certainly undergone name changes. Similarly, 14 of the 160 woods surveyed were present in Wightman's study area (figure 9.1) *c.*AD 1300, and ten of these still bear the same name today. In the Howardian Hills, wood names used by botanists in the late eightheenth and early nineteenth centuries are the same as those in use today (Rob 1963, Gulliver 1990).

Following Peterken (1974), the occurrence or otherwise of each of the 160 woods on the first edition 1 : 63 360 map was noted. For the study area these were surveyed between 1848 and 1854, and a single date of 1850 is used in the text for convenience. The resolution of the 1 : 63 360 maps was sufficiently good to allow even very narrow roadside strips of woodland to be identified; however, for Yorkshire the 1 : 10 560 maps were surveyed at the same time and hence were consulted where necessary.

A special effort was made to ensure that a large sample of post-1850 woods was investigated, but other than this no particular selection procedure was utilized; the woods visited all lay on or near public footpaths or were adjacent to roads. An overlay grid with a 100 dots in a 1 × 1 km grid square was produced, which allowed a size index (which approximated to area in ha) to be worked out. The majority of the woods were broad-leaved throughout, some were mixed, and a few were part coniferized.

## RESULTS

The data were initially sorted three ways: by the nature of the boundaries of the woodland, by the presence or absence of the wood on the first edition OS 1 : 63 360 map, and by the richness of the study species. Woods with rectilinear boundaries had a low count of study species, whether pre- or post-1850. Furthermore, these woods were integral with the surrounding (usually post-enclosure) landscape. Occasionally, the boundary was curved rather than straight (possibly sometimes for landscaping purposes), but always fitted into the matrix of surrounding field boundaries. Woods with irregular boundaries on all or most sides were usually set in a landscape of fields with non-rectilinear boundaries, and fell into three groups. One small group consisted of woods with a very high species count. A second largish group consisted of moderately species-rich woods, while a third group was transitional between the second one and the rectilinear woods. Landscape features and historical sources were then used to elucidate the nature of these three groups of woods, as described subsequently. A further finding was that woods on the chalk wolds and their escarpments always seemed more species-poor than comparable woods in other areas. Accordingly, these woods have been treated in a separate but parallel way in the text.

Fourteen of the 160 woods investigated had been identified by Wightman (1964, 1968) as being present *c.*AD 1300, based on a comprehensive search of all available medieval documents relating to the central part of the current study area (see figure 9.1). Two of the 14 were in the group with a very high count of study species; 11 were in the group of moderately species-rich woods, and one was coniferized. Hence initial indications suggested that these groups corresponded to primary and ancient woods respectively.

The ranking of the study species by abundance (table 9.1), coupled with an analysis of the types of wood in which they occurred, has allowed the plant species to be divided into four categories. The 'Moderate Colonizers' (MC) were the three most abundant overall of the study species, and occurred in at least 12 of the post 1850 woods. Next in the ranking were the 'Frequent' (F) species, with at least 70 occurrences overall and with a certain limited ability to colonize post 1850 woods. Furthermore, most members of this group were occasionally to be found in other habitats, a feature exemplified by the primrose, *Primula vulgaris*. At the opposite end of the scale were the I (Infrequent) species with their relatively low abundance and strong association with the richest woods. All other species were classed as A species.

## THE FEATURES OF THE MAIN KINDS OF WOODLAND

### *Primary woods*

These seven woods which had a extremely high count of study species (21 or more, unless on the chalk wolds), also had several other attributes which led to their categorization as primary woods. All had very steep slopes in sections of the wood. Four had native hybrid limes as trees or as coppice stools. Two were named by Wightman (1964, 1968) as being present in *c.*AD 1300. Each contained at least three I species. For woods not on the chalk wolds and their escarpments, the range of study species and mean occurrence per woodland type is shown in table 9.2, and their locations are plotted in figure 9.2. Primary woods were differentiated completely from all others both by the overall number of study species recorded and by the number of I species present. A parallel situation occurred on the chalk wolds and their escarpments (see table 9.3).

The sloping parts of all the primary woods, which were too steep for arable cultivation, were likely to have only ever been used for timber and underwood production. In three cases they were so steep that fences and hedges would not remain in position, and any stock present would risk death or injury. All

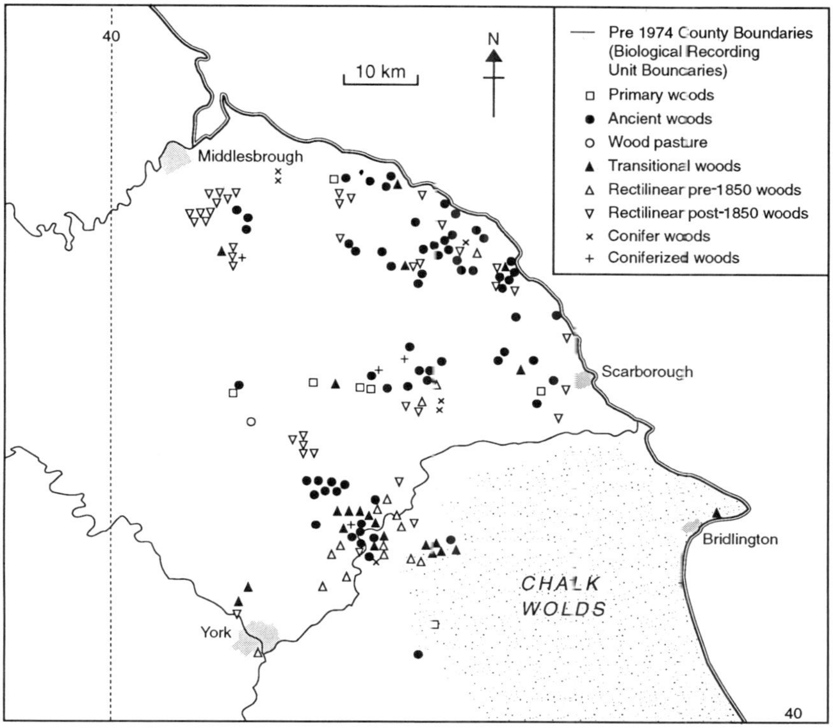


Figure 9.2 The distribution of the 160 woods in the study area.

the primary woods (except the one on the chalk) had watercourses passing through, or acting as a boundary alongside which herbaceous plants could survive during periods of unfavourable management, e.g. heavy grazing or extremely dense canopies.

Hybrid lime (*Tilia cordata* × *platyphyllos*) was present in four primary woods. However, the locations of the limes and their gross morphology (both standards and coppice stools) was strongly suggestive of a natural origin: that is, they are almost certainly descended from fragments of natural hybrid swarms between *Tilia cordata* and *Tilia platyphyllos*, when both parent species were much more common in the landscape, and have survived despite the local extinction of the parent species. It is almost inconceivable that they are introduced hybrid lime (*Tilia* × *europaea*), as they are located in the most inaccessible (and often the most precipitous) parts of the wood. By contrast, anyone planting European lime as a specimen tree would naturally select a high-profile location, such as the entrance to the wood or along the major rides. Native hybrid lime is strongly associated with ancient woodland in Derbyshire (Pigott 1969); it also occurs in the Wye Valley on the English/Welsh border (Pigott 1991).

All but one of the primary woods was located, at least in part, on chalk or limestone. The designation of primary status may be easier to implement for vascular plants in areas with base-rich soils, as several of the potential

**Table 9.2** The number of species in each plant category for five types of wood, except those on the chalk wolds and their escarpments

| *Type of woodland* | *Number of woods* | *Presence of a rectilinear boundary* | *Plant category* | | | | | *Total* | *Study shrubs* |
|---|---|---|---|---|---|---|---|---|---|
| | | | *MC* | *F* | *A* | *A + MC + F*[a] | *I* | | |
| *Range of number of study species* | | | | | | | | | |
| Primary | 6 | – | 3 | 5 | 10–14 | 18–22 | 3–5 | 21–25 | 3–8 |
| Ancient | 62 | – | 1–3 | 2–5 | 4–12 | 9–20 | 0–2 | 9–20 | 0–4 |
| Transitional | 17 | –[b] | 1–3 | 0–5 | 1– 3 | 5–10 | 0 | 4–10 | 0–3 |
| Rectilinear pre-1850 | 13 | + | 0–2 | 0–2 | 0–1 | 0–4 | 0 | 0–4 | 0–4[c] |
| Rectilinear post-1850 | 40 | + | 0–2 | 0–2 | 0–2 | 0–6 | 0 | 0–6 | 0–1 |
| Total | 138 | | | | | | | | |
| Mean species value per category | | | | | | | | | |
| Primary | | | 3.0 | 5.0 | 11.5 | 19.5 | 3.8 | 23.3 | 5.2 |
| Ancient | | | 2.9 | 4.3 | 7.4 | 14.6 | 0.3 | 14.9 | 2.5 |
| Transitional | | | 2.4 | 2.6 | 1.7 | 6.7 | 0.0 | 6.7 | 1.3 |
| Rectilinear pre-1850 | | | 1.1 | 1.0 | 0.2 | 2.3 | 0.0 | 2.3 | 1.1 |
| Rectilinear post-1850 | | | 1.0 | 0.6 | 0.4 | 2.0 | 0.02 | 2.0 | 0.2 |

[a] See table 9.4.
[b] With one exception, believed to be rectilinear by contraction.
[c] The shrub complement in the two rectilinear pre-1850 woods with a score of four species had almost certainly been enriched by planting.

**Table 9.3** The number of species in each plant category for the four types of woods on the chalk wolds and their escarpments

| *Type of woodland* | *Number of woods* | *Presence of a rectilinear boundary* | *Plant category* | | | | | *Total* | *Study shrubs* |
|---|---|---|---|---|---|---|---|---|---|
| | | | *MC* | *F* | *A* | *A + MC + F*[a] | *I* | | |
| *Range of number of study species* | | | | | | | | | |
| Primary | 1[b] | – | 2 | 3 | 8 | 13 | 3 | 16 | 3 |
| Ancient | 2 | – | 2–3 | 3 | 4–5 | 9–11 | 0 | 9–11 | 2–4 |
| Transitional | 6 | – | 1–2 | 0–2 | 1–3 | 4–7 | 0 | 4–7 | 1–2 |
| Rectilinear pre-1850 | 2 | + | 1 | 1–2 | 0–1 | 2–3 | 0 | 2–4 | 1–2 |
| Total | 11 | | | | | | | | |
| *Mean species value per category* | | | | | | | | | |
| Primary | | | 2.0 | 3.0 | 8.0 | 13.0 | 3.0 | 16.0 | 3.0 |
| Ancient | | | 2.5 | 3.0 | 4.5 | 10.0 | 0.0 | 10.0 | 3.0 |
| Transitional | | | 1,7 | 1.7 | 2.3 | 5.7 | 0.0 | 5.7 | 1.3 |
| Rectilinear pre-1850 | | | 1.0 | 1.5 | 0.5 | 3.0 | 0.0 | 3.0 | 1.5 |

[a] See table 9.4.

[b] A considerable part of this wood had been coniferized.

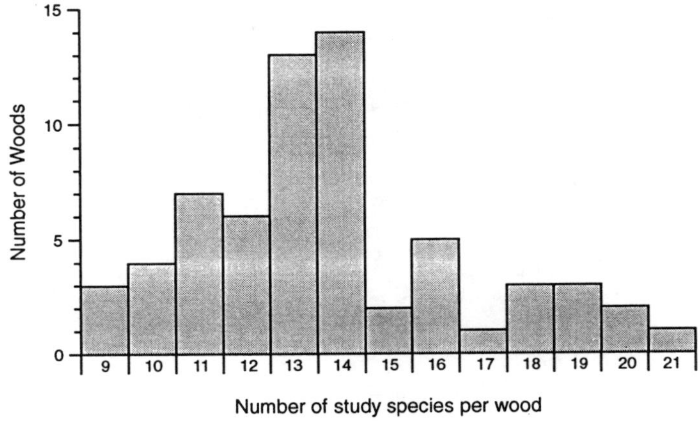


**Figure 9.3** The frequency of occurrence of the study species in the 64 ancient woods.

indicator species are weak calcicoles. By contrast, Peterken and Game (1984) concluded that most of their ancient woodland species were 'calcifuge or associated with moist or poorly drained soils', which suggests that there are important regional differences in the nature of indicator species, as many of their ancient woods are also primary (Peterken 1974).

### *Ancient woods*

The group of 64 woods with a moderately high count of study species was mainly assigned ancient woodland status on the basis that they contained 11 woods shown by Wightman (1968) to have been present in *c.*AD 1300. A further wood in this group was mentioned in a Duchy of Lancaster survey of AD 1608 (NRRS 1894). A second line of evidence was a species count which was lower overall than that for the primary woods, which suggested a somewhat more recent origin for these woods. All of the ancient woods had boundaries that were curved or irregular in shape for part or all of their length.

The woods with a moderately high species count could be distinguished from less species-rich woods by the presence of four or more of the A species, and from primary woods by an absence of three or more of the I species (table 9.2). These differences also held true for the two ancient woods investigated on the chalk wolds and their escarpments (table 9.3).

The ancient woods form a very coherent group, showing a mean value of 14.9 study species per wood (table 9.2), with over 40 per cent of the woods having a count of 13 or 14 of the study species (figure 9.3). Given the likely difference in age of the different woods, the strong possibility of different colonization rates and the differences in management regimes, past and present, the degree of uniformity of woods in this category is remarkable. By contrast the woods varied greatly in size (figure 9.4).

The 'ancient woods' group may include depauperate primary woods; that is, woods which are primary in an absolute sense, but which have passed through

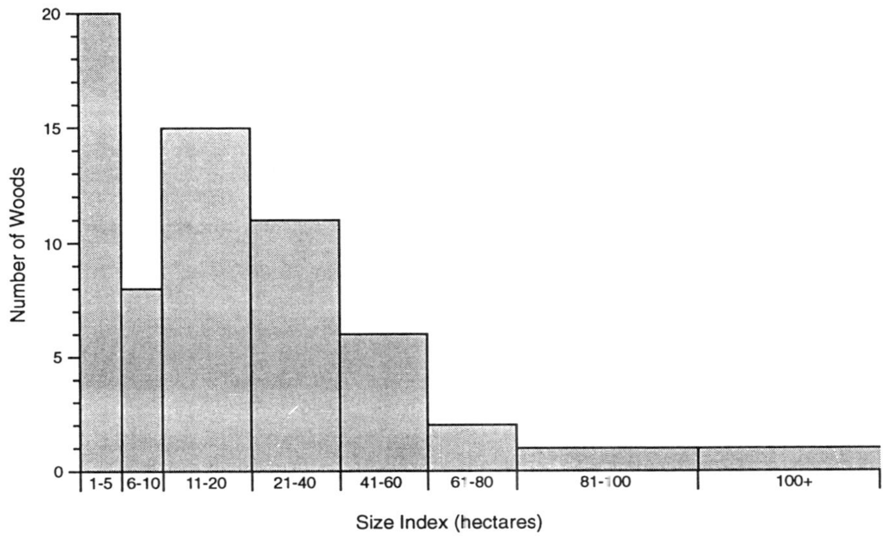


Figure 9.4 The size categories of the 64 ancient woods.

one or more periods of management that have been unfavourable to the survival of herbaceous woodland plants.

*Transitional woods*

The 23 transitional woods were intermediate in characteriztics between ancient woods and rectilinear woods. There were always less than four A species present (see tables 9.2–9.4). Their boundary attributes resembled ancient woods, except for the richest one which had rectilinear boundaries, possibly due to incursions into the wood by the surrounding land parcels. At least three explanations can be advanced to explain the origins of transitional woods:

- They may be ancient woods (in an absolute sense) which have gone through a period(s) of unsympathetic management that has eliminated much of the herbaceous plant and shrub interest; for example, dense shading from the canopy, heavy grazing or dense growth of brambles (*Rubus fruticosus* agg.).
- They could be woods which originated in the seventeenth century, i.e. not ancient woods according to the definition used by Peterken (1981).
- They could be eighteenth century woods (or even recent secondary woods) with a relatively high colonization rate, created in areas where the field boundary pattern was already long established.

It is likely that within the total set of transitional woods, there are some examples of each of these explanations.

**Table 9.4** A summary of the main types of wood and their associated indicator species characteristics

| *Type of wood* | *Indicator species characteristics* | *Number studied* |
|---|---|---|
| Primary | 3 or more I species present | 7 |
| Ancient | 2 or less I species present<br>4 or more A (ancient woodland indicator) species present | 64 |
| Transitional | 3 or less A species present,<br>4–10 A + MC + F species present, curvilinear boundaries[a] | 23 |
| Rectilinear pre-1850 | 0–4 A + MC + F species present,rectilinear boundaries | 15 |
| Rectilinear post-1850 | 0–6 A + MC + F species present, rectilinear boundaries | 40 |
| Total | | 149 |

[a] With one exception, believed to be rectilinear by contraction.

### *Rectilinear pre-1850 woods*

These 14 woods, one of which was on the chalk, had rectilinear boundaries and were sufficiently well-established by *c.*1850 to be shown as woodland on the first edition 1 : 63 360 OS maps. The greatest number of study species present in any rectilinear pre 1850 wood was four, that is, there was no overlap in species-richness with the transitional woods.

### *Rectilinear post-1850 woods*

All of these 40 woods were in their first generation of canopy trees, and were therefore unique amongst the sample of 160 in that they were the only group for which it was possible to gain a feel for the ecological developments that had occurred throughout the life of the wood. The mean number of study species per wood was 2.0, marginally less than than the 2.3 for rectilinear pre-1850 woods, but the single richest example had six study species, compared with a figure of four for the pre-1850 woods. The occurrence of a few woods with five or six study species in this group was unexpected, but the sample size was rather large (40), thereby admitting some woods with several factors favourable to colonization occurring coincidentally. In addition, it is possible that rectilinear woods may lose some of their shade bearing species when the first felling occurs, with the associated vigorous development of light demanding species. A second wave of colonization may occur some time later when the new canopy closes. Five contained a mixture of broad-leaved and coniferous trees, none of them occurred on the chalk wolds and their escarpments.

It is assumed that the woods in this group date from after 1850 (or a little before if the woods were newly planted when the surveyors visited the site); but it is possible that some are earlier, and were not recorded by the Ordnance surveyors as woodland because they had been recently cut over for coppice, and then converted to high forest post 1850. Classification 'errors' do occur from time to time. The June 1954 edition of the 1 : 25 000 OS map does not show a small but well established wood within the study area.

### *Other woods*

The six conifer woods contained 0–2 study species, the mean being 0.8 per wood. The mean number of study species in the four coniferized woods was 4.0, the range being 2–6. Located on a small hill surrounded by commercial forestry, the wood pasture contained five study species. None of these three types occurred on the chalk wolds or their escarpments.

## DISCUSSION

### *Key indicators*

Native hybrid lime occurred in four primary woods, two ancient woods and at the wood pasture (table 9.5). It also occurred in two transitional woods, strongly suggesting that these woods are ancient in an absolute sense, but have lost most of their characteristic flora during a period of unsympathetic management. In none of the woods was the tree abundant, and at one site it occurred as a single specimen. The study area is beyond the margins of the zone in which *Tilia* species (limes) were dominant around 4500 BC (Rackham 1990), but is well within the boundaries of the current distribution of *Tilia cordata*, one of the parents of the hybrid (Pigott 1991). Almost certainly, both parent species were formerly much more abundant. Lime is vulnerable not only to management regimes which favour other species, especially oak, but also because cattle will eat its bast as an emergency food source (Pigott 1991). They may well have been let into lime woods in severe winters, a practice which is not conducive to the survival of large numbers of individual trees.

Spindle (*Euonymus europaeus*), dogwood (*Cornus sanguinea*), toothwort (*Lathraea squamara*), wood barley (*Hordelymus europaeus*), and lily of the valley (*Convallaria majalis*) all showed absolute fidelity to ancient and primary woodlands and were not particularly infrequent (table 9.5). Spurge laurel (*Daphne laureola*) and genuine wild crab apple (i.e. *Malus sylvestris*, not *Malus domestica*) showed on equivalent fidelity, but were not as frequent. Dogwood (*Cornus sanguinea*) was also found planted, along with a variety of other shrubs, in two rectilinear woods.

The cumulative total of the eight study shrubs for ancient woods was very variable (table 9.2) and could not therefore be used as the basis for a collective indicator system for identifying ancient woodlands.

### *Size of wood and species richness*

Peterken (1974) and Hooper (in Peterken 1981) found a significant positive linear correlation between log (total plant species richness) and log (size) in

**Table 9.5** The occurrence of key indicators in various types of woodland

| *Category of woodland* | *Woodbanks*[a] | *Native hybrid lime*[b] | Euonymus europaeus | Cornus sanquinea[c] | Lathraea squamara | Hordelymus europaea | Convallaria majalis |
|---|---|---|---|---|---|---|---|
| Primary ($n$ = 7) | 1 | 4 | 4 | 5 | 7 | 6 | 5 |
| Ancient ($n$ = 64) | 2 | 2 | 9 | 6 | 5 | 3 | 2 |
| Transitional ($n$ = 23) | 1 | 2 | 0 | 0 | 0 | 0 | 0 |
| Coniferized ($n$ = 4) | 1 | 0 | 0 | 0 | 0 | 0 | 0 |
| Total ($n$ = 98) | 5 | 8 | 13 | 11 | 12 | 9 | 7 |

[a] Woodbanks: the transitional wood with a woodbank also has native lime present.

[b] Native hybrid lime: all of the specimens except one were growing near watercourses.

[c] Dogwood (*Cornus sanguinea*): additionally, two rectilinear woods had a variety of shrubs, including dogwood, which were believed to have been planted.

[d] Wood barley (*Hordelymus europaea*): six of the nine occurrences were in woods which also contained spindle (*Euonymus europaeus*).

woods. Peterken and Game (1984) found this relationship to hold for log/log plots, and for linear/log plots. Rackham (1980) found a significant log/log relationship for some of the groups of woods he studied; but in other sets of geographically associated woods no significant relationship was detected. The correlation coefficient for ancient woods between richness of study species used here and size index was $r = 0.24$ (not significant, $n = 62$), and between richness and log (size index) it was $r = 0.18$ (not significant); indeed, two of the richest woods were among the smallest with size indices of 1 and 2. The apparent discrepancy between these different findings is due to the following factors. First, Peterken, Hooper, Peterken and Game, and Rackham all studied the total plant species complement, not just those species indicative of ancient woodlands. As a wood grows in size, the number of generalists and adventives is likely to increase strongly, even though the number of true woodland plants may be increasing very slowly or not at all. However, Peterken and Game did also find a positive correlation between their ancient woodland species and log area. Second, many of the woods investigated were 'patchy', with zones of relative species-richness and zones of relative species poverty. Small woods tended to have good light penetration from the sides. This in turn led to a well-developed canopy and shrub layer, under which a vernal ground flora could also flourish. For many of the larger woods, the situation was less well balanced; for example, some had very dense canopies while others had extensive stands of brambles (*Rubus fruticosus* agg.) and dog's mercury (*Mercurialis perennis*). Third, although colonization rates should rise and extinction rates should fall with increasing woodland size, these factors may not vary greatly for the study species as a group over the range of woodland sizes investigated. Indeed, the fact that a successful dating technique has been developed using data gathered from entire woods of a large range of sizes demonstrates the value and robustness of the indicator species technique.

### *Woodbanks*

Woodbanks are a common feature of ancient woods in East Anglia and elsewhere (Rackham 1980, 1990). They consist of a ditch with an internal bank which surrounds the wood, describing a sinuous course in the process, and which has a characteristic cross-section. Often only fragments of the original bank still survive. They were usually created during or prior to the medieval period (Rackham 1980 1990) and hence can provide useful dating evidence for ancient woods. Their special characteristics make them distinguishable from later (usually straight) boundary banks and from banks created by ditch clearing operations.

One genuine woodbank occurred in a primary wood, two in ancient woods, one in a transitional wood and one in a coniferized wood in the total set of 160 woods. The transitional wood with the woodbank also had native lime, and hence was almost certainly a depauperate ancient wood. Thus although the presence of a woodbank is a useful dating feature in north-east Yorkshire, the incidence of woodbanks is too low to allow them to be widely used as a main indicator of ancient woodland. Indeed the presence of some of the less common shrubs, such as dogwood (*Cornus sanguinea*) proved a much more useful spot indicator of ancient woodland status than woodbanks. In South Yorkshire, by contrast, woodbanks are sufficiently common for them to be very important in the identification of ancient woodland (Jones 1989).

The low incidence of woodbanks was not due to their function being replaced by stone walls, as these were also rarely to be found as woodland boundaries. It therefore seems likely that woods in north-east Yorkshire were surrounded by fences, or by live or dead hedges, in the Middle Ages.

### *Coppice*

The presence of stands of mixed species coppice with well developed coppice stools is a feature of ancient woods in East Anglia (Rackham 1980, 1990), central Lincolnshire (Peterken and Game 1964) and elsewhere (Peterken 1981). Genuine mixed species coppice showing signs of continuous harvesting over a long period of time was however extremely rare in the study area, suggesting that many former coppice woods were converted to high forest so long ago that all traces of the coppice stools have been lost. Whether single- or double-species coppice still survives in north-east Yorkshire, there is sometimes circumstantial evidence that the understorey was once more species-rich (Gulliver 1991).

### *Domesday survey*

The Phillimore edition of the Domesday survey (Faull and Stinson 1986) lists the entries for north-east Yorkshire with great precision; the entries have also been mapped by Darby and Maxwell (1962). However, the Domesday Survey constitutes an imperfect source in this area for the woodland historian. For example, no woodland is recorded for the north-west corner of the North York Moors and the adjacent part of the Vale of the Tees, or from the coastal belt from Falsgrave to Whitby; this latter zone currently includes at least two woods with native limes. The range of types of woods to be found today in these two areas is no different from adjacent regions.

The applicability of Domesday is also restricted by the fact that the survey simply records the existence on an estate or land holding of a wood of given size or value or, in some cases, the mere presence of woodland. As the entries are not precisely located in space, one can never absolutely correlate a Domesday entry with an actual wood on the ground, even though the dimensions of the two may be similar.

Domesday entries were therefore of limited usefulness in determining woodland age. Nevertheless, four woods in the primary and 13 in the ancient categories are in townships in which woodland existed and was quantified on estates in AD 1086. In addition, six ancient woods (from the 64 surveyed) occur at locations that form part of a multiple estate entry with given dimensions, and five have simple woodland presence entries on multiple estate lists. Perhaps a more significant point is that the Domesday Survey does reveal a concentration of woods in the central part of the southern edge of the North

York Moors and in the Howardian Hills, both areas which are rich in ancient woods today. Domesday evidence has therefore been used in this study in a secondary role, to supplement plant and landscape evidence.

## CONCLUSIONS

Of the 160 woods surveyed, only seven were classed as primary. Taking one viewpoint, one might imagine that the landscape of north-east Yorkshire had been devastated by human impact in two main phases, first with the clearance of most of the primary wood during the prehistoric phase, and then with the loss of some of the primary woodland and some ancient woodland in the historical period – and especially in the last 70 years – by processes such as total clearance, coniferization and persistent grazing. In this scenario the negative effects of human activities are only marginally offset by the existence of non-primary woodland in the following categories:

- ancient (i.e. secondary) woodland with its complement of species, which is considerably less than that of primary woodland, and with its absence of continuity with the wildwood;

- recent secondary woodland with its characteristically low numbers of traditional woodland herbaceous and shrub species

- coniferous woodland

Taking the opposite viewpoint, it is possible to envisage the countryside of north-east Yorkshire as one which is enriched by the presence of woods belonging to several different stages in the evolution of the landscape, from fragments of the wildwood to relatively modern woods, reflecting a continuing desire to plant broad-leaved woods for pleasure and profit.

Historical, field archaeological, landscape and botanical evidence have been successfully combined in this study to provide a picture of the interrelationship of people with the woodland environment. The techniques utilized (especially the deployment of suites of indicator species) could be used with advantage for enhancing the Ancient Woodland Inventory by the collection of field data from every wood in the inventory.

The relative rarity of genuine woodbanks, the general absence of surviving mixed species coppice on long established stools, and the importance of some of the shrub species as spot indicators of ancient woodland status all help to give special character to the area. The production of further regional studies will allow a full appreciation of the range of woodland landscape types (and their historical development), building on the excellent work already undertaken elsewhere in the UK by, for example, Peterken (1974) and Peterken and Game (1984) in Lincolnshire; and Rackham (1980, 1990) in East Anglia.

ACKNOWLEDGEMENTS

Sincere thanks are extended to all those who have helped with the various aspects of this study. Particular debts of gratitude are owed to Dr Margaret Atherden, Professor R. A. Butlin, Mrs Jackie Chapman, Mr David Counsell, Mr Robin Edwards, Mr Nigel Fisher, Mr Paul Harris, Miss Anya McCracken, Dr John McDonnell, Mr Tom Medd, Miss Sue Rees, Dr Neil Roberts, Mrs Nan Sykes, Mr Andrew Weston and my wife.

REFERENCES

Ball, D. F. and Stevens, P. A. (1981) 'The role of 'ancient' woodlands in conserving 'undisturbed' soils in Britain', *Biological Conservation* 19: 163–76

Carter, A. (1987) *North Yorkshire Inventory of Ancient Woodlands Part III: Ryedale and Scarborough (Provisional)* (Nature Conservancy Council, Peterborough)

Counsell, D. (1990) *The Royal Forests of Pickering and Spaunton*, unpublished manuscript

Darby, H. C. and Maxwell, I. S. (1962) *The Domesday Geography of Northern England* (Cambridge University Press, Cambridge)

Faull, M. L. and Stinson, M. (1986) *Domesday Book Volume 30, Yorkshire: Parts 1 and 2* (Phillimore, Chicester)

Garland, S. P. (1984) 'Beetles as primary woodland indicators', *Sorby Record* 22: 3–38

Gulliver, R. L. (1990) 'The rare plants of the Howardian Hills, North Yorkshire, 1794–1988', *Watsonia* 18: 69–80

Gulliver, R. L. (1991) 'Report on Howsham Woods' *Annual Report of the Yorkshire Philosophical Society for the Year 1990*, 56–66

Jones, M. (1989) *Sheffield's Woodland Heritage* (Sheffield City Libraries, Sheffield)

Kerney, M. and Stubbs, A. E. (1978) *The Conservation of Snails, Slugs and Freshwater Mussels: Chief Scientist's Team Notes No. 10* (Nature Conservancy Council, London)

Linnard, W. (1982) *Welsh Woods and Forests: History and Utilisation* (National Museum of Wales, Cardiff)

Marren, P. (1990) *Woodland Heritage* (David & Charles, Newton Abbot)

North Riding Record Series (NRRS), Volume 1, N. S. (1894) *The Honour and Forest of Pickering*, ed. R. B. Turton (North Riding Record Society, London)

Packham, J. R. (1983) 'Biological flora of the British Isles No. 155: *Lamiastrum galeobdolon*' *Journal of Ecology* 71: 975–97

Peterken, G. F. (1974) 'A method for assessing woodland flora for conservation using indicator species' *Biological Conservation* 6: 239–45

Peterken, G. F. (1981) *Woodland Conservation and Management* (Chapman and Hall, London)

Peterken, G. F. and Game, M. (1984) 'Historical factors affecting the number and distribution of vascular plant species in the woodland of central Lincolnshire', *Journal of Ecology* 72: 155–82

Pigott, C. D. (1969) 'The status of *Tilia cordata* and *T. platyphyllos* on the Derbyshire limestone', *Journal of Ecology* 57: 491–504

Pigott, C. D. (1991) 'Biological flora of the British Isles No. 174: *Tilia cordata*', *Journal of Ecology* 79: 1147–207

Rackham, O. (1971) 'Historical studies and woodland conservation', *Symposium of the British Ecological Society* 11: 563–80

Rackham, O. (1980) *Ancient Woodland* (Edward Arnold, London)

Rackham, O. (1986) *The History of the Countryside* (Dent, London)

Rackham, O. (1990) *Trees and Woodland in the British landscape*, 2nd edn (Dent, London)

Rob, C. M. (1963) 'An introduction to *A Catalogue of the More Rare Wild Plants of the Castle Howard District by R. Teesdale*; the republished list and comments on the present status of the species', *Annual Report of the Council, Yorkshire Philosophical Society* 1962, York

Rose, F. (1976) 'Lichenological indicators of age and environmental continuity in woodlands', in Brown, D. H., Hawksworth, D. L. and Bailey, R. H. (eds), *Lichenological: Progress and Problems* (Systematics Association Special Volume 8, Academic Press, London), pp. 279–307

Simmons, I. G. and Innes, J. B. (1981) 'Tree remains in a North York Moors peat profile', *Nature* 294: 76–8

Simmons, I. G. and Innes, J. B. (1988) 'Studies in the vegetational history of the North York Moors. X. Investigations on east Bilsdale moor', *Journal of Biogeography* 15: 125–37

Spencer, J. W. and Kirby, K. J. (1992) 'An inventory of ancient woodland for England and Wales', *Biological Conservation* 62: 77–93

Stace, C. A. (1991) *New Flora of the British Isles* (Cambridge University Press, Cambridge)

Steel, R. C. and Welch, R. C. (eds) (1973) *Monks Wood: A Nature Reserve Record* (Natural Environmental Research Council, Huntingdon)

Stubbs, A. E. (1982) 'Hover flies as primary woodland indicators with reference to Wharncliffe Wood', *Sorby Record* 20: 62–7

Watkins, C, (1991) *Woodland Management and Conservation* (David & Charles, Newton Abbot)

Wightman, W. R. (1964) 'Some aspects of the historical geography of the Vale of Pickering area, 1086–1350 AD', unpublished Ph.D. thesis, Department of Geography, University of Durham

Wightman, W. R. (1968) 'The pattern of vegetation in the Vale of Pickering area *c.*1300 AD', *Transactions, Institute of British Geographers* 45: 125–42

# PART III

# The North Atlantic Region

10

# Ancient *Saami* in Finnish Lapland and Their Impact on the Forest Vegetation

Christian Carpelan and Sheila Hicks

ABSTRACT

The social and economic basis of *Saami* life is described together with an outline of the way in which this group occupied the pine forest area of northern Finland. A specific example is provided by a winter village site at Nukkumajoki near Inari, which was seasonally occupied for three periods: in the fifteenth and the seventeenth centuries, and around AD 1800. A pollen diagram is presented from a section through the wall of one of the seasonally occupied *kɔta* within the winter village. This shows which plant species were favoured by human presence and confirms that intensive seasonal occupation can locally transform the pine forest to a more open woodland with much birch. The possibilities of using pollen evidence to confirm the local presence of hunter–gatherers is assessed.

## INTRODUCTION

The study area is situated at the western end of lake Inari in Finnish Lapland. The vegetation, which falls into the northern boreal coniferous forest zone (Hämet-Ahti 1988) is dominated by pine intermixed with varying amounts of birch (figure 10.1). For much of historical time, this region was occupied by Forest-*Saami*. These people lived by hunting and fishing and their way of life can be reconstructed both from written descriptions and from archaeological remains (sources earlier than the seventeenth century are scarce; Tanner 1929, Itkonen 1948, Tegengren 1952).

The Forest-*Saami* of Finnish Lapland traditionally lived in local communities called Lapp villages (Lp. *siida*). Within the territory of the village each family had its hunting grounds and fishing waters and moved between these from one seasonal site to another, following a regular annual cycle. The families of a *siida* spent spring, summer and autumn fishing and hunting by themselves at their individual sites. In the winters the families spent their time together at a common site, the winter village. It was there that social, economic

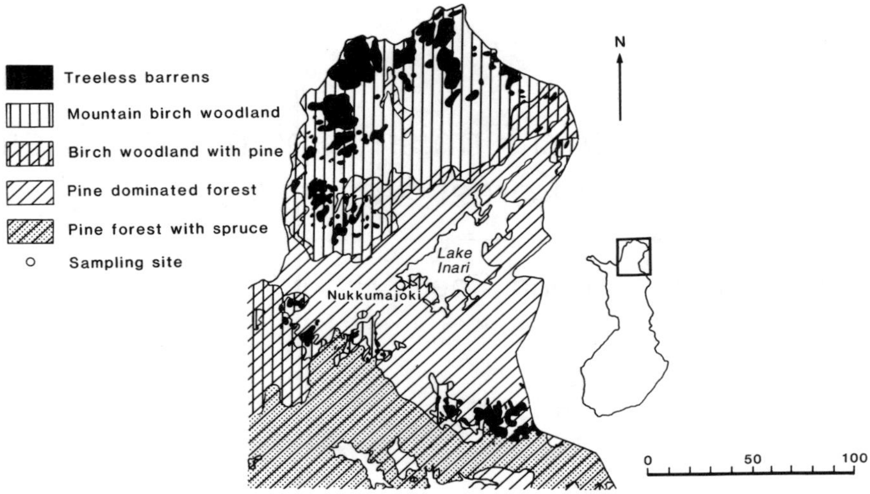


Figure 10.1 Northern Lapland, showing major vegetation belts and the location of Nukkumajoki.

and legal matters were arranged, as well as the relationship with neighbouring *siida* communities. Priests, judges, tax collectors and traders representing remote centres of political power visited the winter villages regularly.

As a physical community, a Forest-*Saami* winter village usually existed for a limited period of time. As the surroundings of a village were exhausted of dry pines for fuel, living pines for bark (i.e. cambium for food) and lichen for feeding reindeer, it was abandoned and a new village was founded at another site. Of course it was possible to return to an ancient village site after several generations, when the forest had regenerated. The turf-covered timber dwellings (Fi. *kota*, Lp. *koahtte*) remained habitable for decades if they were properly maintained and not destroyed by accidental fires.

Near the modern village of Inari at the western end of lake Inari (68° 55′ N, 27° 01′ E) 14 ancient *Saami* winter village sites are known. Ten of them lie by the Nukkumajoki, a small river running in a sheltered valley and discharging into Lake Inari (figure 10.2). Ilmari Itkonen, who discovered most of the sites as early as 1909, interpreted these sites in terms of the *Saami* custom of shifting the location of the winter village. He assumed that the sites probably represented a sequence of winter villages within the Inari *siida* (Itkonen 1913).

In 1966, one of the authors carried out an archaeological survey of *Saami* antiquities of the area. Under the auspices of the University of Helsinki Department of Archaeology he conducted fieldwork in the region between 1978 and 1987. Some new winter village sites were found, each site was surveyed (including phosphate mapping), and excavations were carried out at three sites, viz. Nukkumajoki 2, 4 and 8 (Carpelan 1991).

Except for the river bed, which is rather marshy in places and bordered by birch, the Nukkumajoki valley supports (or did prior to felling) dense pine

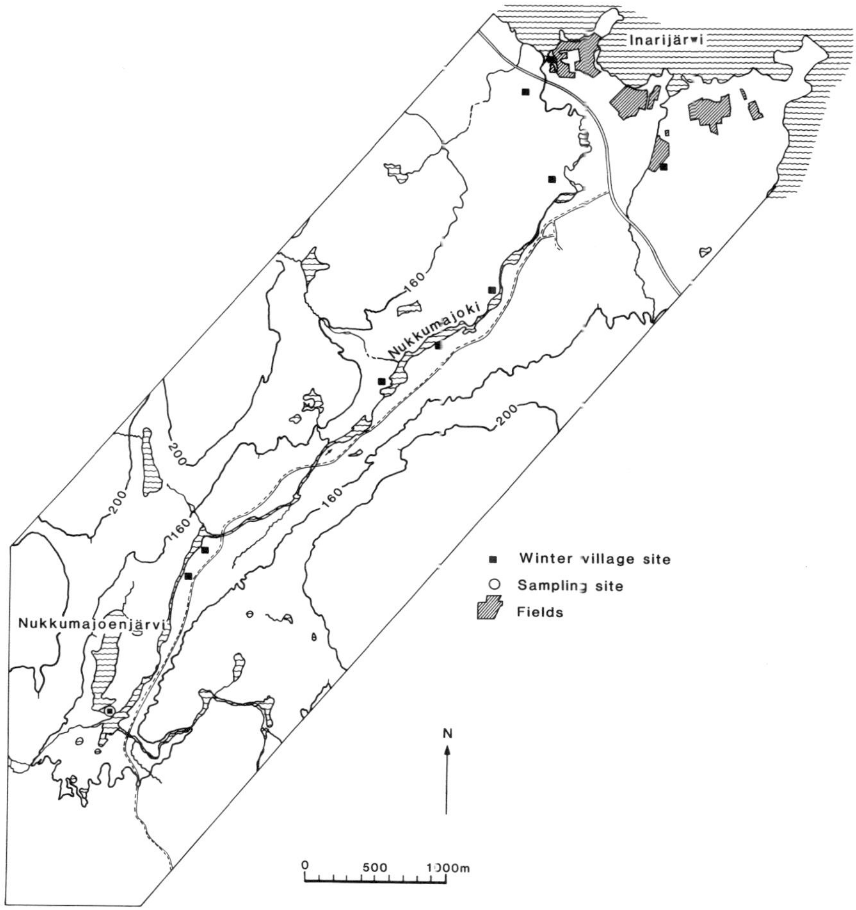


Figure 10.2 The relative locations of the winter villages within the Nukkumajoki valley.

forest with a ground cover of dwarf shrubs and lichen. The village sites, 20–100 m away from the river, are characterized by rich birch stands, with junipers, dwarf shrubs, herbs and mosses. The habitation must have increased the amount of nutrients in the ground, allowing more demanding plants to grow in this otherwise nutrient-poor environment.

This observation led Suominen to carry out a preliminary botanical survey of the sites (Suominen 1975). In the course of the archaeological fieldwork campaign the question of the impact of the *Saami* way of life on the vegetation arose which, in 1991, led to the palynological investigations presented here for the winter village site Nukkumajoki 7.

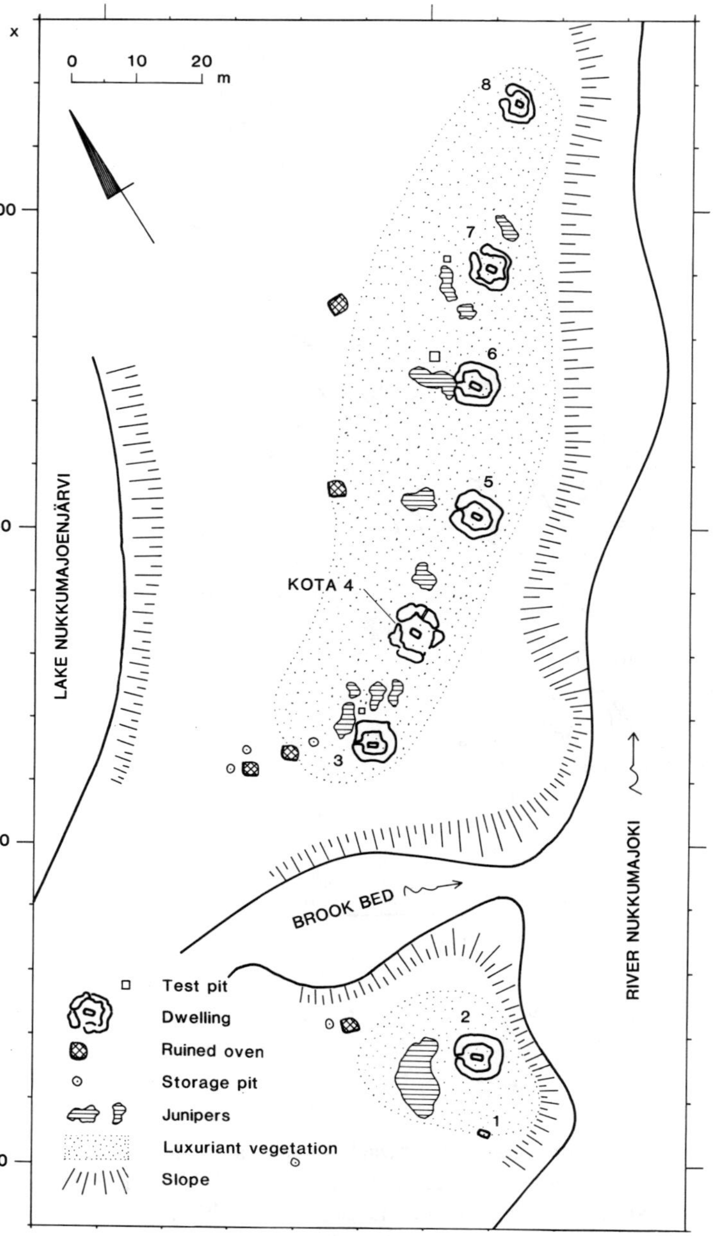


**Figure 10.3** A plan of winter village Nukkumajoki 7.

## ARCHAEOLOGY

The winter village site Nukkumajoki 7 is situated on the isthmus between the river Nukkumajoki and Lake Nukkumajoenjärvi (figure 10.2). The isthmus,

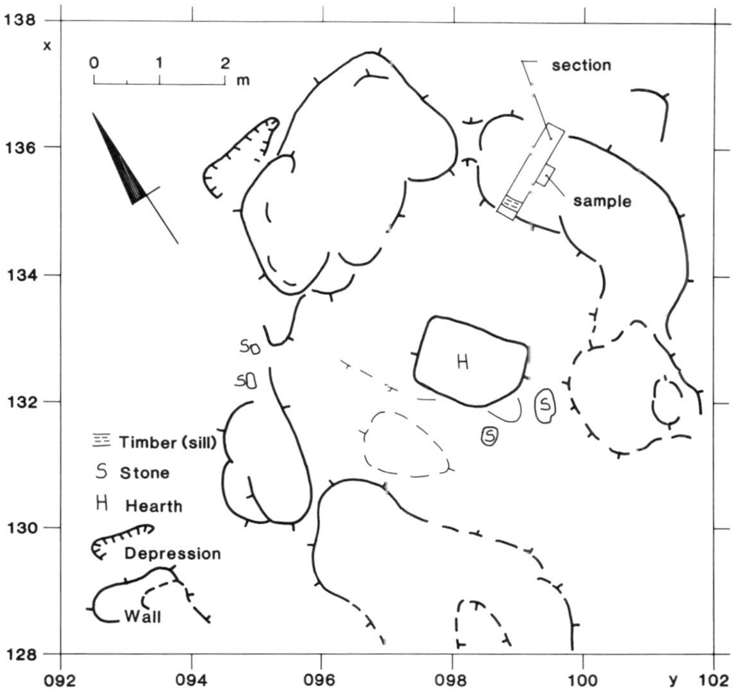


Figure 10.4 A plan of *kota* 4 at Nukkumajoki 7.

of coarse moraine, is *c*.80 m wide and is divided by a brook via which the waters of the lake run into the river. The remains of eight *kota* dwellings form a row, 170 m in length and 5–20 m from the river, along the crest of the isthmus. The remains of at least five small log houses (their ruined ovens) form a parallel row *c*.15 m from the first and closer to the lake. The brook separates two dwellings and one log house from the rest (figure 10.3).

The vegetation follows the general pattern for these winter villages as described above in that a belt of birches, junipers, dwarf shrubs, herbs and mosses occurs along the row of the *kota*, while the surroundings are characterized by pine forest with dwarf shrubs and lichen (figure 10.3). The phosphate values of the soil, which are not dealt with in any more detail here, are clearly higher within this belt of more luxuriant vegetation.

What is left of a *kota* appears as a somewhat irregular square surrounded by a low wall and with a raised area in the middle (figure 10.4). A superficial examination of the remains of 72 dwellings by the Nukkumajoki, as well as the results of excavations carried out at Nukkumajoki 2, 4 and 8, have revealed that the floors were usually hexagonal, or sometimes octagonal, in shape. The central elevation contains the rectangular hearth of stones (now fire cracked).

The foundation of a *kota* was a hexagonal (or octagonal) frame of logs forming sills, against which the supportive structure of the wooden and turf covered tent shaped building leant (figure 10.5). Inside the sills the floor was about 5 × 5 m in size. The sills, in their turn, were pushed up against the

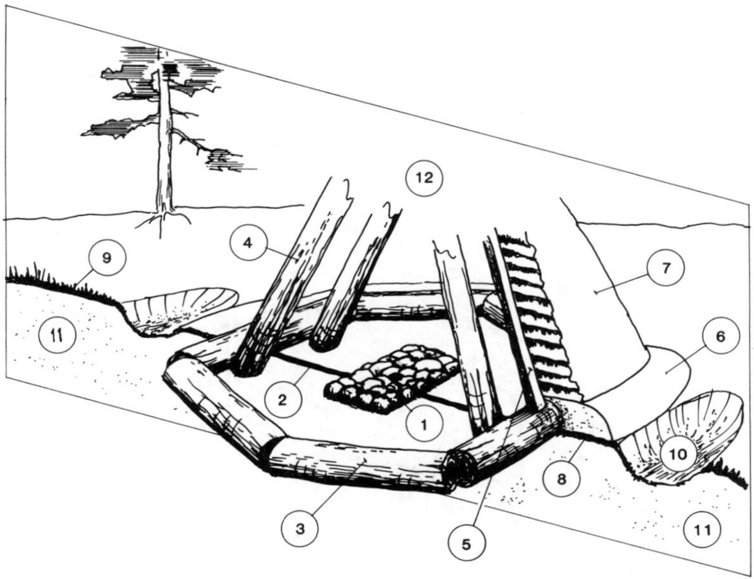


Figure 10.5 A schematic reconstruction of a *kota*. 1, Hearth; 2, floor; 3, sill; 4, supportive structure; 5, plank; 6, earth bank; 7, turf cover; 8, buried black layer; 9, cultural layer with refuse; 10, depression; 11, podsol; 12, top of *kota* (not possible to reconstruct).

earth wall. Oblong depressions often observed immediately outside and parallel with the walls indicate where the earth was taken from. The existence of a 1–2 cm thick black cultural layer beneath many of the excavated walls indicates that the *kota* were built at places which had previously been dwelling sites. The observations made at *kota* 4 of Nukkumajoki 7 are in agreement with this general description (figure 10.6).

Excavations and test pits at all the village sites by the River Nukkumajoki revealed that the cultural layer, below the humus was usually thin, but could be up to 15 cm thick near the dwellings. There it contained large quantities of split bones of (wild) reindeer (*Rangifer tarandus* L.): these formed the main component of the refuse fauna, remains of other game and fish being only of minor importance.

Most of the artefacts (more than 500), collected in the course of excavations (including ornaments, axes, knives, scissors, drills, arrowheads and metal fire-steels, some spoons and other objects of bone and antler as well as flints, whetstones, pieces of pumice, balls of pitch and resin chewing-gums) were found in such bone accumulations. Many of the objects seem to have been unused at the time of deposition. Some objects, imported from Russia and from Central Europe, are datable to the fifteenth and others to the seventeenth centuries. A badly corroded knife, probably from the fifteenth century, was found in a test pit at Nukkumajoki 7.

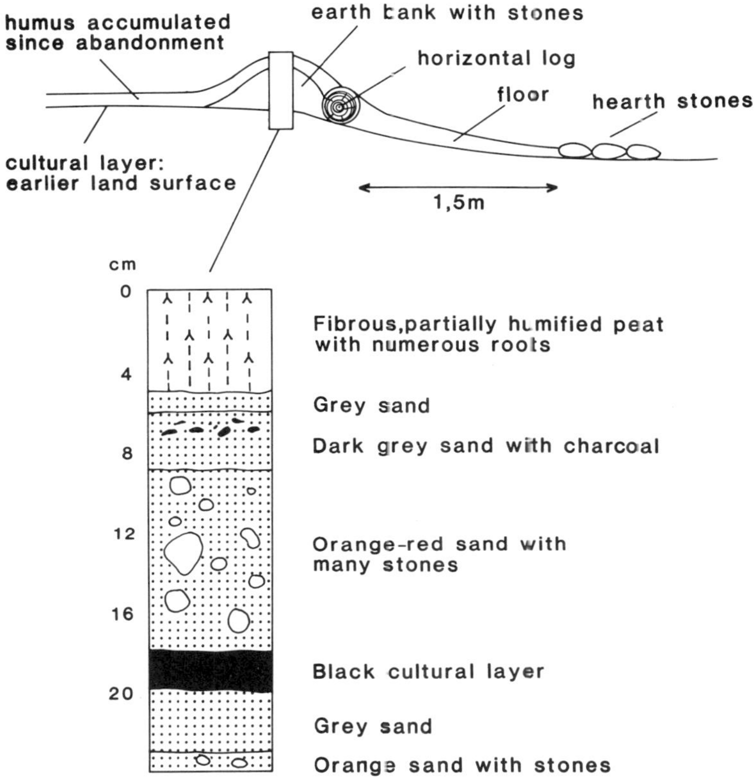


Figure 10.6 A section through the wall of *kota* 4 at Nukkumajoki 7.

Datable objects indicate the chronological position of some of the Nukkumajoki winter village sites. Unfortunately, dendrochronological dating would not be feasible because the remains of the timber structures were in a bad state of decay. Instead, radiocarbon dating of charcoal, timber and bone samples, as well as thermoluminescence dating of hearth stones was used in assessing their chronological positions. Preliminary accounts of the Nukkumajoki dating programme are given in Carpelan and Kankainen (1990) and Carpelan et al. (in press).

The available $^{14}$C and TL dates corroborate the picture given by the artefacts, that several of the Nukkumajoki village sites were in use in the fifteenth and the seventeenth centuries (phases A and B). The dwellings, the walls of which can now be seen, were probably all built in phase B. Apparently the Nukkumajoki winter village sites were inhabited and abandoned simultaneously. The model suggested by Itkonen therefore seems to be erroneous.

No samples from Nukkumajoki 7 itself have been dated so far, but the observations referred to above are consistent with what is known of other village sites at Nukkumajoki. Thus one may assume that phases A and B are also present at Nukkumajoki 7. There, as elsewhere, the log houses represent a still later phase, phase C, datable to around AD 1800. As these latter houses were built some distance away, the activity that they represented did not affect the earlier dwellings or cultural layers in any appreciable way.

## POLLEN ANALYSIS

In order to assess the way in which seasonal occupation can affect vegetation, a section was taken through the wall of *kota* 4 (figure 10.6) and an analysis made of the pollen present in each of the recognizable horizons. The stratigraphy of the section was as follows:

| | |
|---|---|
| 0–5 cm | dark brown humus, fibrous and with numerous woody roots |
| 5–6 cm | grey sand |
| 6–9 cm | dark grey sand with charcoal fragments at 7 cm |
| 9–18 cm | orange-red sand with many stones |
| 18–20 cm | black cultural layer |
| 20–23 cm | grey sand |
| 23– | orange sand with stones |

The black cultural layer represents the land surface on which the *kota* was built, and the grey and orange mineral horizons below it the podsol profile which had formed by that time. The mineral horizons from 5 cm to 18 cm depth are the sand bed that formed the foundation of the *kota* wall. The charcoal in the uppermost part of this mineral horizon must be what remains of the walls, which may eventually have been destroyed by fire, while the grey and orange horizons within this mineral material show the degree of podsolization which has taken place since the sand bed was constructed. The surface humus horizon has accumulated since the *kota* itself was destroyed.

The black cultural layer is associated with the earlier occupation phase, phase A, dated to the fifteenth century; while the mineral horizons of the *kota* sand bed represent the later occupation phase, phase B, dated to the seventeenth century. The humus represents the last 300 years.

The pollen diagram produced from this section is illustrated in figure 10.7. The sand of the lowest podsol was devoid of pollen but contained fragments of charcoal. The two stratigraphic horizons, which can be directly related to the two occupation phases, namely the black cultural layer at 18–20 cm depth and the uppermost horizons of the sand bed at 5–8 cm depth, have pollen assemblages in which *Betula* is more abundant (20–30 per cent and 30–40 per cent of total pollen respectively) and herbaceous species are more frequent. In the lower layer a whole range of pollen types which can be associated with either good light conditions (Ericales, *Lycopodium annotinum*, *L. clavatum*, Poaceae), disturbed ground (Compositae *Solidago* and *Achillea* types *Campanula*, *Ranunculus*, Caryophyllaceae), nitrogen-enriched soils (*Urtica*, Chenopodiaceae) or fire (*Epilobium*, and in some respects *Calluna*) are recorded, while in the upper layer the range of species is more restricted (*Calluna* and other Ericales, *Lycopodium annotinum*, *L. clavatum*, Poaceae, Compositae *Solidago* type,

*Ranunculus* and *Epilobium*). This combination of an increase in *Betula* along with the presence of this particular range of herbs is, therefore, taken as indicative of seasonal occupation.

*Artemisia*, although a pollen type frequently associated with human activity, cannot be deemed indicative here, since the only *Artemisia* species to grow in the region is *Artemisia vulgaris*, and that is a recent introduction, not recorded before 1951 (Hämet-Ahti et. al. 1986). All of the other species are naturally part of the Inari flora but, with the exception of *Lycopodium annotinum* and the various Ericales species, do not normally occur in established pine forest.

The sporadic presence of many of these pollen types in the sand with stones horizon at 9–18 cm depth is because this material was dug from the original land surface in the construction of the *kota* and is, therefore, a mixture of the black cultural layer and its underlying barren podsol.

The occurrence or slightly increased values of *Alnus*, *Salix*, *Rubus chamaemorus*, *Sphagnum*, *Equisetum* and Polypodiaceae in the lower cultural layer suggests that the winter village area was somewhat damper at that stage than subsequently; possibly the water table was higher. The slight peaks in exotic species and in *Picea*, on the other hand, show that the forest was particularly open at this stage. These pollen types, which represent species which do not grow this far north, must all be long-distance transported. Such long-distance transported types can be deposited only when there is an incomplete forest canopy.

The upper humus layer, which has accumulated since the abandonment of the site, is dominated by *Pinus*, while both *Melampyrum* and *Empetrum* are present. The majority of the herbs associated with the occupation phases are absent and those which do persist are present at distinctly lower frequencies. The larger charcoal fragments (> 40 $\mu$m) which are consistently present from the lower cultural layer to the surface of the sand bed are conspicuously absent from this surface humus.

Just at the junction between the mineral material and the surface humus, there is a clear peak in *Epilobium* pollen. Although there are many species of *Epilobium*, it is most probably *Epilobium angustifolium* that is represented. This plant is characteristic of paths and trackways, and even more so of areas which have been burnt. Although it can occur quite widely its pollen is only occasionally found. The peak in this diagram, therefore, is interpreted as indicating local fire, possibly that which destroyed the *kota*.

Suominen (1975) analysed the plant cover at five of the Nukkumajoki winter villages some 20 years ago and noted the most frequently occurring species. In pollen terms (table 10.1), these include *Betula*, Poaceae, *Lycopodium annotinum*, Compositae *Solidago* type and *Epilobium*. The correlation with the pollen types in the second occupation phase confirms that the vegetation that he observed was a remnant from that period of occupation. Obviously the impact on the vegetation is relatively long lasting.

The suggestion that the suite of pollen types present in the black cultural layer is characteristic of an area in the boreal forest where people gather together on a seasonal basis is supported by the fact that a highly similar collection of pollen types is recorded by Aronsson (1991) for reindeer-herding sites in northern Sweden (table 10.2).

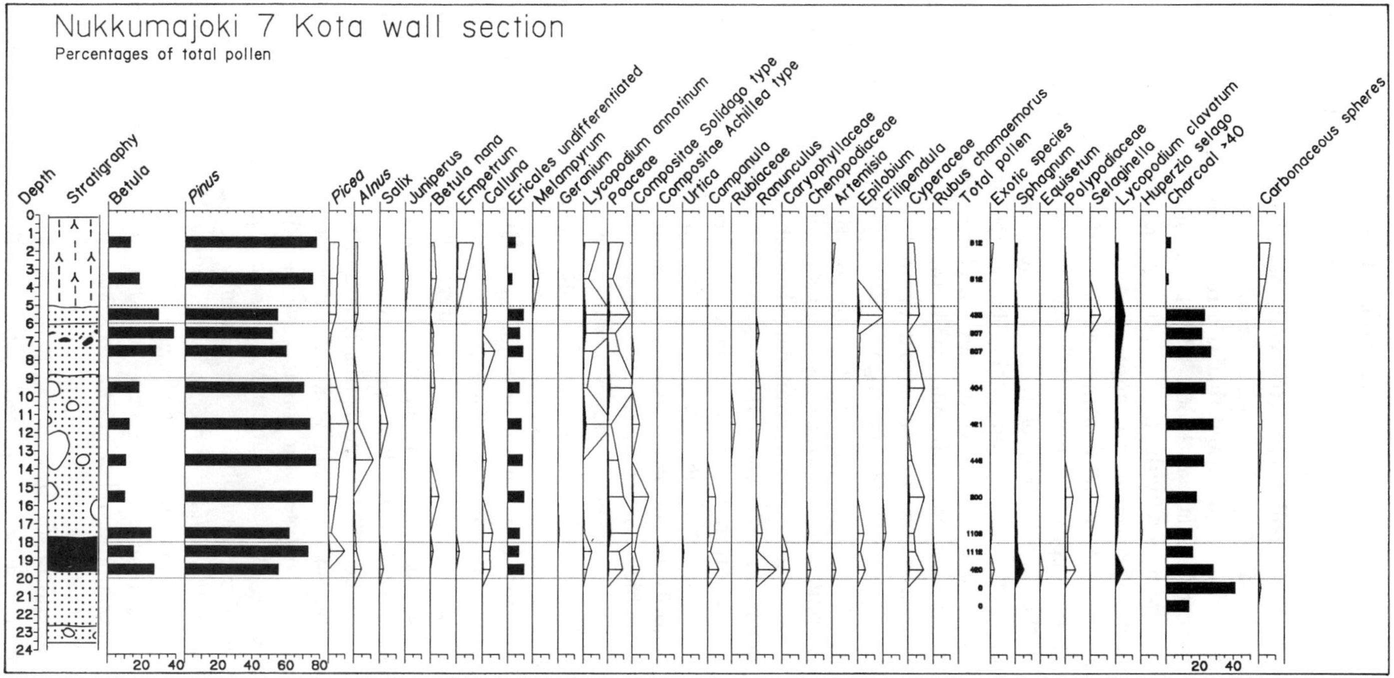


Figure 10.7 A pollen diagram of the wall section.

**Table 10.1** Plants in order of abundance at five winter villages in the Nukkumajoki valley

| *Species* | *Precision of pollen identification* |
|---|---|
| *Betula pubescens* | Tree *Betula*[a] |
| *Juniperus communis* | *Juniperus* |
| *Deschampsia flexuosa* | Poaceae[a] |
| *Lycopodium annotinum* | *Lycopodium annotinum*[a] |
| *Linnaea borealis* | *Linnaea* |
| *Trientalis europaea* | *Trientalis* |
| *Calamagrostis lapponica* | Poaceae[a] |
| *Festuca ovina* | Poaceae[a] |
| *Solidago virgaurea* | Comopositae *Solidago* type[a] |
| *Epilobium angustifolium* | *Epilobium*[a] |
| *Poa rigens* | Poaceae[a] |
| *Diphasiastrum complanatum* | *Diphasiastrum* type |
| Less frequently occurring: | |
| *Melampyrum pratense* | *Melampyrum* |
| *Hierochloë hirta* (*odorata*) | Poaceae[a] |
| *Hieracium vulgatum* | Compositae *Taraxacum* type |
| *Calamagrostis purpurea* | Poaceae[a] |
| *Ramischia secunda* | *Ramischia* |
| *Equisetum sylvaticum* | *Equisetum*[a] |
| *Antennaria dioica* | Compositae *Solidago* type[a] |

[a] Pollen types that are recorded in the cultural assemblages of the Nukkumajoki wall section.

**Table 10.2** Pollen indicators within 50 m of the activity

Increased frequencies of:
- Poaceae[a]

together with:
- Rumex

Presence of some of the following:
- Chenopodiaceae[a]
- *Plantago major/media*
- Caryophyllaceae[a]
- Asteraceae (primarily *Solidago*[a] and *Achillea*[a] types)
- Ranunculaceae[a]
- Apiaceae
- *Epilobium*[a]
- *Juniperus*
- *Urtica*[a]

Possibly also:
- *Rubus* type
- *Melampyrum*

[a] Pollen types that are recorded in the cultural assemblages of the Nukkumajoki wall section.

## CONCLUSIONS

The pollen and archaeological evidence together give a more detailed picture of the extent to which the *Saami* hunters altered their environment. Their influence was felt only on those sites where several families gathered together year after year, such as the winter villages. The necessity to fell standing dead pines, initially for building but later just for repairing the *kotas*, to fell living pines for their bark, and to collect wood for fuel, throughout the long winter months meant that eventually all pines within easily reachable distance of the winter village were removed. This opening up of the forest allowed the dwarf shrub species of the undergrowth to flourish and flower more prolifically, and the *Lycopodium* species to benefit in the same way. The better light conditions, the disturbance of the ground and the increase in available nutrients caused by different types of refuse allowed grasses and other herbs to expand. Once the site was abandoned and the *kotas* fell into disuse or were burnt, then *Epilobium*, and later *Betula* invaded. *Juniperus* also flourished. Eventually, the forest regenerated and *Pinus* was re-established but *Betula*, *Juniperus* and grasses continued to be more common in the winter village sites than elsewhere in the pine forest.

The fact that the pollen evidence for the earlier occupation phase is stronger and more varied than that of the latter may indicate that the earlier phase was of longer duration than the latter or, less probably, that the evidence is the cumulated evidence of all previous occupation phases.

The combination of increased pollen percentages of the Ericales species together with the appearance of a group of pollen types indicating plants which demand good light conditions and/or a nitrogen rich soil, followed by an obvious peak in *Betula*, is sufficiently distinctive that it can be regarded as evidence for long-term seasonal occupation of a boreal forest. The area over which this type of pollen evidence is deposited is certainly restricted. Work on pollen evidence of settlement and agriculture in closed forest (Andersen 1988, Hicks 1985) suggests less than 200 m. It follows, therefore, that if this type of pollen combination is observed in a pollen diagram from some less well known area of boreal forest it can be regarded as evidence of seasonal occupation in the immediate vicinity. It is envisaged that the same evidence is produced whatever the actual age of the settlement site and irrespective of whether the site was a winter village or a summer fishing camp.

## REFERENCES

Andersen, S. T. (1988) 'Changes in agricultural practices in the Holocene indicated in a pollen diagram from a small hollow in Denmark', in Birks, H. H., Birks, H. J. B., Kaland, P. E. and Moe, D. (eds) *The Cultural Landscape Past, Present and Future* (Cambridge University Press, Cambridge) pp. 395–408

Aronsson, K.-Å. (1991) 'Forest reindeer herding A.D. 1–1800', *Archaeology and Environment* 10: 1–125

Carpelan, C. (1991) 'Peuranpyytäjien kylä Inarissa', *Raito* 2/1991: 20–28

Carpelan, C. and Kankainen, T. (1990) 'Radiocarbon dating of a Subrecent *Saami* winter-village site in Inari, Lapland, Finland: a preliminary account', *Pact* 29(1): 357–70

Carpelan, C., Jungner, H. and Mejdahl, V. 'Dating of a Subrecent *Saami* winter-village site near Inari, Finnish Lapland – a preliminary account', *Pact* (in press)

Hicks, S. (1985) 'Problems and possibilities of correlating historical/archaeological and pollen-analytical evidence in a northern boreal environment', *Fennoscandia Archaeologica* II: 51–84

Hämet-Ahti, L. (1988) *Vegetation and Flora* Atlas of Finland 141: 1–2

Hämet-Ahti, L., Suominen, J., Ulvinen, T., Uotila, P. and Vuokko, S. (1986) *Retkeilykasvio* (Suomen Luonnonsujelun Tuki oy)

Itkonen, I. (1913) 'Tietoja Inarin kirkonkylän seudun muinaisuudesta', *Suomen Museo* 20(1): 2–9

Itkonen, T. I. (1948) *Suomen Lappalaiset Vuoteen 1945* I, II (WSOY, Porvoo)

Suominen, J. (1975) 'Kasvipeitteestä saamelaisten muinaisilla talvikylän paikoilla' ('On the plant cover at sites of ancient Lapp winter villages, Finnish Lapland'), *Luonnon Tutkija* 79: 92–4

Tanner, V. (1929) 'Antropogeografiska studier inom Petsamo-området. I. Skolt-lapparna (Résumée. Études de géographie humaine dans la territoire de Petsamo. I. Les Lapons dits "Skolt")', *Fennia* 49(4): 1–518

Tegengren, H. (1952) 'En utdöd lappkultur i Kemi lappmark. Studier i Nordfinlands kolonisationshistoria', *Acta Academiae Aboensis, Humaniora* XIX(4): 1–287

11

# Introductions, Extinctions or Continuity? Faunal Change in the North Atlantic Islands

Jon P. Sadler and Peter Skidmore

ABSTRACT

Many workers have highlighted the Palaearctic nature of the insect fauna of the North Atlantic region, despite the fact that the prevailing ocean currents and winds are westerlies. It has been posited that many, so-called, synanthropic species were introduced to the islands by successive waves of European human migrations. The scale of human impact varies from island to island and the importance apportioned to particular cultural groups in altering the indigenous fauna varies from researcher to researcher. It is evident that before the creation of suitable habitats by humans any synanthropic insect colonist would have found an inhospitable environment, with insufficient continuity of habitat to maintain a viable population. Once permanent settlements were established, the continuity of land management practices ensure the survival of most of the introduced (synanthropic) insects, although occasional extinctions may have occurred. In the case of Greenland, however, there is a period of two hundred or so years between the final demise of the Norse settlements and the island's recolonization by the Norwegian Hans Egede on behalf of the Missionary College in Copenhagen in 1721. What happened to the synanthropic fauna during this period? Does the similarity of the fossil (Norse) and the present-day faunas of Greenland indicate stability and continuity in the synanthropic fauna? Can one argue for extinction followed by reintroduction along with Hans Egede and his colonists? These questions are elucidated by discussing specific examples from the fossil record of Greenland and Iceland.

## INTRODUCTION

The impact of human migrations in the shaping of the biota of a particular region of the Earth has interested biogeographers for many years. Indeed, the similarity of the biota of the North Atlantic region was commented upon by some early scholars (Rink 1877, Saunders 1882). Lindroth (1931, 1957), however, was the first researcher systematically to study the region's biota and he noted that, despite the prevailing westerly winds and ocean currents, the faunas of Faroe, Iceland and Greenland were essentially Palaearctic in nature. Lindroth's work can be viewed in the wider context of a period of Europe's 'biological imperialism' (Crosby 1986) and parallel scenarios can be cited from North America (Buckland et al. ch. 12, this volume, Elton 1558) and the Southern Hemisphere (e.g. Clark 1949). These studies almost exclusively consider change in historical times, utilizing the dual tools of personal accounts and/or documentary records and present-day distributional patterns, to reconstruct past biotal variation. This is clearly illustrated by Lindroth's (1957) study of 'faunal connections' in the North Atlantic. His conclusions regarding the chronology of species movements highlight a perennial biogeographical problem; that of elucidating past human impact from patterns observed in the natural world at the present day. Lindroth (1957) had the added benefit of historical and documentary evidence dating from the late fifteenth century onwards, and this provided for an excellent biohistorical study of the mechanisms of species transport which greatly underestimated the role of earlier Norse (Viking) settlers (Buckland 1988, Sadler 1991a). In the absence of reliable historical records for these early periods one must turn to other sources of data, and for some groups of animals, notably the insects, which preserve well in natural and anthropogenic sediments, the fossil record provides additional data. Despite difficulties relating to the fragmentary nature of the record (Rosen 1988a), these fossil data offer a potential means of re-examining models constructed solely on the basis of the present-day distribution of elements of a region's biota.

This chapter employs the techniques' of the historical biogeographer (*sensu* Rosen 1988b: 34ff) and focuses upon the changing nature of the insect fauna of the North Atlantic in relation to the settlement history of the region. The insects are chosen because they have been extensively researched and therefore the taxonomy and ecology of the species is relatively well known. Emphasis will be placed on the Coleoptera (beetles) and Diptera (true/two-winged flies), and most of the discussion will focus on the Greenlandic and Icelandic fauna because it is only from these islands that there exists an expanding and extensive fossil database of species occurrence.

## BACKGROUND

### *Settlement history*

The first major European settlement of the North Atlantic region was undertaken by Norse or Viking colonists who reached Faroe by AD 850, Iceland by AD 870, Greenland by AD 986 and ultimately America (Newfoundland) by AD 1000 (McGovern 1981, Ingstad 1986, Jones 1986, Wallace 1991). These colonists exported a European, mainly pastoral system, emphasizing the overwintering of domesticates for secondary products such as milk, a diet that they supplemented by the utilization of additional local marine and terrestrial resources. Although not without population fluctuations resulting from epidemics, the cold of the Little Ice Age (Grove 1988) and the lack of attention of increasingly disinterested colonial Norwegian and Danish crowns, viable settlement was maintained on Faroe and Iceland (Thorsteinsson 1981, Sveinbjarnardóttir 1992). In Greenland, however, the Western settlement had been abandoned by *c.*AD 1352 and the Eastern settlement was lost by the sixteenth century (Krogh 1967, McGovern 1980). The island was only recolonized by Europeans some two hundred years later by the Norwegian Hans Egede on behalf of the Missionary College in Copenhagen, in 1721 (Gad 1970, 1973).

Jóhansen (1979, 1985, 1986), on the basis of radiocarbon-dated pollen diagrams from the islands of Streymoy and Mýkines, has consistently argued for pre-Viking settlement of the Faroe islands in *c.*AD 650 by Irish settlers; the presence of Irish monks on the island of Papay, eastern Iceland, has also been considered by Sveinbjarnardóttir (1972; Buckland et al. ch. 13, this volume). Hermanns-Auðardóttir (1991) has presented archaeological evidence from the Vestmanneyjar indicating a phase of early 'Merovingian' settlement of Iceland some 200 years before the generally accepted date for Viking Landnám of *c.*AD 870. In all instances, the evidence is limited to a few sites and further research is necessary to help resolve the archaeological and dating problems (Krogh 1986, Buckland 1989–90, Debes 1989–90, Kaland 1991, Mahler and Malmros 1991). The possibility of pre-Viking European settlement in the region should be borne in mind when the zoogeography of the regions insect fauna is under consideration.

### *Zoogeography*

The faunal characteriztics of the region have been described in detail by Lindroth (1931, 1957) and more recently by Downes (1988) and are summarized here.

1 It is often stated that the North Atlantic insect fauna has no endemics (Lindroth 1931, 1957, Larsson 1959, Downes 1966, 1988, Buckland 1988), a fact that has led many to believe the faunas are relatively young. Böcher (1988a), however, records a putative endemic within the Greenland beetle fauna. In a recent revision of the entire Greenlandic Aleocharinae genus *Atheta*, Mahler (1988) found that the individuals formerly identified as *A. fungi* (Graven.) deviated considerably from this species. He therefore ascribed the individuals to a new species, *A. groenlandica* Mahler, although he indicates that it is closely related to *A. fungi* and is similarly a highly variable species that has records on all of the North Atlantic islands. One could argue, however, that such variability might not indicate a specific rank and indeed, it has parallels with other elements in the fauna. Lindroth (1968) has shown how statistical treatment of the morphs of the carabid, *Carabus problematicus* Herbst indicates subspecific rank. Similarly, *Philonthus varius* var. *shetlandicus*, a variant of *Philonthus varius*, is only recorded on Shetland and Orkney. An equally plausible alternative would be to suggest that if populations of the islands had descended from very few individuals (which seems likely), the reduction in genetic material may lead to the phenotypic form in the isolated population appearing noticeably different from the norm of the parent population. Even if one accepts the specific rank of *A. groenlandica*, the Nearctic *Atheta* fauna is insufficiently researched to state unequivocally that the species

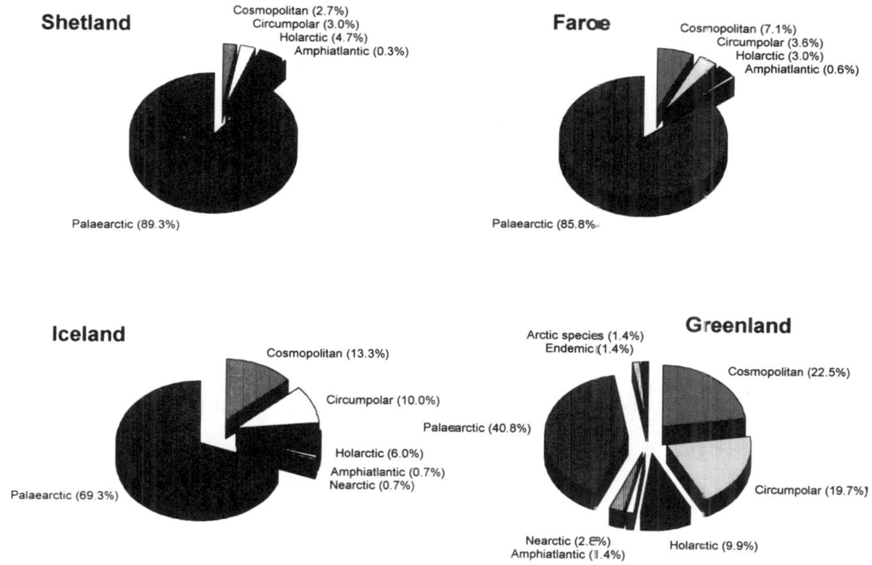


**Figure 11.1** The zoogeography of the Coleoptera of Shetland, Faroe, Iceland and Greenland.

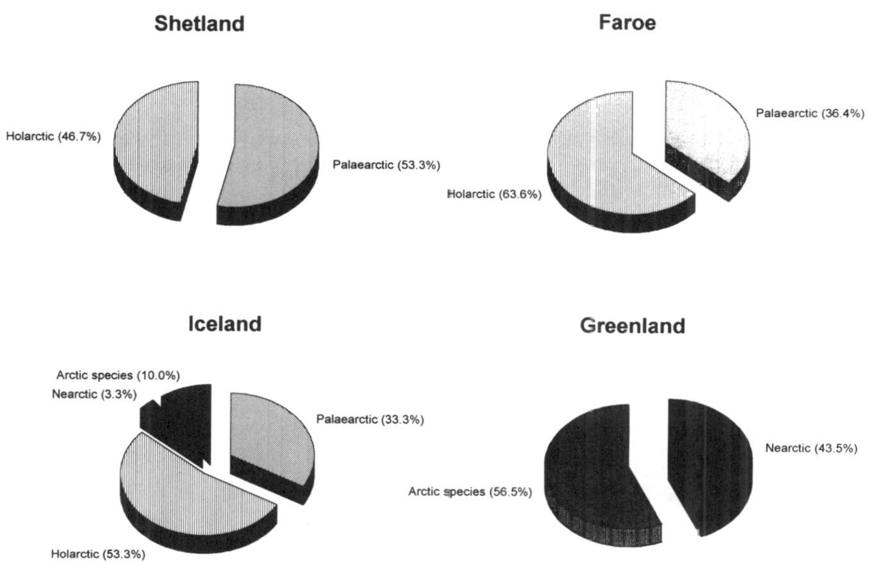


**Figure 11.2** The zoogeography of the Muscidae of Shetland, Faroe, Iceland and Greenland.

**Table 11.1** The zoogeography of the Coleoptera (beetles) of the North Atlantic islands. The figures refer to the number of species recorded from the various islands

| | *Greenland* | *Iceland* | *Faroe* | *Shetland* | *Orkney* |
|---|---|---|---|---|---|
| Cosmopolitan | 16 | 20 | 12 | 9 | 7 |
| Circumpolar | 14 | 15 | 6 | 10 | 8 |
| Holarctic | 7 | 9 | 5 | 16 | 23 |
| Amphiatlantic | 1 | 1 | 1 | 1 | — |
| Nearctic | 2 | 1 | — | — | — |
| Palaearctic | 29 | 104 | 145 | 302 | 326 |
| Endemic | 1 | — | — | — | — |
| Arctic species | 1 | — | — | — | — |
| Totals | 71 | 150 | 169 | 338 | 364 |

*Sources*: Greenland – Bocher (1988), Sadler (1991b); Iceland – Buckland (1988), Buckland et al. (1986b), Dugmore and Buckland (1984), Ingólfsson (1976), Larsson and Gígja (1959), Lindroth (1931, 1957, 1965), Lindroth et al. (1973, 1988), Ólafsson (1979), Sadler and Buckland (unpubl.); Faroe – Bengston (1981), Dinnin (unpubl.), Enckell (1985); Shetland – Berry and Johnston (1980); Orkney – Berry (1985), Sadler (1993).

**Table 11.2** Numbers of North Atlantic Coleoptera common to the British Isles and Scandinavia. Percentage figures are in parentheses. These neighbouring areas are thought to be the most likely sources from where the island's Coleoptera orginated

| *Island* | *British Isles* | *Scandinavia* | *Totals* |
|---|---|---|---|
| Greenland | 19(68.7) | 24(85.7) | 71 |
| Iceland | 82(97.6) | 84(100) | 150 |
| Faroe | 165(97.6) | 166(98.2) | 169 |
| Shetland | 337(99.7) | 330(97.6) | 338 |
| Orkney | 364(100) | 350(96.1) | 364 |

*Sources*: see table 11.1.

**Table 11.3** The zoogeography of the Muscidae (Diptera, or true flies) of the North Atlantic islands. The figures refer to the number of species recorded from the various islands. The Muscidae are presented since they are relatively well researched and effectively mirror the patterns evident in the dipterous fauna

| *Region* | *Palaearctic* | *Holarctic* | *Nearctic* | *Arctic species* | *Totals* |
|---|---|---|---|---|---|
| N. & W. Scandinavia | 250 | 130 | — | 13 | *c*.380 |
| British Isles | 189 | 84 | — | — | 273 |
| Ireland | 100 | 58 | — | — | 158 |
| Scotland | 129 | 82 | — | — | 211 |
| Inner Hebrides | 65 | 46 | — | — | 111 |
| Outer Hebrides | 73 | 28 | — | — | 101 |
| Orkney | 27 | 20 | — | — | 47 |
| Shetland | 16 | 14 | — | — | 30 |
| Faroe | 8 | 14 | — | — | 22 |
| Iceland | 10 | 16 | 1 | 3 | 27 |
| Jan Mayen | — | 2 | — | 1 | 2 |
| Spitzbergen | — | 5 | — | 4 | 5 |
| Greenland | — | 26 | 10 | 13 | 37 |
| N. & E. Canada | — | 83 | 87 | 29 | 170 |

*Sources*: Hennig (1964–7), Huckett (1965), Nielsen et al. (1954), Skidmore (1985, unpubl.).

may not yet be recorded from neighbouring areas such as Arctic Canada. It is notable also that the flies, *Limnophora groenlandica* Malloch, *L. rotundata* Collin and *Dolichopus groenlandicus* Zett. are only recorded from Greenland and may represent endemics, although it must be stressed that their taxonomy is similarly poorly understood and in need of revision.

2 The coleopterous faunas have a dearth of nearctic species (figure 11.1), only the byrrhid *Tylicus subcanus* LeConte, recorded recently from Greenland, and the dermestid, *Reesa vespulae* (Milliron), recently introduced to Greenland, Iceland and Scandinavia (Böcher 1988a, Ólafsson 1979), are nearctic in origin. A further four Greenlandic species, *Micralymma brevilingue* Schiod., *Simplocaria elongata* Sahl., *Coccinella transversoguttata* Fald. and *Dorytomus imbecillus* Faust, are western holarctic and not found in Europe. The faunas of Iceland, Faroe, Shetland and Orkney are predominantly palaearctic and have close affinities with the faunas of the British Isles and Scandinavia (tables 11.1 and 11.2). A similar pattern is evident in the dipterous faunas as one travels westwards from Great Britain, to Faroe and Iceland (table 11.3). The affinities of the flies of Greenland, however, are clearly different and the fauna is markedly nearctic (figure 11.2).

3 The faunas of the islands are cool temperate in character and there is a dearth of arctic forms. Only *Gnypeta cavicollis* Sahl., which is restricted to seven localities in north-east Greenland, has high arctic affinities. This contrasts strongly with the dipterous fauna of Greenland which comprises two groups, a dominant pure high arctic element in the north and a low boreal to subarctic one in the south (table 11.3).

4 There is marked decrease both in the number and diversity of species from east to west across the Atlantic. Perry (1986, and in Coope 1986, fig. 7) has computer modelled the possible carabid (ground beetle) species which might occur on the North Atlantic islands given the local climate and unimpeded access. It is immediately evident that the percentage of the potential carabid fauna which actually occurs on the islands falls off from Shetland (96 per cent), Faroe (49 per cent), Iceland (35 per cent) to Greenland (19 per cent). The only satisfactory view of these data is that the faunas were derived from the east and that there is a progressive loss of species due to operation of a transatlantic sweepstake (*sensu* Simpson 1940).

## *Faunal immigration*

It is clear that, in terms of the diperous fauna, Greenland is very different to the other islands in the region and one must consider its faunal development separately. This had been recognized by Degerbøl (1937), who suggested that the bulk of the flies in Greenland had reached the island via Ellesmere to the north. Downes (1966, 1988) highlighted the near complete harmony of Greenland's northern species of Lepidoptera and Diptera with the faunas of Ellesmere Island and similarly concluded that this was their main route of dispersal. The origins of the southern, nearctic, element are more enigmatic. The harmony of this group with Eastern Canada suggests that it originates from that source, but many of the species would not be able to withstand the

**Table 11.4** Indigenous and introduced Coleoptera of the North Atlantic islands. Species only recorded as fossils are excluded from the totals. Percentage figures are in parentheses

| *Island* | *Indigenous* | *Introduced* | *Totals* |
|---|---|---|---|
| Greenland | 28(39.4) | 43(60.6) | 71 |
| Iceland | 84(56) | 66(44) | 150 |
| Faroe | 140(82.8) | 29(17.2) | 169 |
| Shetland | 289(85.5) | 49(14.5) | 338 |
| Orkney | 328 (90.1) | 36 (9.9) | 364 |

*Sources*: see table 11.1.

high arctic conditions that prevail in the Ellesmere dispersal corridor. Although dispersal in periods of more clement climate is possible (cf. Dansgaard et al. 1975), it is more likely this southern element arrived by aerial drift across the Labrador sea, aided by the prevailing westerly winds.

The European nature of the biota of the islands east of Greenland led several authors to favour an initial early period of faunal immigration over notional land bridges linking the North Atlantic islands to North-west Europe (Löve and Löve 1963, Lindroth 1957), and survival of the biota through the Pleistocene glaciations in areas of ice-free refugia (Lindroth 1965, 1972). With the earliest realistic data for a viable land bridge in the Tertiary, survival in refugia seems increasingly unlikely (Buckland and Dugmore 1991). Buckland et al. (1986a) offer an elegant and controversial model of late-glacial immigration, aided by a reversed North Atlantic Drift and extensive sediment laden ice-floes, which could account for several of the zoogeographical characteriztics hitherto discussed (cf. Böcher 1988a, Coope 1986). Later aerial and hydrochorous dispersal may also be important (Downes 1988), but it cannot explain the presence of species that are clearly 'synanthropic' in the region and cannot maintain breeding populations on an island without the ameliorated conditions provided by human dwellings and aspects of human activity (cf. Tischler 1973). Many of these have been introduced to the islands (table 11.4) and have been described as 'serf species' (Enckell et al. 1987) – the term 'stowaway' is equally applicable, as they were unwittingly dispersed by humans (Sadler 1991a). Ostenfeld (1926) went so far as to suggest that the entire European element in the Greenlandic flora was introduced by the Norse. This is obviously an over-estimate, but the fossil record indicates that many stowaways made their first appearance in Iceland and Greenland with the Norse settlers. It is the nature and development of the synanthropic fauna that the remainder of the chapter will focus upon, and these aspects are best resolved with recourse to the fossil record.

## FOSSIL RECORD

### *Impact of settlement*

Ari the Wise, author of the *Íslendingabók (Book of the Icelanders*), describes the environment of Iceland at the time of first settlement (Landnám). He

**Plate 11.1** The Landnám layer at Kilaarsarfik in the former Norse western settlement of Greenland.

states that 'At that time Iceland was covered with forests between mountain and seashore' (trans. Hermannson 1930: 60). At the time of settlement some 65 per cent of Iceland was vegetated (Árnalds 1987). Birch woodland was common at lower elevations with willow and other dwarf shrubs dominant above 400–500 m a.s.l. The destruction of the predominantly birch and willow forests consequent upon Landnám (Hallsdóttir 1987) was followed by the progressive loss of soils (Thórarinsson 1961). A similar sequence of events can be seen to have occurred on Greenland and Faroe (Fredskild 1973, 1988; Jóhansen 1975, 1979, 1985). In Greenland, at least in the Western Settlement, the predominantly dwarf-shrub vegetation was cleared by burning. A layer of charcoal marking Norse Landnám can be found overlying sterile subsoil in many areas of the former Western settlement (Iversen 1934; Fredskild and Humle 1991; McGovern et al. 1988 – see also plate 11.1) and wind-borne microscopic charcoal recorded from lake sediments markedly increases in concentration in lake basins after settlement (Fredskild, 1983, 1988).

After Landnám, subsequent demands on local timber for uses such as fuel, charcoal for smelting (cf. Ingstad 1974; Roussell 1936, 1941) and dairying processes (Albrethsen and Keller 1986: 100) probably ensured further destruction of woodland. The cutting and removal of turf to construct and repair turf buildings must have laid further land bare. Trampling and over-grazing, occasioned by the intensive system of animal husbandry that was imported with the settlers, had a significant effect on ground cover growth and regeneration (Dugmore and Buckland 1991). The combined impact of these actions is recorded in peat and lake profiles by accelerated deposition caused by soil erosion and deflation. This may have been a deciding factor in the ultimate decline of the Greenland settlements (cf. McGovern et al. 1988), but also appears to have been important in Iceland where it contributed towards the abandonment of several farmsteads (Sveinbjarnardóttir, 1992) and the unconstrained nature of the much of coastal sandur (Buckland et al. 1991a). Although, in Iceland, the large areas of grassland created and modified by the colonists have been further improved by drainage and reseeding during the last century, soil erosion remains a problem today (Friðriksson 1978, Árnalds 1987).

### *Faunal change*

As the settlers modified the environment they not only created new habitats but they altered old ones. The cumulative effect of these activities may not only have led to changes in the distribution and frequency of the indigenous biota, but may have enabled many of the introduced species to establish viable breeding populations on the islands. At the risk of succumbing to what Schoener (1988: 483) terms 'Islomania – the powerful attraction of islands', one can assume that the farm sites exhibited the characteristics of 'habitat islands' (*sensu* MacArthur and Wilson 1967); indeed, the former sites of farmsteads stand out clearly in the landscape (plate 11.2). In the harsh environments of Iceland and Greenland, temperatures would seldom be high enough for the effective dispersal of synanthropic insects associated with the settlement sites. Moreover, the surrounding landscape of heavily grazed shrub and grassland would have provided an impenetrable barrier to most species other than those with a high dispersal potential. The immigration rate would be controlled by some factor of the distance from the source areas, and extinction rates by the size of the farm and the spatial and temporal continuity of the farm habitat. The maintenance of the population of synanthropic species would require either a constant dispersal of species from adjacent farm sites (islands) and/or the continued management of the habitat; a process fuelled by the immigration of stowaways via the movement of fodder, food and other trade items from Europe (Sadler 1991a). Although an obvious simplification, such an approach provides a suitable conceptual framework within which to address several biogeographical questions regarding the nature of the post-Landám faunas of the islands.

### *Introductions and extinctions*

Almost half of the Coleoptera of Iceland and Greenland are anthropochorous, and many other species on Faroe, Shetland and Orkney owe their origins to

Plate 11.2 The abandoned farm site at Reykjarnes, Arnesthreppur, northwest Iceland.

human dispersal (table 11.4). The fossil records from Greenland and Iceland show that many synanthropes appear immediately at Norse Landnám (Buckland et al. 1991a, McGovern et al. 1983) and that most are commonly associated with farm site habitats including stored hay and stored products such as cereals, and waste deposits such as carrion, dung, faeces and rotting plant material. The species diversity away from the archaeological sites increases immediately after Landnám in Iceland (Buckland et al. 1986b: fig. 7), perhaps reflecting the increased habitats created by human activity. This introduced assemblage of stowaways is remarkably consistent and widespread in Norse cultural deposits and has been found in Viking Dublin (Coope 1981), Anglo-Scandinavian York and Oslo (Kenward 1980, Hall et al. 1983), Iceland (Perry et al. 1985) and Greenland. Fewer introduced species have been recorded in the dipterous faunas of the region. Nielsen et al. (1954) record 47 facultative synanthropes from Iceland, but only three species of flies, *Telomerina flavipes* (Meigen), *Melophagus ovinus* L. and *Musca domestica* L., are probable introductions to Greenland. The fields of the Norse farm sites were also rich in introduced species, particularly weeds and apophytes such as chickweed (*Stellaria media*), Shepherd's purse (*Capsella bursa-pastoris*) and knotweed (*Polygonum aviculare*). Pollen and seeds of flax (*Linum usitatissimum*) have been recorded from the Norse middens of two western settlement farms and these may indicate cultivation (Fredskild 1988).

With only one grain of barley pollen recorded from Brattahlid, the home of Eirik the Red, in the Eastern Settlement of Greenland (Fredskild 1978), there is little evidence for the successful cultivation of cereals and, as was noted in *The King's Mirror* (trans. Larsen 1917: 142), they did not form an important part of the Norse economy. Supplies of grain were probably imported from Europe and the *Saga of Eirik the Red* relates that cargoes of 'malt and flour' were also imported (trans. Magnusson and Pálsson 1965: 92). To date, no obligate cereal pests have been recorded from Norse-age deposits in Greenland. In Iceland, the saw-toothed grain beetle, *Oryzaephilus surinamensis*, (L.), and the grain weevil, *Sitophilus granarius* (L.), were introduced as early as the eleventh century (Sadler 1991a,b), and despite finds of the remains of barley at two sites (Friðriksson 1959, 1960), it likely that these pests were imported to Iceland in shipments of grain (Amorosi et al. 1992).

The dispersal corridor provided by the Norse colonists for their stowaway guests also may have operated like a sweepstake. Which stowaway species were transported was dependent upon several factors, including (i) species availability at the port or embarkation point, (ii) species survival during the trans-Atlantic journey, (iii) habitat availability at the destination, and (iv) stochastic mechanisms (or 'biogeographic accidents'). Differences in the faunas of the islands indicate the severity of the dispersal route and this is well exemplified by the coprophile or dung fauna. Dung feeders are rare; only one *Aphodius* beetle occurs in Iceland and none is found in Greenland, despite the fact that 39 and 46 species are found in Great Britain and Scandinavia respectively. Not all introduced species maintain populations on the islands. In Iceland, the scarabaeid beetles, *Aphodius lapponum* Gyll., *A. fimetanus* L., *A.* cf. *prodromus* Brahm and *Oxyomus sylvestris* (Scopoli) all have fossil records, but only the first is recorded at the present day (Ólafsson 1991). Such apparent extinctions are difficult to explain, particularly since their favoured habitat of dung and similar accumulations of rotting plant and animal material was demonstrably widespread. The failure of *O. sylvestris* is probably a result of its thermal requirements. The species has a markedly southern distribution in Britain and Scandinavia (Landin 1961, Jessop 1986), and it is possible that the temperatures in Iceland are not sufficient for it to complete its life cycle.

For many montane dung beetle species, emergence of imagines is dependent upon the duration of snow cover and the length of periods when soil temperatures are above 5 °C (Lumaret and Stiernet 1991). These thermal requirements may well explain why no dung beetles are recorded in Greenland, and it may be significant, similarly, that obligate dipterous coprophiles do not extend into the Arctic.

Notwithstanding the operation of a severe sweepstake, and the few documented stowaways that failed to establish permanent populations, a considerable proportion of the introduced fauna, with the exception of the Norse introductions in Greenland (below), has been remarkably successful on the islands. The reasons for this are not always clear (cf. Simberloff 1981) and when dealing with synanthropic species, it is often difficult to disentangle the effects of transportation and habitat availability, since both are intimately associated (Niemela 1992). Spence (1991) has examined why several species of synanthropic ground beetles introduced into Canada from Europe have been successful. He indicates that the general hypothesis of exotics invading an empty niche in North America is false (cf. Herbold and Moyle 1986) and that their success is a result of their being habitat generalists in the source area (in this case southern England), where most habitats are disturbed to some degree by human activity. This broken ground habitat, allied to the artificially ameliorated habitats provided directly in human settlements, is a prerequisite for the successful introduction of synanthropic species. In the North Atlantic islands suitable habitats were not abundant, or indeed totally absent, prior to Landnám. Thus it is the translocation of habitats from Europe that is important in providing a foothold for potential colonists.

Elements in the indigenous (pre-Landnám) biota may well have benefited from the activities of the settlers. The mould-feeding lathridiid beetle, *Corticaria rubripes* Mannerheim, is a very common fossil in Norse cultural deposits in Greenland, although it is known from only two recorded sites in Greenland at the present day (Böcher 1988a). Initially, *C. rubripes* was assumed to have been introduced to Greenland (Sadler 1991a), but fossil records from a palaeo-Inuit site in the north (Böcher 1988b) indicate that it is native to the island. In Europe, the species is found in forest litter, under bark and in swathes of vegetation, but it also occurs in synanthropic situations (e.g. old barns, manure and so on). It is in these synanthropic habitats that the species was successfully exploiting in the former Norse settlements. Similar examples are evident in the Diptera and the flora of the region. The fly *Heleomyza borealis* (Boh.), accounts for some 53 per cent of the total species recorded from Norse sediments in Greenland. The species is common around human habitations in boreal to arctic areas where accumulations of excrement and decaying animal matter are widespread, but it is also characteriztic of guano deposits around bird cliffs. *H. borealis* is recorded in Greenland at the present day, but may well have been more common in the past. Blinks (*Montia fontana*) is strongly favoured by human activity. A native plant in Greenland, growing in fen areas, often along brooks, it has spread to the nutrient-rich soils among the houses in the former Norse settlements (Fredskild and Humle 1991).

Changing land management practices, widespread erosion and the drainage of wetlands meant that several species have suffered drastic reductions in their former range and there may have been inevitable extinctions. Habitat destruction may well have forced the small hydraenid water beetle, *Hydraena britteni* Joy, into extinction in Iceland, although the impact of reduced temperatures of the Little Ice Age should not be underestimated (Buckland et al. 1983). The introduction of mechanization and the trend towards monoculture of fast-growing grasses has resulted in variations in the abundance of stored hay species in Iceland (Buckland et al. 1991b). The maintenance of settlement since Norse Landnám in Iceland, however, ensured that suitable habitats were sustained for a sufficient length of time for the survival of many introduced insects.

### *Faunal continuity*

In Greenland, there is a period of two hundred years or so between the demise of the former Norse settlements and the European recolonization of

**Table 11.5** Species known only as fossils from the North Atlantic islands

| *Island* | *Coleoptera* | *Diptera* |
|---|---|---|
| Greenland | *Othius angustus* Steph. | *Telomerina flavipes* (Meigen) |
| | *Philonthus politus* (L.) | *Melophagus ovinus* (L.) |
| | *Philonthus cephalotes* (Graven.) | |
| | *Ocalea picata* (Steph.) | |
| | *Atomaria* sp. | |
| Iceland | *Hydraena britteni* Joy. | – |
| | *Aphodius firmetarius* (L.) | |
| | *A. cf. prodromus* (Brahm) | |
| | *Oxyomus sylvestris* (Scopoli) | |
| Faroe | *Calathus micropterus* Dufts. | – |
| | *Ochthephilus omalinus* (Erich.) | |
| Shetland | *Cryptophagus scutellatus* Newm. | – |
| | *Aglenus brunneus* (Gyll.) | |
| | *Mycetaea hirta* (Marsh.) | |
| | *Tipnus unicolor* (Pill. & Mitt.) | |
| Orkney | *Bembidion unicolor* Chaud. | – |
| | *Acritus nigricornis* (Hoff.) | |
| | *Thanatophilus dispar* (Herbst) | |
| | *Micropeplus fulvus* Erich. | |
| | *Xylodromus concinnus* (Marsh.) | |
| | *X. depressus* (Graven.) | |
| | *Carpelimus bilineatus* Steph. | |
| | *Tachinus corticinus* Graven. | |
| | *Aphodius fimetarius* (L.) | |
| | *Dascillus cervinus* (L.) | |
| | *Calyptomerus dubius* (Marsh.) | |
| | *Dryops* spp. | |
| | *Tipnus unicolor* (Pill. & Mitt.) | |
| | *Alophus triguttatus* (F.) | |
| | *Phytobius quadrituberculatus* F. | |

*Sources*: Bengston (1981), Buckland (1988, unpubl.), Dinnin (unpubl.), Sadler (1991b, 1993), Skidmore (unpubl.).

the island by Hans Egede in 1721. The loss of habitat consequent upon this depopulation must have had some effect on the insect fauna, and some of the species in the fossil record are no longer recorded from Greenland at the present day (table 11.5). One must always be cautious of equating absence with extinction, but for strongly synanthropic insects, for which every year would mean one generation, survival for two hundred years, without the shelter provided by buildings and the constant cycling of organic waste materials resulting from the daily pastoral life on the islands, is unlikely. The continuity of habitat management is particularly important. Entomological surveys of abandoned farms in the Arnesthreppur in north-west Iceland (summer 1990) provide corroborating evidence to reinforce this point. At the site of Reykjarnes (plate 11.2), which was finally abandoned in 1956, no obligate synanthropes

**Plate 11.3** A view through the homefield of the abandoned Norse farm site of Sandnaes (V51) in the former western settlement of Greenland.

were recovered, despite the fact that outhouses such as the sheephouse and barn remained standing. The barn, which had been used by a local farmer as recently as ten years ago, had hay residues on the floor.

Böcher (1988a) suggests that *Quedius mesomelinus* (Marsh.) and *Xylodromus concinnus* (Marsh.) might have survived until the present day by switching over to a life associated with Inuit turf huts. Establishing how these insects made the transition from Norse to Inuit structures is problematical, particularly since the archaeological record suggests that contact between the groups was slight (McGhee 1984). Even if one accepts there was enough cultural contact to facilitate the dispersal of these species, it is questionable whether sufficient habitat existed in Inuit settlements, since their resource base was based on hunting, rather than pastoral, activities. It may be significant, however, that both species are were recorded in abundance during a spell of collecting between 1876 and 1899, and since then only a few captures have been made. Böcher (1988a) notes that both species are now probably extinct in Greenland and their demise may be attributable to the transition from turf huts to wooden buildings during the early twentieth century. The bug *Nabis flavomarginatus* (Sch.) and the rove beetle *Omalium excavatum* Stephens may represent the only true Norse relicts. The former is absent in Iceland and has been recovered from Norse sites in Greenland (McGovern et al. 1983). The species is predaceous as a larva and adult and is common on grasses (Southwood and

Leston 1959). *O. excavatum* is widely distributed in the North Atlantic, where it is found in both synanthropic and non-synanthropic habitats. In Greenland today their distributions are closely linked with the former Norse settlement areas and both species were probably introduced by the Norse settlers. Their persistence may be attributable to the longevity of the homefield habitat in the environment. The homefields, which were intensively manured by the settlers, still support a rich flora of introduced species at the present day (plate 11.3).

One is left, therefore, with the suggestion that the majority of the modern Greenland fauna was re-introduced in the post-medieval era with Hans Egede and his colonists. The probabilities of both of these colonization ventures introducing the same few species might at first consideration seem very slight, but the source area in both cases was Scandinavia and many of the species involved are now cosmopolitan pests with global distributions and ideally suited to re-introduction. These problems might be resolved by further study of the fossil record from cultural deposits from Inuit houses, which span the fifteenth to the eighteenth centuries and the later deposits of the European migrants.

As Lindroth (1957) admirably documents, further movement of species of plant and animals took place during the seventeenth and eighteenth centuries as a result of the Newfoundland trade across the North Atlantic, when Danish and Dutch ships were common visitors to North Atlantic waters (Gad 1973). It is since this period that some 50 species of ground beetles appear to have been introduced to, and become established in, North America (Spence 1991, Buckland et al. this volume). Two records of palaearctic species of *Cercyon* hydrophilid beetles from Iceland and Greenland indicate that the dissemination of insects species by human agencies continues apace today. The recent, single, record of *Cercyon obsoletus* Gyllenhali (= *lugubris* Olivier) from Greenland is almost certainly an accidental introduction (Böcher 1988a), as is the chance capture of *Cercyon haemorrhoidalis* F. from Iceland.

## CONCLUSIONS

After an early phase of post-glacial immigration, the insect fauna of the North Atlantic islands has been modified by anthropogenic activities of Norse and later settlers. The first colonists exported a whole 'cultural package' based upon a pastoral system, and it is this translocation of habitat that provided the foothold for a community of stowaway insects. A large proportion of the coleopterous fauna has been passively dispersed by humans, and these introduced species are totally reliant for their continued survival on the islands upon the ameliorated habitats provided by human settlements and their associated activities. Faunal continuity is intimately associated with habitat continuity, and nowhere is this more clearly illustrated than in the case of Greenland, where the loss of habitat consequent upon the demise of the Norse settlements caused the probable extinction of much of the synanthropic fauna.

ACKNOWLEDGEMENTS

Support for this work has been provided by the Science and Engineering Research Council, the Scottish Development Department and the National Science Foundation of America. Interest in the faunas of the North Atlantic was stimulated by discussions with Dr P. C. Buckland, as both a teacher and a colleague. Thanks are due to the two referees for their helpful suggestions and to our British, American and Icelandic colleagues, too numerous to mention here, without who kindly assisted in the field, seminar room and laboratory.

REFERENCES

Albrethsen, S. E. and Keller, C. (1986) 'The use of the saeter in medieval Norse farming in Greenland', *Arctic Anthropology* 23(1–2): 91–107

Amorosi, T., Buckland, P. C., Ólafsson, G., Sadler, J. P. and Skidmore, P. (1992) 'Site status and the palaeoecological record: a discussion of the results from Bessastaðir, Iceland', in Morris, C. and Rackham, J. (eds) *Norse and Later Settlement and Subsistence in the North Atlantic* (Glasgow University Press, Glasgow) pp. 169–192

Árnalds, A. (1987) Ecosystem disturbance in Iceland', *Arctic and Alpine Research* 19: 508–13

Bengston, S.-A. (1981) 'Terrestrial invertebrates of the Faroe Isles III. Beetle (Coleoptera): checklist, distribution and habitats', *Fauna Norvegica* B28: 52–82

Berry, R. J. (1985) *The Natural History of Orkney* (New Naturalist, Collins, London)

Berry, R. J. and Johnston, J. L. (1980) *The Natural History of Shetland (New Naturalist, Collins, London)*

Böcher, J. (1988a) 'The Coleoptera of Greenland', *Meddelelser om Grønland, Bioscience* 26: 1–100

Böcher, J. (1988b) 'Insektundersogelserne. Boplads i dybfrost', *Naturens Verden* (1988): 423–4

Buckland, P. C. (1988) 'North Atlantic connections: introductions or endemics', *Entomologica Scandinavica Supplement* 32: 7–29

Buckland, P. C. (1989–90) 'Insects, man and the earliest settlement of the Faroe Islands: a case not proven', *Fróðskaparrit* 38–9: 107–15

Buckland, P. C. and Dugmore, A. J. (1991) 'If this is a Refugium, why are my feet so bloody cold?', in Maizels, J. and Caseldine, C. (eds) *Environmental Change in Iceland* (Kluwer, Dordrecht) pp. 107–26

Buckland, P. C., Dugmore, A. J. and Sadler, J. P. (1991b) 'Faunal change or taphonomic problem? A comparison of modern and fossil insect faunas from southeast Iceland', in Maizels, J. and Caseldine, C. (eds) *Environmental Change in Iceland* (Kluwer, Dordrecht) pp. 127–46

Buckland, P. C., Perry, D. W. and Sveinbjarnardóttir, G. (1983) '*Hydraena briffeni* Joy (Coleoptera; Hydraenidae) fundin í setlögum á Íslandi, frá því seint á nútíma', *Náttúrufræðingurinn* 52: 37–44

Buckland, P. C., Perry, D. W., Gíslasson, G. M. and Dugmore, A. J. (1986a) 'The pre-Landnám fauna of Iceland: a palaeontological contribution', *Boreas* 15: 173–84

Buckland, P. C., Dugmore, A. J., Perry, D. W., Savory, D. and Sveinbjarnardóttir, G. (1991a) 'Holt in Eyjafjallasveit, Iceland: a palaeoecological study of the impact of Landnám', *Acta Archaeologica* 61: 252–71

Buckland, P. C., Gerrard, A. J., Larsen, G., Perry, D. W., Savory, D. R. and

Sveinbjarnardóttir, G. (1986b) 'Late Holocene palaeoecology at Ketilssta∂ir in Mýrdalur, South Iceland', *Jökull* 36: 41–55

Clark, A. H. (1949) *The Invasion Of New Zealand By People, Plants And Animals* (Greenwood Press, Westport)

Coope, G. R. (1981) 'Report on the Coleoptera from an eleventh-century house at Christ Church Place, Dublin', in Bekker-Nielson, H., Foote, P. and Olsen, O. (eds) *Proceedings of the Eighth Viking Congress (1977)* (Odense University Press) pp. 51–6

Coope, G. R. (1986) 'The invasion and colonisation of the North Atlantic Islands: a palaeoecological solution to a biogeographic problem', *Philosophical Transactions of the Royal Society, London B* 314: 619–35

Crosby, A. W. (1986) *Ecological Imperialism: the Biological Expansion of Europe, 900–1900 AD* (Cambridge University Press, Cambridge)

Dansgaard, W., Johnsen, S. J., Reek, N., Clansen, H. B. and Hammer, E. V. (1975) 'Climatic change, Norsemen and modern man', *Nature* 255: 24–8

Debes, H. J. (1989–90) 'Problems concerning the earliest settlement in the Faroe Islands', *Fró∂skaparrit* 38–39: 23–34

Degerbøl, M. (1937) 'The contribution to the investigation of the fauna of the Bosseville coast, with special reference to zoogeography', *Meddelelser om Grønland* 104(19): 1–35

Downes, J. A. (1966) 'The Lepidoptera of Greenland: some geographic considerations', *Canadian Entomologist* 98: 1134–44

Downes, J. A. (1988) 'The post-glacial colonization of the North Atlantic Islands', *Memoirs, Entomological Society of Canadian* 144: 55–92

Dugmore, A. J. and Buckland, P. C. (1984) 'Vatnabjallan Oreodytes sanmarki (Sahl.) (Col., Dytiscidae) fundin á Íslandi', *Náttúrufræ∂ingurinn* 53: 7–12

Dugmore, A. J. and Buckland, P. C. (1991) 'Erosion in Iceland', in Maizels, J. and Caseldine, C. (eds) *Environmental change in Iceland* (Kluwer, Dordrecht)

Elton, C. S. (1958) *The Ecology of Invasions by Plants and Animals* (Chapman and Hall, London)

Enckell, P. (1985) *Island Life: Agency of Man upon Dispersal, Distribution and Genetic Variation in Faroese Populations of Terrestrial Invertebrates* (Department of Animal Ecology, University of Lund, Sweden)

Enckell, P. H., Bengston, S-A. and Wimon, B. (1987) 'Serf and waif colonization: distribution and dispersal of invertebrate species in Faroe island settlement areas', *Journal of Biogeography* 14: 89–104

Fredskild, B. (1973) 'Studies in the vegetational history of Greenland', *Meddelelser om Grønland* 194(4): 1–245

Fredskild, B. (1978) 'Palaeobotanical investigations of some peat deposits of Norse age at Qagssiarssuk, South Greenland', *Meddelelser om Grønland* 204(5): 1–41

Fredskild, B. (1983) 'The Holocene vegetational development of the Godthaabsfjord area, West Greenland', *Meddelelser om Grønland, Geoscience* 10: 3–28

Fredskild, B. (1988) 'Agriculture in a marginal area – South Greenland from the Norse Landnám (985 AD) to the present (1985 AD). In Birks, H. H., Birks, H. J. B., Kaland, P. E. and Moe, D. (eds) *The Cultural Landscape: Past, Present and Future* (Cambridge University Press, Cambridge), pp. 381–93

Fredskild, B. and Humle, L. (1991) 'Plant remains from the Norse farm Sandnes in the Western Settlement Greenland', *Acta Boreal* 8: 69–81

Fri∂riksson, S. (1959) 'Korn frá Gröf í Öraefum', *Árbók hins íslenzka Fornleifafélags* (1959): 88–91

Fri∂riksson, S. (1960) 'Jurtaleifar frá Bergpórshvoli á söguöld', *Árbók hins íslenzka Fornleifafélags* (1960): 64–75

Friðriksson, S. (1978) 'The degradation of Icelandic ecosystems', in Holdgate, M. W. and Woodman, M. J. (eds) *The Breakdown and Restoration of Ecosystems* (Plenum Press, New York) pp. 145–56

Gad, F. (1970) *A History of Greenland*, vol. 1 (David Hurst, London)

Gad, F. (1973) *A History of Greenland*, vol. 2. (David Hurst, London)

Grove, J. M. (1988) *The Little Ice Age* (Methuen, London)

Hall, A. R., Kenward, H. K., Williams, D. and Greig, J. R. A. (1983) *Environment and Living Conditions at Two Anglo-Scandinavian sites* Archaeology of York, 14/4 (Council for British Archaeology, London)

Hallsdóttir, M. (1987) 'Pollen analytical studies of human influence on vegetation in relation to the Landnám layer tephra in south west Iceland', Lundqua thesis, 18, University of Lund)

Hennig, W. (1964) *Die Fliegen der Palarctischer Region*, vols 1 and 2 (Nägler and Obermiller, Stuttgart)

Herbold, B. and Moyle, P. B. (1986) 'Introduced species and vacant niches', *American Naturalist* 128(5): 751–60

Hermanns-Auðardóttir, M. (1991) 'The early settlement of Iceland', *Norwegian Archaeological Review* 24(1): 1–9

Hermannson, H. (trans.) (1930) *The Book of the Icelanders (Íslendingabók)* (Ithaca, New York)

Huckett, H. C. (1965) 'The Muscidae of Northern Canada, Alaska and Greenland (Diptera)', *Memoirs, Entomological Society of Canadian* 42: 1–370

Ingólfsson, A. (1976) 'Smádýralif og gróður á sjávarfitjunn við Gálgahraun', *Náttúrufræðingurinn* 46: 223–37

Ingstad, A. S. (1986) *The Discovery of America*, vol. 1: *Excavations at L'Anse aux Meadows, Newfoundland, 1961–68* (Norwegian University Press)

Ingstad, H. (1974) *Westward to Vinland* (Book Club Associates, London)

Iversen, J. (1934) 'Moorgeologische Untersuchungen auf Grönland', *Meddelelser Dansk Geologiske Forening* 8: 341–58

Jessop, L. (1986) *Dung Beetles and Chafers Coleoptera: Scarabaeoidea* Handbooks for the Identification of British Insects 5(2) (Royal Entomological Society, London)

Jóhansen, J. (1975) 'Pollen diagrams from Shetland and Faroe Islands', *New Phytologist* 75: 369–87

Jóhansen, J. (1979) 'Cereal cultivation in Mykines, Faroe Islands AD 600', *Danmarks Geologiske Undersogelse, Årbok* (1981): 31–7

Jóhansen, J. (1985) *Studies in the Vegetational History of the Faroe and Shetland Islands* (Foroya Fróðskaparrit, Tórshavn)

Jóhansen, J. (1986) 'Foroya fyrstu búseting – eina ferð enn', *Möndul* 12: 3–6

Jones, G. (1986) *A History of the Vikings* (2nd edn) (Oxford University Press, Oxford)

Kaland, S. (1991) 'Comments on the early settlement of Iceland', *Norwegian Archaeological Review* 24(1): 10–12

Kenward, H. K. (1980) 'Insect remains', in Schia, E. (ed.), Feltene 'Olsogate 3 og 7', *Die Arkeologiske Utgravningeri Gamlebyen* (Oslo, Olso 2) pp. 134–7

Krogh, K. J. (1967) *Viking Greenland* (Nationalmusset, Copenhagen)

Krogh, K. J. (1986) 'Um foroya fyrstu búseting – eina ferð enn', *Möndul* 12: 33–36

Landin, B.-O. (1961) 'Ecological studies on dung beetles', *Opuscula Entomologica Supplement* 19

Larsen, L. M. (trans.) (1917) *The King's Mirror* (American–Scandinavian Foundation, New York)

Larsson, S. G. (1959) 'Coleoptera 2. General remarks', *The Zoology of Iceland* (Munksgaard, Copenhagen) 3(46b)

Larsson, S. G., and Gígja, G. (1959) 'Coleoptera 1. Synopsis', *The Zoology of Iceland* (Munksgaard, Copenhagen) 3(46a)

Lindroth, C. H. (1931) 'Die insektenfauna Islands und ihre probleme', *Zoologiske Bidrag Fran Uppsala* 13: 105–600

Lindroth, C. H. (1957) *Faunal Connections between Europe and North America* (Almqvist and Wiksell, Stockholm)

Lindroth, C. H. (1965) 'Skaftafell, Iceland; a living glacial refugium', *Oikos, Supplement* 6

Lindroth, C. H. (1968) 'The Icelandic form of *Carabus problematicus* Hbst.: a statistical treatment of subspecies', *Opuscula Entomologica* 33: 157–182

Lindroth, C. H. (1972) 'Reflections on glacial refugia', *Ambio*, Special Report No. 2

Lindroth, C. H., Bengston, S.-A. and Enckell, P. H. (1988) 'Terrestrial faunas of far isolated areas: a study in tracing old faunal centres. *Entomologica Scandinavica* 32: 31–65

Lindroth, C. H., Böðvarsson, H., Sigurður, H. and Richter, S. H. (1973) 'Surtsey, Iceland – the development of a new fauna, 1963–1970: terrestrial invertebrates', *Entomologica Scandinavica Supplement* No. 5

Löve, A. and Löve, D. (1963) *North Atlantic Biota and Their History* (Pergamon Press, London)

Lumaret, J-P. and Stiernet, N. (1991) 'Montane dung beetles', in Hanski, I. and Cambefort, Y. (eds) *Dung Beetle Ecology* (Princeton University Press, Princeton, New Jersey) pp. 242–55

MacArthur, R. H. and Wilson, E. O. (1967) *The Theory of Island Biogeography* (Princeton University Press, Princeton, New Jersey)

Magnusson, M. and Pálsson, H. (1965) *The Vinland Sagas: The Norse Discovery of America* (Penguin, Harmondsworth)

Mahler, D. and Malmros, C. (1991) 'Comments on the early settlement of Iceland', *Norwegian Archaeological Review* 24(1): 14–18

Mahler, V. (1988) 'A new species of *Atheta* Thomsom, 1858, from Greenland (Coleoptera, Staphylinidae, Aleocharinae)', *Steenstrupia* 14: 93–7

McGhee, R. (1984) 'Contact between native north Americans and the medieval Norse: a review of the evidence', *American Antiquity* 49(1): 4–28

McGovern, T. H. (1980) 'Cows, harpseals and churchbells: adaptation and extinction in Norse Greenland', *Human Ecology* 8(1): 245–77

McGovern, T. H. (1981) 'The Vinland adventure: a North Atlantic perspective', *North American Archaeologist* 24(4): 285–08

McGovern, T. H., Bigelow, G. F., Amorosi, T. and Russell, D. (1988) 'Northern islands, human error and environmental change in the medieval North Atlantic', *Human Ecology* 16: 225–70

McGovern, T. H., Buckland, P. C., Savory, D., Sveinbjarnardóttir, G., Andeasen, C. and Skidmore, P. (1983) 'A study of the faunal and floral remains from two Norse farms in the Western Settlement, Greenland', *Arctic Anthropology* 20(2): 93–120

Nielsen, P., Ringdahl, O. and Tuxen, S. L. (1954) 'Diptera I', *The Zoology of Iceland*, 111(48a) (Ejnar Munksgaard, Copenhagen)

Niemela, I. (1992) 'Carabid beetles in the Åland Archipelago', in Noonan, G. R., Stork, N. E. and Ball, G. E. (eds) *The Biogeography of Ground Beetles of Mountains and Islands* (Intercept, Andover)

Ólafsson, E. (1979) 'Hambjalla, Ressa vespulae (Mill.) (Coleoptera, Dermestidae), nýtt meindýr á Íslandi', *Náttúrufræðingurinn* 49: 155–62

Ólafsson, E. (1991) *Fjölrit Náttúrufrædistofnunar* 17 (Íslenskt Skordýratal, Reykjavík)

Ostenfeld, C. H. (1926) 'The flora of Greenland and its origins', *Biologiske Meddelelser* B6: 1–87

Perry, D. W (1986) The analysis of subfossil insect remains: a numerical approach, unpubl. Ph.D. thesis, Department of Geology, University of Birmingham

Perry, D. W., Buckland, P. C. and Snæsdóttir, M. (1985) 'The application of numerical techniques to insect assemblages from the site of Stóraborg, Iceland', *Journal of Archaeological Science* 12: 335–45

Rink, H. (1877) *Danish Greenland – its people and products* (H. S. King, London)

Rosen, B. R. (1988a) 'From fossils to earth history: applied historical biogeography', in Myers, A. A. and Giller, P. S. (eds) *Analytical Biogeography* (Chapman and Hall, London) pp. 437–78

Rosen, B. R. (1988b) 'Biogeographhic pattern: a perceptual overview', in Myers, A. A. and Giller, P. S. (eds) *Analytical Biogeography* (Chapman and Hall, London) pp. 23–53

Roussell, A (1936) 'Sandnes and the neighbouring farms', *Meddelelser om Grønland* 88(2): 1–232

Roussell, A. (1941) 'Farms and churches of the Medieval Norse settlements of Greenland', *Meddelelser om Grønland* 89(1): 1–354

Sadler, J. P. (1991a) 'Beetles, boats and biogeography: insect invaders of the North Atlantic Islands', *Acta Archaeologica* 61: 199–212

Sadler, J. P. (1991b) 'Archaeological and biogeographical implications of palaeoentomological studies in Orkney and Iceland', unpubl. Ph.D. thesis, Department of Archaeology and Prehistory, University of Sheffield

Sadler, J. P. (1993) 'Records of modern and fossil Coleoptera from the Orkney Islands', *Entomologist's Monthly Magazine* 129: 21–3

Saunders, W. (1882) 'On the introduction and dissemination of noxious insects', *Proceedings and Transactions of the Royal Society of Canada, Section 4. Geological and Biological Sciences* 1: 77–9

Schoener, A. (1988) 'Experimental island biogeography', in Myers, A. A. and Giller, P. S. (eds) *Analytical Biogeography* (Chapman and Hall, London) pp. 483–512

Simberloff, D. H. (1981) 'Community effects of introduced species', in Nitecki, T. H. (ed.) *Biological Crises in Ecological and Evolutionary Time* (Academic Press, New York) pp. 53–81

Simpson, G. G. (1940) *Mammals and landbridges Journal of the Washington Academy of Science* 30: 137–63

Skidmore, P. (1985) *The Biology of the Muscidae of the World* (Junk, Dordrecht)

Southwood, T. R. E. and Leston, D. (1959) *Land and Water Bugs of the British Isles* (Wayside and Woodland, London)

Spence, J. R. (1991) 'Success of European Carabid species in Western Canada: preadaptation for synanthropy?', in Stork, N. (eds.) *The Role of Ground Beetles in Ecological and Environmental Studies* (Intercept, Andover) pp. 129–41

Sveinbjarnardóttir, G. (1972) 'Papar', *Mimir* 11: 2–20

Sveinbjarnardóttir, G. (1992) *Farm abandonment in Medieval and Post-medieval Iceland: an Interdisciplinary Study (Oxbow, Oxford)*

Tischler, W. (1973) 'Ecology of Arthropod fauna in man-made habitats: the problem of synanthropy', *Zoologischer Anzeiger, Leipzig* 191: 157–61

Thórarinsson, S. (1961) 'Population change in Iceland', *Geographical Review* 51: 519–33

Thorsteinsson, A. (1981) 'On the development of Faroese settlements', in Bekker-Nielsen, H., Foote, P. and Olsen, O. (eds) *Proceedings of the Eighth Viking Congress (1977)* (Odense University Press) pp. 189–202

Wallace, B. L. (1991) 'L'Anse Aux Meadows: gateway to Vinland', *Acta Archaeologica* 61: 166–97

12

# By-passing Ellis Island: Insect Immigration to North America

Paul C. Buckland, Allan C. Ashworth and Don W. Schwert

ABSTRACT

The fauna of North America contains many immigrants, largely of European origin; yet, despite five hundred years of connections, few Americans have been successful in Europe. This is particularly apparent with the terrestrial invertebrate faunas, where North America contains many species of Old World origin and few New World species have successfully made transition to Europe. The reasons for this are not immediately apparent. New World crops are grown widely in Europe and a number of pests have established themselves, but successful colonization has been predominantly the other way. It is argued that the key to successful immigration has not been the existence of so-called vacant niches in New World ecosystems but the progressive Europeanization of agricultural and other systems, providing a home from home for most of the casual introductions. In much of New England, where Late Woodland Indian exploitation had produced a pasture woodland landscape, rather than forest, the introduction of domestic herbivores and arable by European settlers provided numerous opportunities for synanthropic and open-ground insects, which were not available before, and it was into these that the bulk of introduced species were able to move.

## INTRODUCTION

The biology and history of plant and animal invasions represents one of the more important areas of co-operation between biogeographers and historical geographers. The works of Elton (1958) and Lindroth (1957) remain classic examples of such integrated approaches, the former presenting a world view on invasions and the latter examining the connections between Europe and North America. Crosby (1986), in line with several historians (e.g. Wolf 1982), has emphasized that biological impact on the world scale has led to an essential Europeanization of landscape to the extent that native systems of tenure and apparently sustained utilization of resource have been all but abandoned. Even the more marginal environments, of less direct interest to European exploitative systems, are not without their Palaearctic immigrants. Much of the success of deliberately and accidentally imported European species has inevitably been viewed, subconsciously if not overtly, as if the introductions had an inert superiority over indigenous taxa, an inevitable consequence of the fate of native culture at the hands of the European – what Gould (1982), in a similar context, has termed 'the usual smug assumption that different from us is worse'. Yet, even attempting a dispassionate view from the current rash of remorse amongst writers in the European cultural milieu (see, for example, Sale 1991), there can be no doubt that, beyond the many deliberate introductions, many more species have made a successful transition from Old to New World than the converse. In the Coleoptera alone, over 300 species of Palaearctic origin have found suitable habitats in North America, particularly in the Maritime Provinces and New England, while the return lists less than 20. When those species associated with either stored products or introduced crops are extracted, the dominance of Europe is even more profound. Over 50 European ground beetles appear to be established in North America (Bousquet 1990); none seems to have come the other way and established itself in the wild.

If the North American continent is viewed as an island (cf. Carlquist 1974), it could be argued that the random nature of dispersal and the biogeographical impact of the Pleistocene left many vacant niches to be exploited by immigrant species; an argument which might be sustained for large vertebrates such as the horse, because of the nature of Late Quaternary extinctions (Martin and Klein 1984). The concept of the 'vacant niche' itself is not without problems (Herbold and Moyle 1986). While it may be possible to define an organism's requirements and hence its potential range, to itemize what is missing from an ecosystem depends upon one's perception of its component habitats. Eventually, resources are utilized, even if by saprophytes rather than by something perceived as absent, and one is left with purely theoretical questions (cf. Simberloff 1981) of interest but doubtful practicality. The approach here therefore is unashamedly historical, both natural and geographical; it is indeed doubtful whether island biogeographical theory can be applied effectively to continents with geological histories of coevolving populations. The biota of the American continent is at least as likely to be 'harmonic' as that of Eurasia. Indeed, the orientation of mountain chains in the latter has had considerable impact on the selection of the surviving biota through the Pleistocene, with a progressive impoverishment of West European tree diversity over the last two million years compared with North America, where hickory and tulip tree, for example, still survive (cf. Nilsson 1983). The degree of resistance to invasion (*sensu* Simberloff 1986) by the native biota should therefore at least be equal, and the question has to be posed in terms of whether the majority of introduced species are able to exploit principally habitats which are at least novel if not new to the American landscape. Lindroth (1957) showed that the majority of successful early immigrants, particularly amongst the Coleoptera, were likely to have come from south-west England, a consequence of sailing contacts from the sixteenth century onwards. Exceptions, coming from further south in

Europe, include several dung beetles, particularly in the genus *Onthophagus* (e.g. Glaser 1986), which are either rare or absent in Britain, the elm leaf beetle, *Galerucella xanthomelaena* (Schr.), not currently recorded in Britain, and the alder leaf beetle, *Agelastica alni* (L.), probably extinct in the British Isles (Hyman 1992). It is this northern temperate connection, principally between England and the Maritimes and New England and largely concerning the Coleoptera, which will form the core of the discussion in this chapter.

## TAXONOMIC AND BIOGEOGRAPHICAL PROBLEMS

Any attempt to examine those invertebrate species which have been accidentally transported from Europe to North America is immediately beset by problems both of taxonomy and of isolating species which have a disjunct Holarctic distribution. The problem is typified by the arguments over the status of the common amphiatlantic land snail, *Cepaea hortensis* (Mull.), regarded as introduced by biogeographers (Lindroth 1957) and native by archaeologists, who find it in the pre-Columbian middens of coastal New England (Barber 1988). To the north, the intermittent freeway provided by Beringia has meant that many arctic to boreal elements are common to both North America and Eurasia (Hopkins et al. 1982), and only in more temperate latitudes do the faunas apparently diverge sufficiently to become taxonomically distinct. In these areas and southwards, species have been isolated since the early Tertiary (Matthews 1978); yet some taxa do not appear to have evolved either morphologically or physiologically (Coope 1978). Such apparent taxonomic stability can make the problem one of species definition. The fungus-feeding Mycetophagid *Typhaea stercorea* (L.) has been introduced from Europe to all the Atlantic islands (Sadler 1991) and – now cosmopolitan (Hinton 1945) – must have been introduced to North America on many occasions. Its native status there, however, is indicated by fossils from packrat middens of mid-Wisconsinan age in Texas (Elias 1992). An American origin is precluded by several Roman fossil records from Britain (e.g. Osborne 1971, Kenward and Williams 1979, de Moulins 1990). The similarity between the urine-soaked vegetation in packrat nests and the species' synanthropic habitats in hay residues and mouldy stored products is considerable, and habitat stability has no doubt played an important part, but the evidence for species constancy in this warm temperate beetle is nonetheless remarkable; however, DNA studies, might suggest otherwise.

Origins are particularly difficult to trace where animals have become cosmopolitan, and taxonomists have frequently been misled. The grain pests, *Oryzaephilus surinamensis* (L) and *Rhyzopertha dominica* (F), were first described from New World specimens, and this has led many (e.g. Gígja 1944; Munro 1966) to assume that this is their origin; both have a Palaearctic fossil record extending back at least to the Bronze Age Aegean (Buckland 1981, 1990; Panagiotakopulu and Buckland 1991). Such problems also extend to purely anthropochorous taxa. The small ground beetle, *Bembidion tetracolum*, and the dung beetle, *Aphodius fasciatus* Ol., were first described from North America by Say, in the early nineteenth century, before their European origin became apparent (Lindroth 1957); to reinforce the point, both have a European Holocene fossil record (Kelly and Osborne 1965; Chowne et al. 1986). As Lindroth once perceptively remarked, the fossil record provides a means of checking our wilder biogeographical speculations. It is unfortunate that little work has yet been done with post-Columbian archaeological insect assemblages. The mechanisms of transport, however, have been the subject of detailed study (Lindroth 1957), although much remains to be gleaned from cargo manifests, port records and accounts of early voyages.

## SHIPS: FILTERS OR SWEEPSTAKES?

Where the fossil record for the last half millennium has not been examined, through the provision of samples from suitably dated archaeological contexts, the progress of introductions can only be charted on the basis of a slight historical record and theoretical considerations. Even at the present day, animals only tend to become noticed when they approach pest proportions. It is not surprising, therefore, that the earliest record of introduced insects in North America concern two field pests of cereals. The Angoumois grain moth, *Sitotroga cerealella* (Ol.), was first recorded in Europe in 1734, perhaps having been introduced there from South Africa (Sigaut, personal communication, in Buckland 1990); by 1743, it was to be found in New England (Lindroth 1957). The Hessian fly, *Mayetiola destructor* (Say), gained its vernacular name from the Hessian soldiers in whose straw bedding it was allegedly introduced to Long Island in 1779 (Jones and Jones 1974). In the early seventeenth century, the Jamestown colonists were troubled not only by ship's rats destroying their stores (Barbour 1969: 446), but also by grain that 'having fryed some 26 weeks in the ships hold, contained as many wormes as grains' (ibid., 384). The principal pest is likely to have been the grain weevil, *Sitophilus granarius* L., although there are several other possibilities. Much of the insect fauna which is characteristic of stored products and associated habitats had become widespread in Europe by the Roman period (Buckland 1981). As trade expanded further through the medieval period, many species were again taken on board ships, reaching Iceland by the eleventh century (Amorosi et al. 1992) and, although the confirmatory fossil evidence is lacking, the New World by the sixteenth. The earliest confirmation of movement of an American pest eastwards across the Atlantic comes from an unlikely location. In seventeenth century stores for Dutch whalers on Spitzbergen was found the American bean weevil, *Acanthoscelides obtectus* (Say), already established in a native European crop, the horse bean, *Vicia faba* (Wijngaarden-Bakker and Pals 1981); the beetle is now a cosmopolitan pest (Dillon and Dillon 1972).

The development of ships that tied up at quays rather than being beached increased the probability of faunas which were essentially self-sustaining on the boats, finding suitable habitats or refuges between edible cargoes in everything from food residues through dunnage and ballast to the ship's structure itself. The increasingly large ships of the post-medieval period, at sea for extended periods of time, provided floating hotels for a wide range of invertebrates.

Occasionally, the infestation got out of hand. In 1593, Thomas Cavendish, engaged in a Thatcherite excursion in the South Atlantic, took on a cargo of penguins as victualing:

> But after we came neere unto the sun, our dried Penguins began to corrupt, and there bred in them a most lothsome & ugly worme of an inch long. This worme did so mightily increase, and devour our victuals, that there was in reason no hope how we should avoid famine, but be devoured of these wicked creatures: there was nothing that they did not devour, only yron excepted: our clothes, boots, shooes, hats, shirts, stockings: and for the ship they did so eat the timbers, as that we greatly feared they would undoe us, by gnawing through the ships side. Great was the care and diligence of our captaine, master, and company to consume these vermine, but the more we laboured to kill them, the more they increased; so that at the last we could not sleepe for them, but they would eate our flesh and bite like Mosquitos. (Hakluyt 1906: XI, 415)

Such an eclectic diet and a propensity for boring into wood suggests that the culprit was the larva of a Dermestid beetle. Hinton (1945) goes so far as to suggest the leather beetle, *Dermestes maculatus* Deg., but there are several other possibilities amongst the Dermestidae and the origins of the pest species in question are debatable. Only *D. laniarius* Ill. has a European fossil record, from a Bronze Age well in England, where it is not now native (Osborne 1969, 1989), and this is not commonly a pest of stored animal foodstuffs. Fauvel (1889), in a study of species common to the Old and New Worlds, favoured a European origin for *D. maculatus*, as well as *D. lardarius* L., while *D. ater* Deg., *D. carnivorus* F. and *D. peruvianus* Cast. were American. If this is the case, then it is surprising that the extensive studies of urban archaeological insect assemblages, particularly in Britain (cf. Hall and Kenward 1990, Hall et al. 1983; Kenward and Allison, in press) have failed to provide examples. *D. ater* has been recorded from a Chinese tomb of *c.*2100 BP (Chu and Wang 1975) but, if this example and records from Egyptian mummies (Hope 1834, Lesne 1930) reflect exploitation of a long-stored product, the pest species could all be of New World origin, occurring naturally in penguin rookeries and similar habitats. The shipboard fauna must rapidly have achieved cosmopolitan status, and tracking origins is particularly difficult. Hakbijl (1986, 1987) has recently examined insect remains from the wreck of the Dutch ship *Amsterdam*, lost off Hastings in 1749, on an outward voyage to the East Indies (Marsden 1985). As well as a cargo, including one of the few commercially valuable beetles of the period, *Lytta vesicatoria* (L.), imported for the drug cantharidin, sometime used as an aphrodisiac, was a range of pests, presumably loaded with tobacco from the West Indies, including the weevils *Sitophilus linearis* (Hbst.) and *Catolethrus* sp., a longhorn beetle, *Acanthinus* sp. and the Cucujid *Cathartus quadricollis* (Guer.) It is evident that by the eighteenth century the fauna of the Dutch East India Company's warehouses in Amsterdam had become thoroughly cosmopolitan.

The problems of the resident biota on board shipping remained considerable throughout the post-medieval period and were not effectively curbed until the advent of modern insecticides and strict quarantine and import legislation. From Drake's discovery of 'a most wondrous collection of moths called Blattae' on an Eastindiaman off the Azores in 1587 (Buckland 1981) onwards, the cockroaches were constant travelling companions, defying all attempts to eradicate them. Charles Wilkes, in charge of the United States Exploring Expedition of 1838–42, after vainly attempting to sell his weevil infested bread in Rio de Janiero, tried with a brimstone and charcoal fire to smoke out the *Vincennes*' particularly noxious population of what polite ladies were beginning to call 'roaches', but without success (Stanton 1975: 88); the problem continued to dog this unhappy ship. It is singularly unfortunate that the naturalist on the voyage, Charles Pickering, chose to ignore the pests in his expansive study, 'Animals and Plants Removed from their Native Localities by the Hand of Man' (Pickering 1854), choosing instead to present a quixotic account of the role of Egypt in the dispersal of plants.

As well as this largely synanthropic assemblage, there was also ample opportunity for other travellers, in what Sadler (1991) has called the 'stowaway' fauna. Dunnage around any cargo, sometimes sacking or other cloth, but equally likely to be twigs and hay, must have provided many suitable occasions for early introductions. Cattle, sheep, goats and horses were premium cargoes on the North Atlantic run to colonists in need of additional stock, and these required the carrying of both hay and fresh water. Recent authors have tended to underestimate the probability of insect dispersal with the early stages of human colonization. Simberloff (1986) has suggested that of the 1554 successful introductions to North America, about 1.7 per cent of the fauna, only 36 pre-dated 1800, consisting of those species which could survive and reproduce on humans, domestic animals, stored products and in ballast. Of these, the fossil record shows that the so-called human flea, *Pulex irritans* L., was already widespread, apparently a prehistoric emigrant from North America (Buckland and Sadler 1989), and human lice, *Pediculus humanus* Deg., with their closer association with their host, were equally early cosmopolitan (Ewing 1924, Sveinbjarnardóttir and Buckland 1982). The ectoparasites of sheep at least had reached Iceland and Greenland with the first Norse settlers (Sadler 1991) and the majority of the present synanthropous fauna of these islands were similarly early travellers (Buckland et al. 1991).

The importance of ballast in the early transport of plants and animals formed the core of Lindroth's (1957) study of North Atlantic faunal connections. Any ship with superstructure requires ballast to maintain its position in the water, and ballast was equally essential when sailing with or without cargo, requiring trimming whenever goods were either off-loaded or taken on. The occasional captain used the ballast to his advantage in growing fresh vegetables on it (Buckland 1981), but others allowed it to become so foul as to endanger the health of the crew:

> Our ship's company is sickly, one great occasion of it is, our ballast being so bad, stinking and all of a quagmire, and sandy that it soaks the

> timbers, that the water has no course to the pump. Had we that ballast out, and shingle in, I doubt not but our ship would be healthy and in good condition. (Kiggins 1689, in Martin 1978)

Such conditions perhaps account for the introduction of the coffin beetle, *Rhizophagus parallelocollis* Gyll., to North America (Bousquet 1990). Lindroth (1957) identified the annual fishing cycle, carried on from the sixteenth century onwards, largely between south-west England and Newfoundland, as the principal agent in the dispersal of both flora and fauna. A virtual ban on settlement in Newfoundland, imposed by English merchants, and an elusive indigenous population there (Grant Head 1976), meant that boats travelled out in ballast, dumping this before fishing from long boats, similar to the later dory fisheries. Port records are inevitably full of complaints, either about illegal ballast loading at the European end, or of dumping into harbours at the other. Before the advent of pig iron ingots, rock provided the most useful ballast, occasionally finding a market at the end of a voyage (Buckland and Sadler 1990). Despite the dangers of a potentially thixotropic sand load, captains, particularly of smaller fishing boats, often took the cheaper alternative of loading off the beach and, perhaps as a result, the littoral fauna tends to be amphi-Atlantic. The Hydrophilid *Cercyon littoralis* (Gyll.) and the small rove beetle *Micralymma marina* (Strom.) are common to both sides of the ocean. Lindroth (1931) favoured distribution by the North Atlantic Drift from the New World for the latter species, but most later workers have regarded it as a typical ballast passenger from Europe (Larsson and Gígja 1958, Buckland 1988). Dune species have also been similarly moved. Amongst the Scarabaeids, *Aegialia arenaria* (F.) has arrived on the coast of Massachusetts (Darlington 1927) and *A. rufa* (F.), a Red Data Book species in Britain (Shirt 1987), has travelled, probably with French or Breton fishermen to the Nearctic offshore island of St. Pierre (Lindroth 1957). This group also provides one of the few examples of movement in the opposite direction, Johnson (1976) noting a West Coast American species, *Psammodius caelatus* LeC., on the Lancashire coast.

Lindroth's (1955, 1957) studies, although later extended to the ground beetles of Canada and Alaska (Lindroth 1961–9), concentrated on introductions to Newfoundland, where continual off-loading of ballast left many of the outports with an essentially European flora on their disturbed ground habitats (Cooper 1981), containing a ground beetle population of largely similar origin (Lindroth 1957, Larson and Langor 1982). These habitats were new to the island, initially created around the nutrient-rich temporary fish processing stands and later moving into the settlements of the Europeans, largely of a similar south-west English origin (Mannion 1977). On the mainland, the more permanent pattern of early settlement perhaps led to less ballast dispersal and the New England fisheries developed ties with the slave plantations of Virginia and the Caribbean, rather than with Europe. Native weeds, such as species of *Ambrosia*, were more successful as invaders of the Europeans' fields (Bazzaz 1986), although again many of the invaders of disturbed ground, both plant and animal, are of Palaearctic origin (Pimental 1986).

COMPARISON AND CONTRAST

The deep seated image of the pre-Columbian New England shore and its hinterland is of a forested landscape with native peoples having a minimal, localized impact; the great clearance of the woods did not begin until the late eighteenth century (Delcourt and Delcourt 1987). There is a certain irony in an image, owing more to nineteenth-century novels, such as Nathaniel Hawthorne's *The Scarlet Letter* (1850), than to the earliest accounts of the New World (cf. McManis 1972, Denevan 1992). The 'Great Dying', a consequence of the introduction of Eurasian diseases, which reduced the indigenous population from an estimated four million to less than a quarter of this figure (Butzer 1992), had an inevitable impact on the landscape. There was more forest in 1750 than in 1500 (Denevan 1992). This has had significant impact on the native insect fauna. While some problems with taxonomy and quantification of data are currently insuperable, it is evident that, where species pairs can be considered, the Old Forest (*Urwald*) element in New England remains much more widespread than in western Europe. This could be seen as support for the late survival of natural forest, yet it may be more a reflection of the nature of woodland management. In much of Europe, and particularly England, where many of the earliest colonists came from, management strategies had seriously depleted a once renewable resource and dead wood insect habitats had become progressively rarer. The relative fates of two weevils, *Dryophthorus corticalis* (Payk.) and *D. americanus* Bed. are instructive. The latter, recorded from old logs and beneath bark of pine, is included in the Dillons' manual of common beetles of eastern North America (Dillon and Dillon 1972). *D. corticalis* is a Red Data Book species in Britain, restricted to a few old oaks in Windsor Great Park (Shirt 1987), where it was first found in 1925. It is not an infrequent fossil in British early to mid-Holocene sediments (Buckland 1979) and forms one of an increasing number of *Urwaldrelikt* beetles, now having very restricted ranges in Europe (Buckland and Dinnin 1993). Their survival is largely due not to the relicts of forest, long gone from most of Europe, but to the maintenance of what is perhaps the oldest surviving form of landscape management, that of pasture–woodland (cf. Harding and Rose 1986). One European Urwaldtier, the Cucujid *Prostomis mandibularis* (F.), has been accidentally introduced to British Columbia (Campbell 1978). In Europe, it is a rare and rapidly disappearing species of undisturbed forest (Palm 1959), living between the loose fungoid annual rings of dead wood; in Britain, its most recent record is from the Bronze Age (Buckland 1979). Importation from mainland Europe to Western Canada, presumably in unbarked timber, has provided a new, albeit temporary, refuge in what Williams (1989) has described as the last American frontier, the unlogged forests of the Northwest. Such apparent successes amongst the forest fauna are very much in the minority, despite the fact that the opportunities for movement in the opposite direction have been much the greater: one of the material aspects drawing settlers to New England was the availability of timber and the first return trip of the *Fortune* to England from Plymouth, Massachusetts, in 1621, was loaded with 'good clapboard' (Cronon 1983: 109). The Virginian colonists were also at pains to stress the value of the timber to a homeland rapidly running out of renewable resources. Thereafter, the possibilities of introductions to Europe were considerable and many species, particularly Cerambycids, which bore deep into the timber, have been frequently introduced, but appear to have failed to establish themselves (Duffy 1953). Hickin (1968) suggests that the powder post beetle *Lyctus brunneus* (Steph.), widespread in western Europe, was introduced from North America during the early nineteenth century, although the late Maureen Girling's (1989) record of a fossil from a late Roman well in Chichester cast doubt upon this hypothesis, which only further palaeoentomological research can clarify. Exploiting dry cut timber, the Lyctids, both native and introduced, have experienced a massive increase in available habitat as a result of human activity and an exchange of species in synanthropic situations is not surprising. Most of the Old Forest fauna has a low dispersal potential, a result of evolution through the Tertiary in a landscape of continuous forest (Buckland and Dinnin 1993), where dead wood would always be available; chance introductions, therefore, have little chance of finding suitable habitats in a wholly cleared and managed landscape, even where unoccupied niches – if such could be claimed to exist as a result of local extinction – were available. Westwards, across the Atlantic, both the death watch beetle, *Xestobium rufovillosum* (Deg.) and the furniture beetle, *Anobium punctatum* (Deg.), have succeeded in establishing themselves in the wild, having presumably been introduced in infested woodwork. The former is likely to have been carried in oak ship's timbers from Europe, since this wood forms the normal synanthropic host in England (Fisher 1940), although, in New England, it is principally found in pine (Dillon and Dillon 1972). Another likely pest of ship's timbers, the wharf borer, *Nacerdes melanura* (L.), mainly found in wood soaked with salt water, was introduced to North America before 1910 and now extends westwards to North Dakota, as well as in locations on the West Coast (Balsbaugh et al. 1979).

The relationship between the New England Late Woodland Indians and their forests was one of several thousand years standing and, at least in its southern part, the woods were as much a managed landscape as in Europe, although the interference in the natural succession had a differing set of priorities (Cronon 1983, Denevan 1992), and therefore a different impact on the insect fauna. The extent of controlled modification of the forest has been the source of much debate, from the minimalist views of Russell (Russell 1983, Forman and Russell 1983) to the major impact envisioned by Denevan (1992) and others (Hammitt 1992, Patterson and Sassaman 1988). The arguments range around the scale of the deliberate use of fire, a debate paralleled in Europe by that over the nature of the mid-Holocene mesolithic and neolithic landscape (cf. Bennett et al. 1990, Edwards 1990, 1993). Contemporary accounts of the earliest contacts are revealing. The French expedition under Verrazzano in 1524 noted woods in New England so open that an army might march through them (Sauer 1971), and Champlain's map of the later site of the Plymouth colony in 1613 shows a cleared landscape of settlements, fields and scattered trees (McManis 1972: fig. 16). This contrasts with the much more influential image of a wilderness described by the Puritan William Bradford (ibid., 121), although the latter, aside from his wish to seek biblical parallels, was seeing neglect rather than the intended landscape. European diseases had by then more than decimated the Late Woodland Indians, completely disrupting patterns of land use several thousand years in the making. With a potential colonist's eye, Rosier wrote of exploration in 1605:

> ...over very good ground, pleasant and fertile, fit for pasture for the space of some three miles, having but little wood and that Oke like stands left in our pastures in England, good and great, fit timber for any use. Some small Birch, Hazle, and Brake, which might in small time with few men by cleansed and made good arable land; but as it now is will feed cattell of all kindes with fodder enough for Summer and Winter... . And surely it did all resemble a stately Parke, wherein appeare some old trees with high withered tops, and others flourishing with living greene boughs. (McManis 1972: 101)

This description of pasture–woodland shows an awareness of the value of leaf fodder. In much of western Europe, such landscapes had largely become the preserve of the nobility, and park and forest maintenance relied more on grazing pressure and fodder collection than fire. In New England, burning for game and agriculture was widely practised, although the reasons for it were not always grasped by the incoming Europeans:

> The Salvages are accostomed to set fire to the Country in all places where they come, and to burn it twize a yeare, viz: at the Spring, and the fall of the leafe. (Thomas Morton, in McManis 1972: 127)

Both Palaearctic and Nearctic insect faunas include elements adapted to the exploitation of fire-damaged woodland (Evans 1972), and several of these species are Holarctic in distribution. The survival of such taxa clearly relates to the balance between fire frequency and species' dispersal potential. In a study of the insects of the boreal forest zone of Canada, Danks and Foottit (1989) suggest an average fire rotation of 60–100 years. The circumpolar Buprestid *Melanophila acuminata* Deg. is able to seek out charred timber by use of infrared detection (Evans 1966) and is regarded as the characteristic species of this habitat. It is included in the manual of common beetles of eastern North America (Dillon and Dillon 1972); in the British Isles, it is restricted to heathlands, such as the New Forest in southern England, where suppression of fire has no doubt contributed to its recent decline (Levey 1977). Much of the remainder of the burnt ground fauna has suffered major retractions in range, in Britain – some, such as *Agonum quadripunctatum* Deg. and *Zimioma grossum* (L.) (Buckland 1979), to the point of extinction. Opportunities for transfer were, or have become, very limited.

The extent of pasture–woodland in New England may have been sufficient to provide refuges for saproxylic elements in the native fauna. Similarly the old grassland fauna also consists of native species, even to the Elaterids, the larvae of which, the wireworms, were to become pests of the roots of introduced wheat crops. In southern New England and southwards into Virginia, Indian fire maintenance of parkland was not wholly for the management of game, and the spaded agriculture practised was described by the earliest European visitors:

> ...with the corn they put in each hill three or four Brazilian beans (Kidney beans), which are of different colours. When these grow up, they interlace with the corn, which reaches to the height of from five to six feet; and they keep the ground very free from weeds. We saw there many squashes, and pumpkins, and tobacco, which they likewise cultivate. (Champlain 1613, quoted in Cronon 1983: 43)

The combination of intercropped maize, beans and squash, noted throughout eastern North America, from Mexico to southern New England (Sauer 1971), had the advantage not only of insurance against crop failure but also of maintaining soil fertility. On the poor soils of New England, later largely abandoned to secondary woodland by European settlers (Lewis 1990), the system had to be modified to include fallowing and burning as part of a shifting pattern of land use, which also relied on the natural plant and animal resources of the forest; the contrasts between this and European systems and their consequences are expertly drawn out by Cronon (1983). The extensive areas of bare ground which characterize much of the European arable tradition were largely absent from pre-Columbian North America and, at least in the East, the dispersed nature of the large mammal population also precluded the establishment of extensive tracts of short turf grazed grassland. The indigenous East Coast fauna therefore lacked that typical European beetle assemblage, that of *culturesteppe* (cf. Hammond 1974), and several elements, including ground beetles of the genera *Amara* and *Harpalus*, have been introduced.

It is, however, in the insect fauna directly or indirectly associated with vertebrates and their utilization that the most profound changes have been unleashed on the American fauna. Apart from the dog and perhaps the turkey, native societies had no domestic animals. One catastrophic consequence of this was a lack of resistance to European epidemic diseases, most of the origins of which lie in coevolution with ailments of domestic stock (Butzer 1992). As well as the ectoparasites directly and inevitably introduced with the stock, the greater part of the insect fauna associated with domesticated animals has also been transported and has become successfully established The immediacy of this process has recently been examined on the North Atlantic islands (Buckland et al. 1991, Sadler 1991), but similar work on fossil assemblages has only recently begun in North America. The few samples from the Norse site at L'Anse aux Meadows in Newfoundland, examined by Anne Morgan (personal communication) produced no evidence for European insects, although the transitory nature of the occupation (Wallace 1991) and the northern location would have militated against dispersal. In the absence of fossil assemblages from post-Columbian archaeological contexts, the insect fauna can only be examined on the basis of existing records. While the North American fauna contains an abundance of forest coprophages, the pasture element is restricted to a minority of generalist species, and this habitat is largely taken up by introduced European animals (Hanskii 1991), including not only 12 species of the Scarabaeid genus *Aphodius*, but also the associated dung fauna of *Cercyon* and *Sphaeridium* Hydrophilid beetles, and a range of Carabid and Staphylinid predators. The contrast in the way in which the landscapes have evolved under human management again lies at the root of these changes. Pasture–woodland was seriously in decline in Europe before the colonists began their expansion from their tenuous coastal settlements in New England, and the characteristic Palaearctic forest dung beetle *A. zenkeri* Germ., like the American examples largely restricted to deer dung (Landin 1961), was not introduced; it is not known whether this reflects the unlikelihood of accidental loading or flight to a ship, or competitive exclusion in the New World. With the opening up of the Great Plains to European farming, the replacement of vast herds of bison with cattle, and the removal of any biogeographical barrier that the Eastern Forest Belt might offer, it is curious that the indigenous dung fauna of the Great Plain was not able to move eastwards. Gordon (in Hanskii 1991), however, has pointed out that the climate of the Plains is not ideal for the dung community, with rapid drying out of available habitat, and many members of otherwise coprophilous genera are restricted to rodent burrows. European animal husbandry, the intensive rather than extensive use of range and irrigation, therefore provides habitat for European species.

The keeping of cattle in significant numbers demands a varied pattern of ancillary activities, each of which has an associated insect fauna. During the primary phases of settlement, and particularly in the more northerly latitudes, the acquisition of sufficient fodder to overwinter stock would have been a prime consideration. The introduced beetle fauna includes leaf-feeders, which, where not recent introductions with nursery stock, might suggest a familiarity with leaf fodder on the part of French if not English colonists. The Chrysomelids, *Agelastica alni*, *Galerucella xanthomelaena* and *Plagiodera versicolora* Laich. skeletonize the leaves of alder, elm and willow respectively and are likely transportees from France rather than England, where the first is extinct and the second unknown. The branch feeding weevils, *Cryptorhynchus lapathi* (L.) and *Trachodes hispidus* (L.), the latter only known in America from the French offshore island of St. Pierre (Lindroth 1957) may have a similar origin, but the extent of trade in exotic and familiar plants, which has introduced many phytophages (LeSage 1988), makes conclusions purely speculative without a fossil record. The more familiar fodder fauna, that of grass hay, has been shown to be widely spread across the North Atlantic, at least to Greenland by the Norse settlers (Sadler 1991) and its early arrival in New England is to be expected. The origin of much of this fauna of stored hay, in rodent and bird nests, may have allowed some species, such as *Typhaea stercorea*, to have been Holarctic, but the bulk was translated wholesale with the agricultural system. The assemblages include a range of Cryptophagid and Lathridiid beetles which have since become cosmopolitan, and these, with other species, inevitably show an overlap with the granary and stored product fauna, which were early immigrants, and one, otherwise characteristic of hay residues, *Lathridius minutus* (L.), is also one of the most common species in the granaries of the Canadian

Prairies (Barker and Smith 1990). Research upon fossil assemblages from post-medieval Iceland (Buckland et al. 1991) has shown major changes in the composition of the stored hay fauna, resulting from the move towards controlled leys of grasses and clover, and similar changes are to be expected through the progressive improvement of American pasture and hay sources. Introduced clover weevils of the genera *Sitona*, *Hypera* and *Tychius* form a significant proportion of the introduced species, some creating serious pest problems on their introduced crops (Jones and Jones 1974). Decline in elements of the native fauna are also evident, although only one species, the Rocky Mountain locust, *Melanoplus spretus* (Walsh), is currently known to have apparently become extinct (Lockwood 1989) as a result of recent human interference in its floodplain habitat.

## CONCLUSIONS

The importance of synanthropy in the successful establishment of introduced insects has been the focus of much discussion (cf. Spence and Spence 1988). While some of the argument revolves around the definition of synanthropy, Tischler (1973) and others have pointed out the importance in breeding of the protected thermal regimes that are provided by human accumulations of waste. These allow the extension of the range of many species, of varying degrees of synanthropy, beyond their natural limits, something which is particularly marked in the fauna of stored hay and manure (Buckland et al. 1991). This applies as much to groups such as the Muscid flies (Skidmore 1986) as to the Coleoptera. Spence (1991) has argued that the success of exotic ground beetles in Canada is more a reflection of 'their flexibility in habitat use and competitive ability than to strict habitat matching'; yet there can be little doubt that Europeans created a whole series of disturbed ground and detrital habitats, which had previously been either absent or very rare in the North American landscape, and it is into these that the newcomers predominantly moved. It cannot be doubted that biogeographical accident, both in terms of Late Wisconsinan and Early Holocene climatic and dispersal events (Morgan and Morgan 1980, Schwert and Ashworth 1988) and availability of shipboard transport opportunities, played a significant role in the development of the fauna. Without a detailed fossil record, particularly from those archaeological contexts which document the arrival and dispersal of that ultimate weed, the European, the details of the processes remain uncertain.

## ACKNOWLEDGEMENTS

After many years' work in our respective Wallacian realms, this chapter grew from a visit to Fargo by one of us (P. C. B.) and is intended to spur a practical, palaeontological approach to the problems of Atlantic faunal connections. Thanks are due to the University of Sheffield for funding the visit, further to a National Science Foundation (Division of Polar Programs) sponsored meeting of NABO, the North Atlantic Biocultural Organisation, of whose brief this research forms part. The chapter has benefited much from discussion with T. Amorosi, G. R. Coope, R. Crawford, K. J. Edwards, D. D. Gilbertson, P. Halstead, T. H. McGovern, A. V. Morgan, N. Roberts and J. P. Sadler.

## REFERENCES

Amorosi, T., Buckland, P. C., Ólafsson, G., Sadler, J. P. and Skidmore, P. (1992) 'Site status and the palaeoecological record: a discussion of the results from Bessastaðir, Iceland', in Morris, C. D. and Rackham, D. J. (eds) *Norse and Later Settlement and Subsistence in the North Atlantic* (Dept of Archaeology, University of Glasgow) pp. 169–92

Balsbaugh, E. V. Jr., Kopp, D. D. and Scholl, C. (1979) 'The wharf borer *Nacerdes melanura* L., (Coleoptera: Oedemeridae) in North Dakota', *Coleopterists Bulletin* 33: 455–6

Barber, R. J. (1988) 'The use of land snails from prehistoric sites for paleoenvironmental reconstruction', in Nicholas, G. P. (ed) *Holocene Human Ecology in North-eastern North America* (Plenum Press, New York) pp. 11–28

Barbour, P. L. (ed. ) (1969) *The Jamestown Voyages Under the First Charter, 1606–1609* (The Hakluyt Society, Cambridge)

Barker, P. S. and Smith, L. B. (1990) 'Influence of granary type and farming practices on the relative abundance of insects in granary residues', *Canadian Entomologist* 122: 393–400

Bazzaz, F. A. (1986) 'Life history of colonizing plants: some demographic, genetic, and physiological features', in Mooney, H. A. and Drake, J. A. (eds) *Ecology of Biological Invasions of North America and Hawaii* (Springer-Verlag, New York) pp. 96–110

Bennett, K. D., Simonson, W. D. and Peglar, S. M. (1990) 'Fire and man in post-glacial woodlands of eastern England', *Journal of Archaeological Science* 17: 635–42

Bousquet, Y. (1990) 'A review of the North American Species of *Rhizophagus* Herbst and a revision of the Nearctic members of the Subgenus *Anomophagus* Reitter (Coleoptera: Rhizophagidae), *Canadian Entomologist* 122: 131–71

Buckland, P. C. (1979) *Thorne Moors: a Palaeoecological Study of a Bronze Age Site (a Contribution to the History of the British Insect Fauna)* (Department of Geography, University of Birmingham, Occasional Publication 8)

Buckland, P. C. (1981) 'The early dispersal of insect pests of stored products as indicated by archaeological records', *Journal of Stored Produce Research* 17: 1–17

Buckland, P. C. (1988) 'North Atlantic faunal connections – introductions or endemics? *Entomologica Scandinavica, Supplement* 32: 7–29

Buckland, P. C. (1990) 'Granaries stores and insects: the archaeology of insect synanthropy', in Fournier, D. and Sigaut, F. (eds) *Le Preparation Alimentaire des Cereales*, (PACT, Rixensart) pp. 69–81

Buckland, P. C. and Dinnin, M. (1993) Holocene woodlands, the fossil insect evidence', in Kirby, K. J. and Drake, C. M. L. (eds) *Dead Wood Matters: the Ecology and Conservation of Saproxylic Invertebrates in Britain* (English Nature, Peterborough) pp. 6–20

Buckland, P. C. and Sadler, J. (1989) 'A biogeography of the human flea, *Pulex irritans* L. (Siphonaptera: Pulicidae)', *Journal of Biogeography* 16: 115–20

Buckland, P. C. and Sadler, J. (1990) 'Ballast and building stone: a discussion', in

Parsons, D. (ed. ) *Stone: Quarrying and Building in England AD 43–1525*, (Phillimore, Chichester) pp. 114–25

Buckland, P. C., Dugmore, A. J. and Sadler, J. P. (1991) 'Faunal change or taphonomic problem? A comparison of modern and fossil insect faunas from southeast Iceland', in Maizels, J. K. and Caseldine, C. (eds) *Environmental Change in Iceland* (Kluwer Academic, Dordrecht) pp. 127–46

Buckland, P. C., Dugmore, A. J., Perry, D., Savory, D. and Sveinbjarnardóttir, G. (1991) 'Holt in Eyjafjallasveit, Iceland: a palaeoecological study of the impact of Landnám', *Acta Archaeologica* 61: 252–71

Butzer, K. A. (1992) 'The Americas before and after 1492: an introduction to current geographical research', *Annals of the Association of American Geographers* 82: 345–68

Campbell, J. M. (1978) 'Cucujoidea'. in Danks, H. V. (ed. ) *Canada and its Insect Fauna, Memoirs of the Entomological Society of Canada* 108: 377–80

Carlquist, S. (1974) *Island Biology* (Columbia University Press, New York)

Chowne, P., Girling, M. and Greig, J. (1986) 'Excavations at an Iron Age defended enclosure at Tattershall Thorpe, Lincolnshire', *Proceedings of the Prehistoric Society* 52: 159–88

Chu, H. F. and Wang, L. -Y. (1975) 'Insect carcasses unearthed from Chinese antique tombs', *Acta Entomologica Sinica* 18: 333–37 (in Chinese).

Coope, G. R. (197B) 'Constancy of insect species versus inconstancy of Quaternary environments', in Mound, L. A. and Waloff, N. (eds) *Diversity of Insect Faunas* (Blackwell, Oxford) pp. 176–87

Cooper, K. (1981) 'Alien anthropophytic vegetation of the Avalon Peninsula', in Macpherson, A. G. and Macpherson, J. G. (eds) *The Natural Environment of Newfoundland Past and Present* (Department of Geography, University of Newfoundland, St. Johns) pp. 251–65

Cronon, W. (1983) *Changes in the Land. Indians, Colonists and the Ecology of New England* (Hill and Wang, New York)

Crosby, A. W. (1986) *Ecological Imperialism: the Biological Expansion of Europe 900–1900* (Cambridge University Press, Cambridge)

Danks, H. V. and Foottit, R. G. (1989) 'Insects of the boreal zone of Canada', *Canadian Entomologist* 121: 625–90

Darlington, P. J. JR. (1927) '*Aegialia arenaria* Muls. in New England, with local records for other species', *Psyche* 34: 98–9

Delcourt, P. A. and Delcourt, H. R. (1987) *Long-term Forest Dynamics of the Temperate Zone* (Springer-Verlag, New York)

de Moulins, D. (1990) 'Environmental analysis', in Mahony, C and Moulins, D. *The Archaeology of Roman London*, vol. 1: *The Upper Walbrook in the Roman Period* (Council for British Archaeology Research Report 69, London) pp. 85–152

Denevan, W. M. (1992) 'The pristine myth: the landscape of the Americas in 1492', *Annals of the Association of American Geographers* 82: 369–85

Dillon, E. S. and Dillon, L. S. (1972) *A Manual of Common Beetles of Eastern North America* (Dover, New York)

Duffy, E. A. J. (1953) *A Monograph of the Immature Stages of British and Imported Timber Beetles* (British Museum (Natural History), London)

Edwards, K. J. (1990) 'Fire and the Scottish Mesolithic: evidence from microscopic charcoal', in Vermeersch, P. M. and van Peer, P. (eds) *Contributions to the Mesolithic of Europe* (Leuven University Press, Leuven) pp. 71–9

Edwards, K. J. (1993) 'Models of mid-Holocene forest farming for north-west Europe', in Chambers, F. M. (ed. ) *Climatic Change and Human Impact on the Landscape* (Chapman and Hall, London) pp. 133–45

Elias, S. A. (1992) 'Late Quaternary Zoogeography of the Chihauhaun Desert insect fauna, based on fossil records from packrat middens', *Journal of Biogeography* 19: 285–98

Elton, C. S. (1958) *The Ecology of Invasions by Animals and Plants* (Chapman and Hall, London)

Evans, W. G. (1966) 'Perception of infrared from forest fires by *Melanophila acuminata* DeGeer (Buprestidae, Coleoptera)', *Ecology* 47: 1061–5

Evans, W. G. (1972) 'The attraction of insects to forest fires', in *Proceedings of the Tall Timbers Conference on Ecological Animal Control by Habitat Management* (Tallahassee, Florida) pp. 115–27

Ewing, H. E. (1924) 'Lice from human mummies', *Science* 60: 389–90

Fauvel, A. (1889) 'Liste des coleoptérès communes á Europe et á l'Amerique du Nord', *Revue d'Entomologie (Caen)* 22: 203–378

Fisher, R. C. (1940) 'Studies of the biology of the death-watch beetle, *Xestobium rufovillosum* Deg. III: fungal decay in timber in relation to the occurrence and rate of development of this Insect', *Annals of Applied Biology* 27: 545–57

Forman, R. T. T. and Russell, E. W. B. (1983) 'Evaluation of historical data in ecology', *Bulletin of the Ecological Society of America* 64: 5–7

Gígja, G. (1944) *Meindýr í Húsum og Gróðri og varnir gegn þeim* (Jens Guðbjörnsson, Reykjavík)

Girling, M. A. (1989) 'The insect fauna of the Roman Well at the Cattlemarket', in Down, A., *Chichester Excavations* 6 (Philimore, Chichester) pp. 234–41

Glaser, J. D. (1986) 'Notes on some Maryland Scarabaeidae and Cerambycidae (Coleoptera)', *Coleopterists Bulletin* 40: 355–6

Gould, S. J. (1982) 'Sticking up for marsupials', in Gould, S. J. *The Panda's Thumb* (W. W. Norton, New York) pp. 289–95

Grant Head, C. (1976) *Eighteenth Century Newfoundland* (MacClellan & Stewart, Toronto)

Hakbijl, T. (1986) 'Remains of insects', in Gawronski, J. H. G. (ed.) *Annual Report of the VOC-Ship 'Amsterdam' Foundation 1985* (Amsterdam) pp. 72–3

Hakbijl, T. (1987) 'Insect remains: unadulterated *Cantaridum* and tobacco from the West Indies', in Gawronski, J. H. G. (ed.) *Annual Report of the VOC-Ship 'Amsterdam' Foundation 1985* (Amsterdam) pp. 93–4

Hakluyt, R. (1906) *The Principal Navigations Voyages Traffiques & Discoveries of the English Nation* (James MacLehose & Sons, Glasgow)

Hall, A. R. and Kenward, H. K. (1990) *Environmental Evidence from the Colonia* Archaeology of York, 14/6 (Council for British Archaeology for York Archaeological Trust, London)

Hall, A. R. and Kenward, H. K., Williams, D. and Greig, J. R. A. (1983) *Environment and Living Conditions at Two Anglo-Scandinavian Sites*. Archaeology of York, 14/4. (Council for British Archaeology, for York Archaeological Trust, London)

Hammitt, J. E. (1992) 'The shape of adaptation: historical ecology of anthropogenic landscapes in the Southeastern United States', *Landscape Ecology* 7: 121–35

Hammond, P. M. (1974) 'Changes in the British Coleopterous fauna', in Hawksworth, D. L. (ed.) *The Changing Flora and Fauna of Britain*, (Academic Press, London) pp. 323–69

Hanski, I. (1991) 'North temperate dung beetles', in Hanski, I. and Cambefort, Y. (eds) *Dung Beetle Ecology* (Princeton University Press, Princeton) pp. 75–96

Harding, P. T. and Rose, F. (1986) *Pasture–Woodlands in Lowland Britain* (Institute of Terrestrial Ecology, Huntingdon)

Hawthorne, N. (1850) *The Scarlet Letter* (Ticknor, Reed & Fields, Boston)

Herbold, B. and Moyle, P. B. (1986) 'Introduced species and vacant niches', *American Naturalist* 128: 751–60

Hickin, N. E. (1968) *The Insect Factor in Wood Decay* (Hutchinson, London)

Hinton, H. E. (1945) *A Monograph of the Beetles Associated with Stored Products* (British Museum, (Natural History) London)

Hope, F. W. (1834) 'Notice of several species of insects found in the heads of Egyptian mummies', *Proceedings of the Royal Entomological Society London* 1834: 11–13

Hopkins, D. M., Matthews, J. V. JR., Schweger, C. E. and Young, S. B. (1982) *Paleoecology of Beringia* (Academic Press, New York)

Hyman, P. S. (1992) *A Review of the Scarce and Threatened Coleoptera of Great Britain* (revised and updated by M. S. Parsons) (UK Nature Conservation, Peterborough)

Johnson, C. (1976) 'Nine species of Coleoptera new to Britain', *Entomologist's Monthly Magazine* 111: 177–83

Jones, F. G. W. and Jones, M. G. (1974) *Pests of Field Crops* (Edward Arnold, London)

Kelly, M. and Osborne, P. J. (1965) 'Two faunas and floras from the alluvium at Shustoke, Warwickshire', *Proceedings of the Linnean Society, London* 176: 37–65

Kenward, H. K. and Allison, E. P. in press 'Rural origins of the urban insect fauna', in Hall, A. R. and Kenward, H. K. (eds) *Urban – Rural Connexions: Perspectives from Environmental Archaeology* (Oxbow, Oxford)

Kenward, H. K. and Williams, D. (1979) *Biological Evidence from the Roman Warehouses in Coney Sheet*, Archaeology of York, 14/2 (Council for British Archaeology for York Archaeological Trust, London)

Landin, B. (1961) 'Ecological Studies on Dung Beetles', *Opuscula Entomologica Supplement* 19

Larson, D. J. and Langor, D. W. (1982) 'The Carabid beetles of insular Newfoundland (Coleoptera: Carabidae: Cicindellidae) – 30 years after Lindroth', *Canadian Entomologist* 114: 591–7

Larsson, S. G. and Gígja, G. (1958) *Coleoptera: Zoology of Iceland* 46 (Eijnar Munksgaard, Copenhagen)

Lesage L. (1988) 'Notes on European *Longitarsus* species introduced in North America (Coleoptera: Chrysomelidae: Alticinae)', *Canadian Entomologist* 120: 1133–45

Lesne, P. (1930) 'Le Dermeste des cadavres (*Dermestes fiischii* Kug. ) dans les tombes de l'Egypte ancienne', *Bulletin de la Société d'Entomologie d'Egypte* 1930: 21–4

Levey, B. (1977) 'Coleoptera: Buprestidae', *Handbooks for the Identification of the British Insect Fauna* V, 1(b) (Royal Entomological Society of London)

Lewis, P. F. (1990) 'The Northeast and the making of American geographical habits', in Conzen, M. P. (ed.) *The Making of the American Landscape* (Unwin Hyman, Boston) pp. 80–103

Lindroth, C. H. (1931) 'Die Insektenfauna Islands und ihre Probleme', *Zoologiske Bidrag fran Uppsala* 13: 105–600

Lindroth, C. H. (1955) 'The Carabid beetles of Newfoundland, including the French islands St. Pierre and Micquelon', *Opuscula Entomologica Supplement* 12

Lindroth, C. H. (1957) *The Faunal Connections between Europe and North America* (John Wiley, New York)

Lindroth, C. H. (1961–9) 'The Ground-beetles (Carabidae excl. Cicindellinae) of Canada and Alaska, parts 1–6, *Opuscula Entomologica Supplement* 20, 24, 29, 33–5

Lockwood, J. A. (1989) 'Taxonomic status of the Rocky Mountain locust: comparisons of *Melanoplus spretus* (Walsh) with solitary and migratory *Melanoplus sanguinipes* (F.)', *Canadian Entomologist* 121: 1103–9

Mannion, J. J. (1977) *The Peopling of Newfoundland* (Memorial University of Newfoundland, St Johns)

Marsden, P. (1985) *The Wreck of the* Amsterdam (Hutchinson, London)

Martin, C. J. M. (1978) 'The Dartmouth, a British frigate wrecked off Mull, 1690. 5: The ship', *International Journal of Nautical Archaeology and Underwater Exploration* 7: 29–58

Martin, P. S. and Klein, R. G. (eds) (1984) *Quaternary Extinctions, a Prehistoric Revolution* (University of Arizona Press, Tucson)

McManis, D. R. (1972) *European Impressions of the New England Coast 1497–1620* (University of Chicago, Department of Geography, Research Paper 139)

Matthews, J. V. JR. (1978) 'Tertiary and quaternary environments: historical background for an analysis of the Canadian Insect Fauna', in Danks, J. V. (ed.) *Canada and its Insect Fauna Memoirs of the Entomological Society of Canada*, vol. 108, pp. 31–86

Morgan, A. V. and Morgan, A. (1980) Faunal assemblages and distributional shifts of Coleoptera during the Late Pleistocene in Canada and the northern United States. *Canadian Entomologist* 112: 1105–28

Munro, J. W. (1966) *Pests of Stored Products* (Hutchinson, London)

Nilsson, T. (1983) *The Pleistocene* (D. Reidel, Dordrecht)

Osborne, P. J. (1969) 'An insect fauna of Late Bronze Age date from Wilsford, Wiltshire', *Journal Animal Ecology* 38: 555–66

Osborne, P. J. (1971) 'An insect fauna from the Roman site at Alcester, Warwickshire', *Britannia* 2: 156–65

Osborne, P. J. (1989) 'Insects', in Ashbee, P., Bell, M. and Proudfoot, E., *Wilsford Shaft. Excavations 1960–62* (English Heritage, London) pp. 96–9

Palm, T. (1959) 'Die Holz und Rindenkäfer der süd- und mittel-Schwedischen Laubbaume', *Opuscula Entomologica Supplement* 16

Panagiotakopulu, E. and Buckland, P. C. (1991) 'Insect pests of stored products from Late Bronze Age Santorini, Greece', *Journal of Stored Produce Research* 27: 179–84

Patterson, W. A. and Sassaman, K. E. (1988) 'Indian fires in the prehistory of New England', in Nicholas, G. P. (ed.) *Holocene Human Ecology in Northeastern North America* (Plenum Press, New York) pp. 107–36

Pickering, C. (1854) *The Geographical Distribution of Animals and Plants* (Little, Brown, Boston)

Pimental, D. (1986) 'Biological invasions of plants and animals in agriculture and forestry', in Mooney, H. A. and Drake, J. A. (eds) *Ecology of Biological Invasions of North America and Hawaii* (Springer-Verlag, New York) pp. 149–62

Russell, E. W. B. (1983) 'Indian-set fires in the forests of the northeastern United States', *Ecology* 64: 78–88

Sadler, J. (1991) 'Beetles, boats and biogeography: insect invaders of the North Atlantic', *Acta Archaeologica* 61: 199–211

Sale, K. (1991) *The Conquest of Paradise: Christopher Columbus and the Columbian Legacy* (Hodder & Stoughton, London)

Sauer, C. O. (1971) *Sixteenth Century North America* (University of California Press, Berkeley)

Schwert, D. P. and Ashworth, A. C. (1988) 'Late Quaternary history of the northern beetle fauna of North America: a synthesis of fossil and distributional data', *Memoirs, Entomological Society of Canada* 144: 93–107

Shirt, D. B. (1987) *British Red Data Books: 2. Insects* (Nature Conservancy Council, Peterborough)

Simberloff, D. (1981) 'Community effects of introduced species', in Nitecki, T. H. (ed.) *Biotic Crises in Ecological and Evolutionary Time* (Academic Press, New York) pp. 53–81

Simberloff, D. (1986) 'Introduced insects: a biogeographic and systematic perspective', in Mooney, H. A. and Drake, J. A. (eds) *Ecology of Biological Invasions of North America and Hawaii* (Springer-Verlag, New York) pp. 1–26

Skidmore, P. (1986) *The Biology of the Muscidae* (Junk, The Hague)

Spence, J. R. (1991) 'Success of European Carabid species in Western Canada: preadaptation for synanthropy?', in *The Role of Ground Beetles in Ecological and Environmental Studies*, (Intercept, Andover) pp. 129–41

Spence, J. R. and Spence, (1988) 'Of ground-beetles and men: introduced species and the synanthropic fauna of western Canada', *Memoirs, Entomological Society of Canada* 144: 151–68

Stanton, W. (1975) *The Great United States Exploring Expedition of 1838–1842* (University of California Press, Berkeley)

Sveinbjarnardóttir, G. and Buckland, P. C. (1983) 'An uninvited guest', *Antiquity* 58: 127–30

Tischler, W. (1973) 'Ecology of arthropod fauna in man-made habitats: the problem of synanthropy', *Zoologischer Anzeiger, Leipzig* 191: 157–61

Wallace, B. L. (1991) 'L'Anse aux Meadows: gateway to Vinland', *Acta Archaeologica* 61: 166–98

Wijngaarden-Bakker, L. H. and Pals, J. P. (1981) 'Life and work in Smeerenburg: the bio-archaeological aspects', *Proceedings of an International Symposium, Early Exploitation of the North Atlantic, 800–1700 (University of Groningen)* pp. 133–51

Williams, M. (1989) *Americans and their Forests: an Historical Geography* (Cambridge University Press, Cambridge)

Wolf, E. R. (1982) *Europe and the People without History* (University of California Press, Berkeley)

# 13

# A question of Landnám: Pollen, Charcoal and Insect Studies on Papey, Eastern Iceland

Paul C. Buckland, Kevin J. Edwards, J. J. Blackford, Andrew J. Dugmore, Jon P. Sadler and Guðrún Sveinbjarnardóttir

ABSTRACT

'Papey' is one of a small group of Icelandic place names which appear to refer to settlement by Irish monks before Norse settlement (Landnám) in the ninth century AD. In a similar situation in the Faroe Islands, Jóhansen has proposed that there is palynological evidence in support. This has been doubted on both historical and palynological grounds. Recent archaeological research on Papey has failed to find traces of the Irish papar or culdees. Environmental research, assisted by tephrochronology and radiocarbon dating, has provided insights into environment and landscape change both on the island and the Icelandic mainland. Evidence from pollen and insects provides a detailed picture of the landscape before and after settlement and helps to resolve the problems of human impact on a remote island ecosystem.

## INTRODUCTION

The earliest convincing reference to settlement on islands in the mid-North Atlantic occurs in a description of the world by an eighth-century Irish monk, Dicuil, writing at the court of Charlemagne's successors (Tierney 1967). He refers to islands two days' sailing from the outer Scottish islands, once occupied by monks but, by his time, *c*.825, abandoned to sheep and seabirds because of attacks by pirates. The islands are usually taken to be the Faroes (Buckland 1990). His further comment, 'in that island... no darkness occurs during that very brief period of time, but whatever task a man wishes to perform, even to picking lice out of his shirt, he can manage it precisely as in broad daylight', may refer to Iceland (Jones 1986). The Norse settlement of

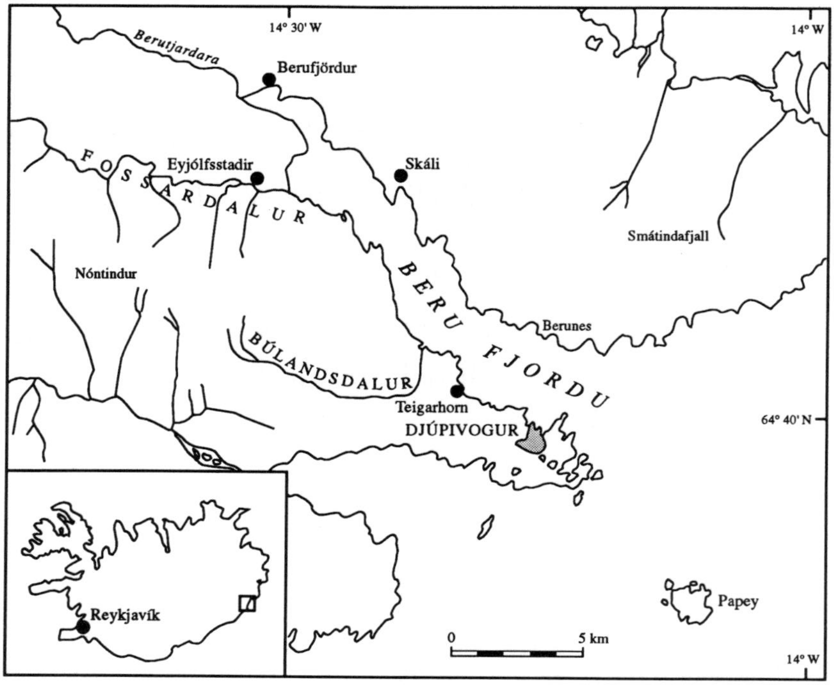


Figure 13.1 A location map of south-east Iceland and the offshore island of Papey.

Iceland (Landnám), from the mid-ninth century AD onwards, is recorded in the book of the Icelanders (*Íslendingabók*), written in the twelfth century. Here mention is made of previous Christian *Irskmen*, called *Papar* by the Norse, who occupied the country when they arrived (*Íslenzk fornrit* 1, 1968: 5).

The small island of Papey (latitude 64°36′ N, longitude 14°11′ W) lies off the south-east coast of Iceland, at the mouth of Berufjörður, some 6 km from the small fishing port of Djúpivogur (figure 13.1). The low basalt island, approximately 2 km × 2 km in extent, reaches a maximum height of 58 m at Hellisbjarg and is bounded by cliffs. Hoppe (1982) noted evidence for Papey having been overridden by the mainland ice of the last glaciation and the present biota owes its origins to postglacial immigration (Buckland 1988). In a survey in 1962, Steindór Steindórsson recorded 124 species of plant (Steindórsson 1962a) and, between the rocky knolls, the surface is extensively covered with *Carex*-dominated bog over deep peat deposits. The island is no longer permanently occupied, although the old farm and small church are still maintained. Until abandonment some 30 years ago, the farmer on Papey maintained some 70 ewes with their lambs, three cows and a horse on the island, and it has since continued to be used as summer grazing for a declining number of sheep.

The place name, Papey, has inevitably been associated with the supposed Celtic monks (Sveinbjarnardóttir 1972). This and additional place-name

**Table 13.1** Radiocarbon dates from Papey, Iceland

| *Lab. no.* | *$^{14}C$ age (BP)* | *Calibrated age (AD)[a]* | *Material* | *Site* |
|---|---|---|---|---|
| St.4191 | 1665 ± 100 | 130 – 600 | *Pinus pumila* | Undir Hellisbjargi |
| St.4192 | 260 ± 245 | 1270 – 1955 | *Picea* sp. | Undir Hellisbjargi |
| U-4014 | 1090 ± 80 | 770 – 1113 | *Larix* and *Betula* spp. | Undir Hellisbjargi |
| St.3605 | 935 ± 100 | 890 – 1270 | *Betula* charcoal | Goðatættur I |
| St.3368 | 1440 ± 100 | 410 – 780 | Driftwood charcoal | Goðatættur II |
| St.3604 | 980 ± 100 | 880 – 1260 | *Betula* charcoal | Goðatættur III |
| St.8348 | 925 ± 80 | 970 – 1260 | Charcoal | Áttahringsvogur |
| Birm-1128 | 940 ± 60 | 980 – 1230 | *Betual* Charcoal | Áttahringsvogur |

[a] After Stuiver and Reimer (1986). Calibration ranges to 2δ levels.

evidence, with evident traces of former structures, led to a series of archaeological excavations, directed between 1967 and 1981 by the late Kristján Eldjárn (1989) and completed for publication by Guðrún Sveinbjarnardóttir. Both the artefactual record and radiocarbon dates (table 13.1) imply occupation of several sites by the late ninth or tenth century, and the tephrochronological sequence further indicates that many had been abandoned before the eruption of Öræfajökull in south-east Iceland in 1362; no site produced convincing evidence for pre-Norse occupation.

Discussions with Kristján Eldjárn suggested that a palaeoecological approach might not only find trace of the elusive papar but also provide evidence for the use of individual structures, as research elsewhere in Iceland was beginning to indicate (Sveinbjarnardóttir et al. 1981). Unfortunately, by this stage in the archaeological research, most suitable deposits had already been excavated and the small samples taken in 1980 and 1981 added nothing to the interpretation. In 1982, therefore, a visit was made to Papey for the express purpose of sampling for macrofossil sequences. Of the two sites with adjacent peat deposits, that at Undir Hellisbjargi was rejected because of the possibility of relatively recent peat cutting, and a small pit was dug 25 m south of the excavated structures at Goðatættur, where the slight rock ridge on which the farm had stood gave way to a small peat-filled basin (figure 13.2). Although conditions restricted the pit to 1 m depth and sampling could only be carried out by constant bailing, it was possible to locate the Ö 1362 tephra horizon, and eleven 2.5 litre samples (P1–P11) were taken in 5 cm slices below it (figure 13.4). In addition, three further tephra horizons were noted. In the Faroes, Jóhannes Jóhansen (1971) had adopted a palynological approach to the search for the earliest inhabitants of the islands and it was evident that a similar study would provide data on Papey. In 1991, therefore, as part of a study of the impact of tephra on terrestrial biota, the opportunity was taken to supplement the macrofossil samples with cores for palynological as well as further tephrochronological study from both Undir Hellisbjargi and Goðatættur.

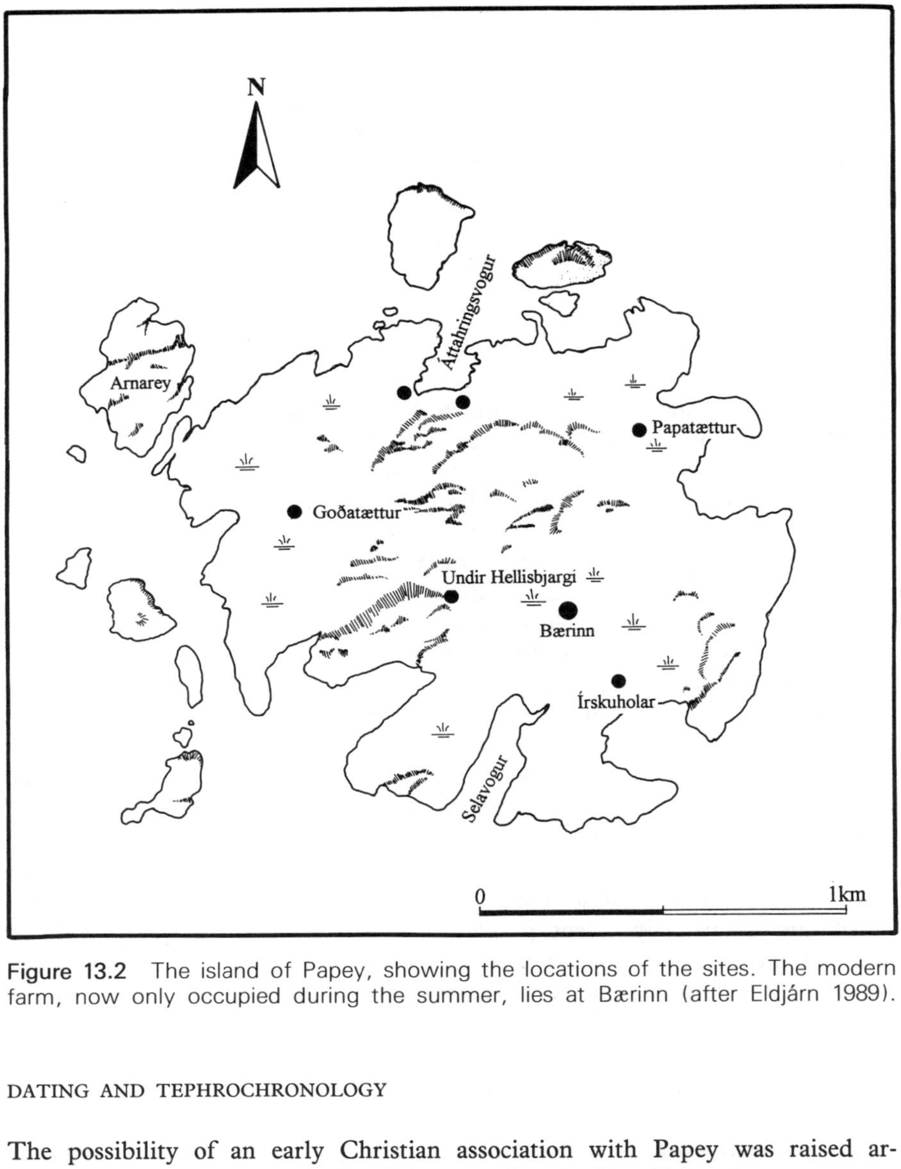


**Figure 13.2** The island of Papey, showing the locations of the sites. The modern farm, now only occupied during the summer, lies at Bærinn (after Eldjárn 1989).

## DATING AND TEPHROCHRONOLOGY

The possibility of an early Christian association with Papey was raised archaeologically in 1928 with the discovery of a number of fragmentary crosses during the construction of a wellhead at Undir Hellisbjargi. Similar pieces are known from medieval burials at Herjólfsnes in Greenland (Nørlund 1924), but crude crosses continued to be placed at springs as votive deposits into the present century in Norway (Eldjárn 1989). The radiocarbon dates from the Undir Hellisbjargi pieces (table 13.1) provide the first of several cautionary tales in the use of driftwood for dating; both dates are on wood of probable Siberian origin. The first calibrated date of AD 130–600, from a structural timber, might be taken to support a pre-Landnám occupation but the second,

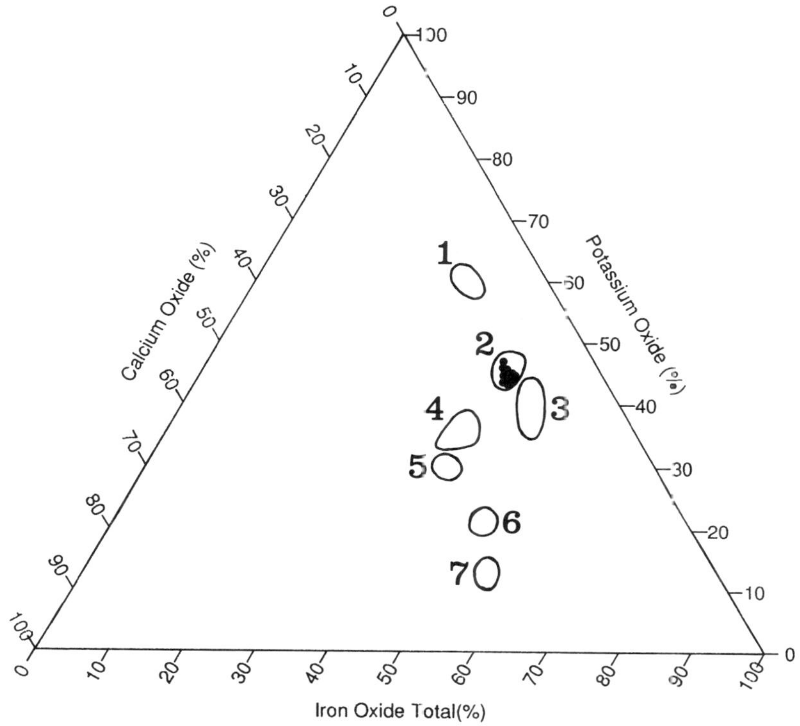


**Figure 13.3** Geochemical characteristics of tephras from various eruptions: (1) Vatnaöldur in AD 900; (2) Öræfajökull in 1362; (3) Eyjafjallajökull in 1821; (4) Hekla in 1104; (5) Askja in 1875; (6) Hekla in 1158; (7) Hekla in 1510 (data from Larsen et al. in prep.). The points derived from table 13.2 confirm the identification of Öræfajökull 1362 on Papey.

of AD 1270–1955, is contradictory. Although driftwood may only be trapped in the Arctic pack ice for a few years, it may remain in fossil deposits for much longer before it is utilized by people. The third date from Undir Hellisbjargi, of AD 770–1113, straddles the period of Landnám but, as the material is a mix of *Betula* and *Larix* charcoal from the structure floor, it must remain suspect. Similarly, a date on unidentified charcoal from the site at Áttahringsvogur, while it is in close agreement with a birch twig charcoal date from the same site, must also be doubted.

Although native birch and willow charcoal might seem to provide the most reliable materials for site dating, recent results, particularly claims for a pre-Norse presence on Vestmannaeyjar (Hermannsdóttir 1986, Hermanns-Auðardóttir 1989), urge caution in the use of the results. The problems have recently been addressed by Vilhjálmsson (1991). Birch may live far in excess of 100 years and, more critically, in areas with little or no timber resources, charcoal, particularly for smithing, could be made from peat (cf. Woolner 1967). In addition, ethnographic and entomological evidence for the widespread use of peat for litter (Amorosi et al. 1992) presents further problems

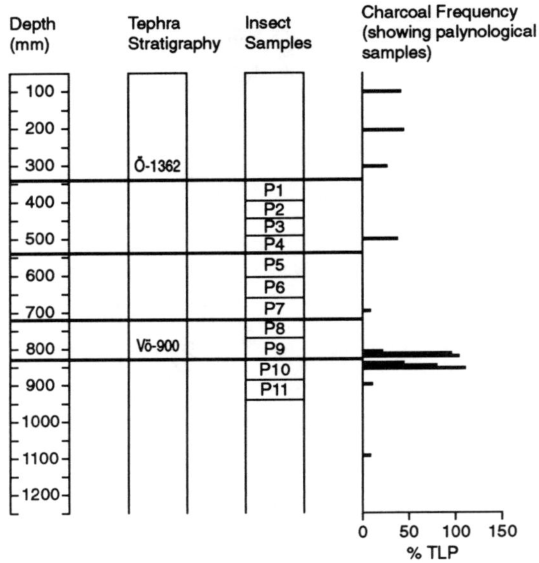


**Figure 13.4** Stratigraphy and samples from Goðatættur I, Iceland.

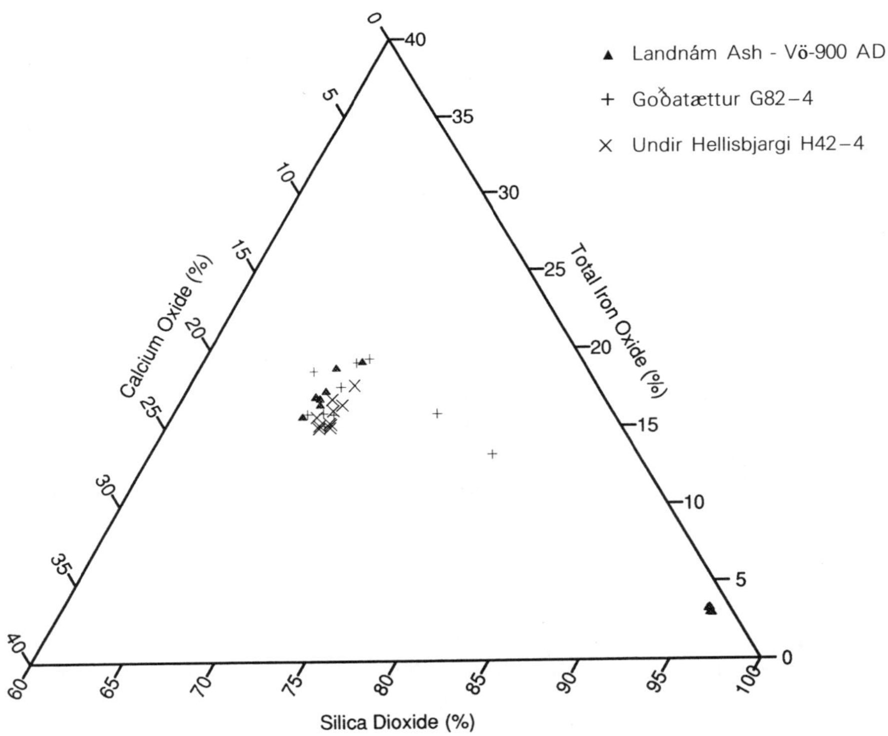


**Figure 13.5** A comparison of three key major elements that indicate that the Landnám Ash (Vö-900) is present at Goðatættur and Undir Hellisbjargi.

**Table 13.2** Major element chemistry of individual glass shards from the Öræfajökull tephra of AD 1362 from Goðatættur. Analyses were performed on thin section using a Cambridge Instruments Microscan V electron microprobe. A standard WDS technique was used with an accelerating voltage of 20 kV and a beam current of 15 nA. Peak counts were collected over a 10 s period, and total analysis time was 50 s. Calibration was based on a mixture of pure elements, oxides and simple silica compounds. Corrections were made for counter dead-time, atomic numbers effects, absorptions and fluorescence using a ZAF correction programme. Total iron is calculated as FeO

| $SiO_2$ | $TiO_2$ | $Al_2O_3$ | *FeOt* | *MnO* | *MgO* | *CaO* | $Na_2O$ | $K_2O$ | *Total* |
|---|---|---|---|---|---|---|---|---|---|
| 73.88 | 0.20 | 13.30 | 3.01 | 0.12 | 0.05 | 0.96 | 4.45 | 3.40 | 99.39 |
| 73.83 | 0.15 | 13.26 | 2.98 | 0.16 | 0.03 | 0.90 | 5.01 | 3.54 | 99.87 |
| 73.66 | 0.26 | 13.15 | 3.11 | 0.14 | 0.02 | 0.95 | 5.15 | 3.40 | 99.84 |
| 73.53 | 0.21 | 13.20 | 3.30 | 0.17 | 0.02 | 1.04 | 4.72 | 3.37 | 99.57 |
| 73.30 | 0.19 | 13.04 | 3.16 | 0.17 | 0.06 | 1.01 | 5.04 | 3.52 | 99.50 |
| 73.27 | 0.26 | 13.27 | 3.26 | 0.14 | 0.02 | 0.95 | 5.29 | 3.48 | 99.95 |
| 73.22 | 0.19 | 13.49 | 3.32 | 0.15 | 0.03 | 0.96 | 4.98 | 3.41 | 99.74 |
| 73.12 | 0.20 | 13.28 | 3.28 | 0.12 | 0.03 | 1.03 | 5.19 | 3.32 | 99.56 |
| 73.02 | 0.22 | 12.79 | 3.14 | 0.17 | 0.06 | 0.96 | 3.74 | 3.23 | 97.32 |
| 72.94 | 0.20 | 13.19 | 3.20 | 0.14 | 0.03 | 0.90 | 5.03 | 3.21 | 99.02 |
| 72.89 | 0.23 | 13.35 | 3.22 | 0.13 | 0.01 | 0.92 | 5.18 | 3.25 | 99.17 |
| 71.89 | 0.22 | 13.12 | 3.16 | 0.18 | 0.03 | 0.89 | 4.96 | 3.30 | 97.75 |
| 71.79 | 0.25 | 13.24 | 3.40 | 0.19 | 0.05 | 1.04 | 4.79 | 3.55 | 98.30 |
| 71.73 | 0.18 | 12.92 | 3.26 | 0.16 | 0.02 | 1.07 | 5.16 | 3.10 | 97.89 |
| 71.52 | 0.25 | 13.30 | 3.20 | 0.19 | 0.06 | 0.96 | 5.05 | 3.40 | 97.92 |

**Table 13.3** Geochemical analyses of grains from the tephra layers at 82–4 cm depth at Goðatættur (a) and 42–4 cm depth at Undir Hellisbjargi (b). The composition indicates an origin in the Veiðivötn volcanic system, probably during the *c.*AD 900 eruption

| *$SiO_2$* | *$TiO_2$* | *$Al_2O_3$* | *FeOt* | *MnO* | *MgO* | *CaO* | *$Na_2O$* | *$K_2O$* | *Total* |
|---|---|---|---|---|---|---|---|---|---|
| (a) | | | | | | | | | |
| 57.25 | 1.59 | 14.88 | 9.610 | 0.23 | 2.49 | 5.82 | 3.54 | 1.04 | 96.45 |
| 54.17 | 2.37 | 14.50 | 11.53 | 0.27 | 3.37 | 7.01 | 3.41 | 1.16 | 98.39 |
| 51.86 | 3.27 | 13.00 | 14.55 | 0.28 | 3.92 | 8.66 | 2.51 | 0.86 | 98.92 |
| 50.07 | 2.24 | 13.77 | 11.51 | 0.19 | 6.44 | 11.14 | 2.48 | 0.3 | 98.14 |
| 49.95 | 1.71 | 12.33 | 11.75 | 0.18 | 8.85 | 12.42 | 2.3 | 0.24 | 99.71 |
| 49.75 | 2.73 | 12.85 | 12.81 | 0.25 | 5.91 | 10.21 | 2.66 | 0.4 | 97.59 |
| 49.61 | 1.95 | 13.63 | 13.88 | 0.24 | 6.26 | 11.22 | 1.37 | 0.29 | 98.45 |
| 49.53 | 2.23 | 13.62 | 11.55 | 0.22 | 7.03 | 11.50 | 2.51 | 0.35 | 98.55 |
| 49.26 | 3.26 | 12.89 | 13.76 | 0.26 | 4.93 | 8.9 | 2.31 | 0.48 | 96.17 |
| 49.17 | 2.04 | 13.65 | 10.80 | 0.26 | 7.01 | 11.61 | 2.4 | 0.31 | 97.27 |
| (b) | | | | | | | | | |
| 51.33 | 1.82 | 14.05 | 12.23 | 0.22 | 6.21 | 10.83 | 2.82 | 0.28 | 99.80 |
| 50.52 | 2.08 | 13.89 | 11.19 | 0.16 | 7.25 | 11.62 | 2.58 | 0.31 | 99.61 |
| 50.28 | 2.76 | 13.21 | 12.89 | 0.22 | 5.48 | 9.64 | 2.96 | 0.54 | 97.98 |
| 49.98 | 2.00 | 14.00 | 11.00 | 0.16 | 7.06 | 11.47 | 2.51 | 0.28 | 98.45 |
| 49.92 | 2.14 | 13.68 | 10.84 | 0.21 | 7.26 | 11.55 | 2.57 | 0.27 | 98.44 |
| 49.38 | 2.32 | 13.73 | 12.15 | 0.22 | 6.43 | 10.8 | 2.48 | 0.29 | 97.81 |
| 49.21 | 1.82 | 14.03 | 10.87 | 0.20 | 7.21 | 11.85 | 2.44 | 0.32 | 97.94 |
| 49.09 | 2.11 | 13.68 | 11.49 | 0.14 | 6.42 | 10.89 | 2.56 | 0.35 | 96.72 |
| 49.02 | 2.04 | 13.46 | 10.69 | 0.18 | 7.16 | 11.88 | 2.51 | 0.30 | 97.25 |
| 48.82 | 2.05 | 13.96 | 11.24 | 0.16 | 6.84 | 11.74 | 2.62 | 0.31 | 97.75 |

in selection of dating materials. At Goðatættur, one date from each of the sites implies a post-Norse landnám context, while a third, on driftwood from site I, is clearly much older (table 13.1).

Despite the reservations over the radiocarbon dates, the archaeological evidence, particularly the artefacts from Goðatættur, suggest an early date for the majority of the excavated sites on Papey, and this is confirmed by the tephra sequence, which also provides a means of correlation with the cores and bulk samples from close to both Goðatættur, and Undir Hellisbjargi. During excavation it was noted that the Ö 1362 ash extended over both sites and that a considerable amount of silt accumulation separated the tephra from the house floors. Geochemical analysis confirms the identity of this horizon across the island (figure 13.3, table 13.2) but, in the excavation at Undir Hellisbjargi, two further tephra layers intervene between it and the floor of the structure, although there is some doubt over the location of this section. The tephrochronologist Guðrún Larsen did not date these directly but examination of the stratigraphy in Rafnsson's study of Hrafnkelsdalur, some 80 km to the west, recorded by the late Sigurður Thórarinsson (in Rafnsson 1990), and Larsen's (1982) work in Jökuldalur, suggests that the upper of these layers might be from the Katla eruption of 1262, and the lower possibly from Hekla in 1158. The comparative scarcity of historical tephras older than 1362 in the area is shown by the work of both Thórarinsson (1958) and Larsen (1982). The implication is that the sediment at Undir Hellisbjargi containing the pre-1362 tephras probably represents at least two centuries of accumulation; this suggests that the structure was abandoned before the middle of the twelfth century.

The excavated site at Goðatættur is more clearly a farm and associated byre and enclosure, with extensive parallels in the Norse world. The Ö 1362 tephra sealed all structures but other ash layers were not noted during excavation and it is only in the pit dug for bulk samples and the adjacent core that further tephra horizons were evident (figure 13.4). The third tephra below the Ö 1362 layer is interpreted as part of the so-called Landnám Ash (Larsen 1984) of *c.*AD 900 (Vö-900). Tephra grains from this horizon at both Goðatættur and Undir Hellisbjargi were geochemically analysed (table 13.3). The layer at the latter site produced the more coherent group of results but, when both data sets are compared with material known to originate from the basaltic part of the Landnám Ash, a good match is produced (figure 13.5). These results support the interpretation that the Landnám Ash is present at both sites although, at this stage, the possibility that the tephra might be from other eruptions within the Veiðivötn volcanic system, if unlikely, cannot be completely excluded (Larsen 1982).

## PALAEOECOLOGY

### *Insects*

The column of 2.5 litre samples from Goðatættur was taken in 1982 sufficiently close to the archaeological site to pick up evidence for activity on the site without disturbance of the stratigraphy by human or animal action. This condition seems to be met by the integrity of the tephra layers in the excavated section, although the faunal remains do suggest some problems in this direction. Samples were sieved over a 300 μm sieve and subjected to paraffin (kerosene) flotation (Coope and Osborne 1968). The insect faunas recovered are of low abundance (table 13.4) but contain sufficient diagnostic taxa to indicate human activity. The two lowest samples (P10–11) contain no anthropogenic indicators, but the appearance of ectoparasites of humans and sheep in P9, over the putative Landnám tephra, provides clear indication of the occupation of the farm. Ectoparasites only leave their hosts at death, relying on direct contact for dispersal. The human louse has been widely recognized in medieval and later contexts in Iceland and Greenland, although only in one case, at Sandnes in the Western Settlement of Greenland, has it proved possible to differentiate between body (*Pediculus humanus humanus*) and the head (*P. h. capitis*) louse (Sveinbjarnardóttir and Buckland 1983): both find not infrequent literary reference, from Thomas Becket's hair shirt to Stigandi in Laxdæla Saga, betrayed by his female slave as she picked lice from his hair (op. cit). The fleece louse, *Damalinia ovis*, and the ked, *Melophagus ovinus*, a blood sucking wingless fly, found only on sheep, are less easily tracked in the literary record and the latter are usually subsumed under the general term 'tick'. Both, however, have a fossil record in the North Atlantic (Sadler 1991) and their importance in archaeological interpretation has recently been considered (Buckland and Perry 1989). The ked, either as fragments of its distinctive puparium, which is glued to the wool of the living animal (Askew 1971) or as individual tergites of exoskeleton, is present in samples P6–9 inclusive and it is supplemented in P6 and P8 by *D. ovis*.

Two other species in the assemblage are also strongly synanthropous and, in the context of Papey, also anthropochorous. Essential to the long-term survival of any pastoral community in the northern temperate to subarctic zone is the acquisition of sufficient fodder to overwinter the core animals, on whose secondary products the farm is maintained. While seaweed was extensively used in coastal locations in Iceland (Amorosi et al. 1994), the essential fodder remained hay and it is the insect fauna associated with the decay of stored hay which appears in the earliest anthropogenic deposits examined (Buckland et al. 1991). *Lathridius minutus* occurs in samples P6–8 at Papey and another synanthrope, *Xylodromus concinnus*, joins it in P6. Their need for the artificially warmed habitats created by humans in order to overwinter in the north makes them certain indicators of a permanent farm, rather than a shieling, an argument which has been advanced in the interpretation of the site at Engihlíð in adjacent Berufjörður (Buckland and Sadler 1991). Their principal habitat in Iceland is in hay barns (Lindroth et al. 1973) but, as both species are flighted, they are likely to be found away from the farm in decaying plant debris, in, for example, cut hay lying in the fields, and their presence can only indicate the proximity of an occupation site.

The period of occupation of Goðatættur is clearly constrained to that during which samples P6–9 inclusive accumulated, with their obligate synanthropic faunas. The insect faunas suggest accumulation on the distal edge of the farm's

**Table 13.4** Insect remains from Papey: taxonomy follows Ólafsson (1991)

| Taxa | Samples | | | | | | | | | | |
|---|---|---|---|---|---|---|---|---|---|---|---|
| | *P1* | *P2* | *P3* | *P4* | *P5* | *P6* | *P7* | *P8* | *P9* | *P10* | *P11* |
| Anoplura | | | | | | | | | | | |
| *Pediculus humanus* sp. | 0 | 0 | 0 | 0 | 0 | 0 | 0 | 0 | 1 | 0 | 0 |
| *Damalinia ovis* (L.) | 0 | 0 | 0 | 0 | 0 | 2 | 0 | 2 | 0 | 0 | 0 |
| Trichoptera | | | | | | | | | | | |
| Limnephilidae indet. | 0 | 0 | 1 | 0 | 0 | 1 | 12 | 27 | 3 | 0 | 16 |
| Coleoptera | | | | | | | | | | | |
| *Patrobus septentrionis* (Dej.) | 1 | 0 | 0 | 0 | 0 | 0 | 0 | 3 | 0 | 1 | 3 |
| *Pterostichus adstrictus* Esch. | 0 | 0 | 0 | 0 | 0 | 0 | 1 | 0 | 0 | 0 | 0 |
| *Calathus melanocephalus* (L.) | 0 | 0 | 0 | 0 | 0 | 0 | 1 | 0 | 0 | 0 | 0 |
| *Agabus solieri* Aubé | 0 | 0 | 0 | 0 | 0 | 0 | 0 | 1 | 0 | 0 | 0 |
| *Xylodromus concinnus* (Marsh) | 0 | 0 | 0 | 1 | 0 | 0 | 0 | 0 | 0 | 0 | 0 |
| *Geodromicus longipes* (Mann.) | 1 | 0 | 1 | 0 | 0 | 0 | 0 | 0 | 0 | 0 | 0 |
| *Stenus carbonarius* Gyll. | 2 | 0 | 1 | 0 | 1 | 1 | 2 | 3 | 4 | 2 | 6 |
| *Quedius umbrinus* Er. | 0 | 0 | 0 | 0 | 0 | 0 | 2 | 1 | 0 | 1 | 0 |
| Aleacharinae gen. indet. | 2 | 0 | 5 | 3 | 6 | 2 | 17 | 9 | 1 | 3 | 12 |
| *Bryaxis puncticollis* (Denny) | 0 | 0 | 0 | 0 | 0 | 0 | 0 | 0 | 1 | 0 | 0 |
| *Lathridius minutus* (L.) grp. | 0 | 0 | 0 | 0 | 0 | 1 | 1 | 1 | 0 | 0 | 0 |
| *Otiorhynchus nodosus* (Mull.) | 0 | 0 | 0 | 0 | 0 | 0 | 0 | 0 | 0 | 0 | 1 |
| *Otiorhynchus* sp. | 0 | 0 | 1 | 0 | 0 | 0 | 0 | 0 | 0 | 0 | 0 |
| Diptera | | | | | | | | | | | |
| *Melophagus ovinus* L. | 0 | 0 | 0 | 0 | 0 | 1 | 1 | 2 | 1 | 0 | 0 |

midden, otherwise only indicated by the occasional fragments of charcoal. This, with the tephra stratigraphy, further indicates that the farm was settled after the deposition of the Landnám ash in *c.*AD 900 and had been abandoned between the deposition of succeeding tephra falls.

The remainder of the invertebrate fauna consists of elements of the natural biota of the island and most were either taken during the fieldwork in 1982 and 1991 or have a pre-Landnám fossil record in Iceland (Buckland 1988). The small rove beetle, *Geodromicus longipes*, however, has not previously been recorded as a fossil, and the Pselaphid *Bryaxis puncticollis* is much more common fossil than its present few captures (Lindroth et al. 1973) might suggest. However, its habitat, at the root of grass tussocks, is difficult to collect from and it may have been overlooked.

Apart from the anthropochorous elements in the fauna, the insects provide no direct evidence of human impact on Papey and a better regional picture is provided by the palynology. The invertebrate remains, however, suggest that some caution is necessary in the interpretation of any results. Although the integrity of the tephra layers implies that the soft peat had not been extensively bioturbated by the farm's animals, the presence of ectoparasite remains in particular indicate that material, cleared out of the farm on to its midden, was accumulating at least intermittently at the sampling location. While charcoal and some faunal elements clearly reflect human activity, biota introduced in peat residues are less clearly defined. In both floor and midden deposits from several Icelandic farm sites, the presence of larvae of Trichoptera, Chironomidae and larval water beetles provides evidence of the use of peat (Amorosi et al. 1992), probably as both animal and human bedding. In the edge of the bog it is impossible to differentiate between the autochthonous and allochthonous elements; some elements may have been several thousand years old before reincorporation in the fossil record.

### *Pollen analysis*

A monolith, 4.0 m in depth and 5 m from the site sampled earlier for insect remains, was obtained in 1991. Samples beneath 400 mm were taken with a Russian corer of 50 mm chamber diameter (Jowsey 1966). Standard NaOH, HF, HCl and acetolysis pretreatments were employed (Faegri and Iversen 1989) with samples embedded in silicone oil. Samples were prepared volumetrically with the addition of *Lycopodium clavatum* spore tablets for the determination of palynomorph concentrations (Stockmarr 1971).

Pollen data are presented as a skeleton diagram covering the whole profile from Goðatættur (figure 13.6), and a more detailed diagram from contiguous 5 mm thick samples either side of the *c.*AD 900 'Landnám' tephra (figure 13.7). This enables the vegetation changes shown by the detailed work to be seen in a context of longer-term changes, and allows a comparison of pre- and post-occupation assemblages. Pollen diagrams were constructed with the aid of the computer programs TILIA and TILIA.GRAPH (Grimm 1991).

Figure 13.6 shows selected taxa from the whole profile. The earliest samples, of unknown age, show an aquatic environment with high percentages of

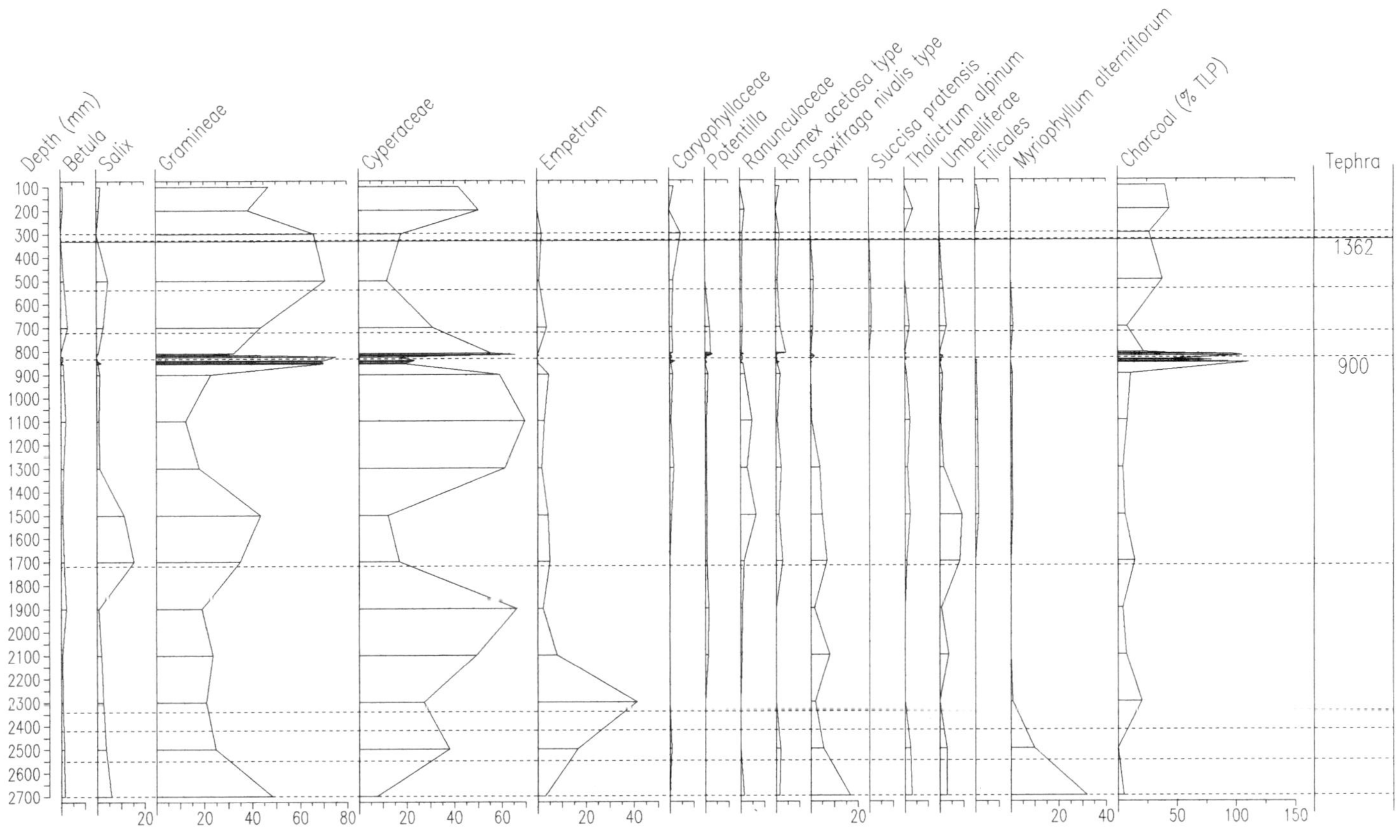


**Figure 13.6** A 'skeleton' pollen diagram: selected pollen percentages for the whole profile. Pollen sum = total land pollen (TLP > 300). Charcoal curve shows particles > 15 $\mu$m as per cent TLP. Tephra deposits are shown as horizontal broken lines, with the Ö 1362 ash as a solid line.

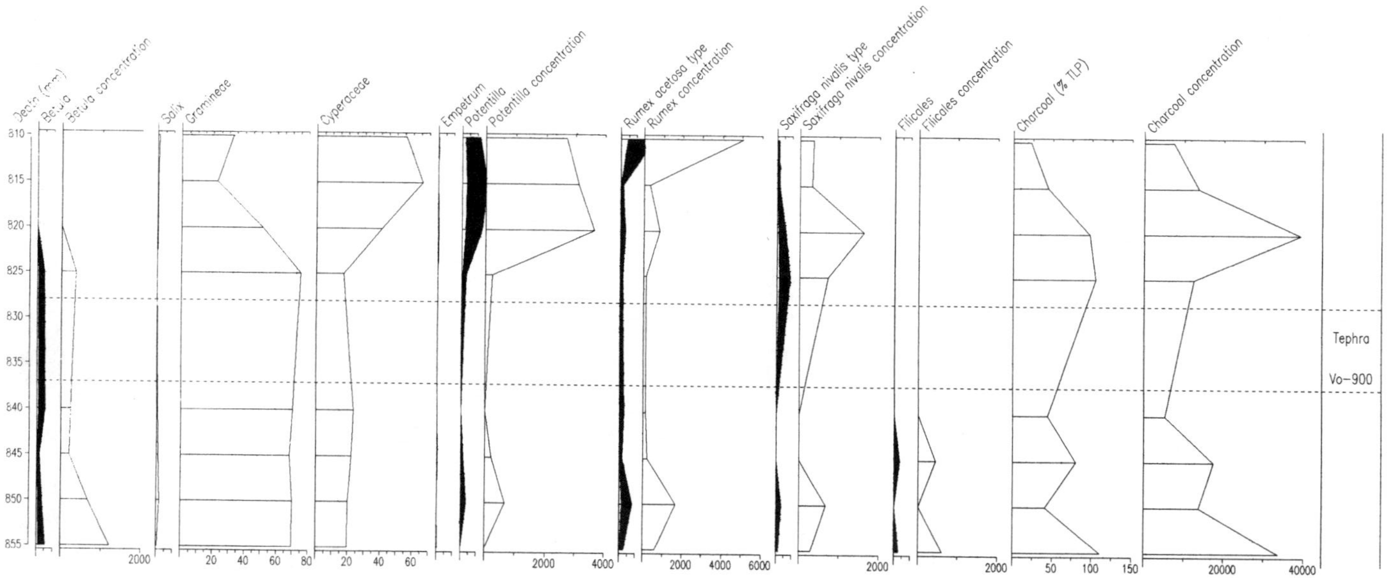


**Figure 13.7** Pollen and charcoal data from 0.5 cm thick contiguous samples either side of the Vö-900 tephra. Pollen sum = total land pollen, TLP > 350. Concentrations are in terms of grains $cm^{-3}$, and charcoal particles $cm^{-3}$. Shaded curves show × 5 exaggeration. Broken horizontal lines show the upper and lower boundaries of the tephra.

*Myriophyllum alterniflorum*, replaced by an *Empetrum*–Cyperaceae–Gramineae 'heath'. Thereafter, the dominant taxa are Gramineae and Cyperaceae, with phases in which these types dominate alternately. Arboreal taxa are rare throughout, as *Betula pubescens*, *Salix*, *Populus* and *Sorbus aucuparia* are the only trees native to Iceland. *Betula* is most abundant in the pollen diagram, but it never exceeds 3.5 per cent of the pollen sum. Studies of surface pollen samples at similar latitudes have shown that tree species of *Betula* are well represented in the pollen rain, especially when other trees are absent (Hicks 1977a,b, 1985) and are recorded as a constant background component when the trees are on a regional scale (Tyldesley 1973). Although *Betula* must have been present regionally, there is no clear evidence for it being present on Papey. *Salix* reaches 12 per cent of the pollen sum at depths of 1500 mm and 1700 mm, but then declines again to less than 3 per cent. The present-day treeless and shrubless vegetation appears to have dominated throughout the period of sediment accumulation.

### *Vegetation changes after the Landnám tephra*

Just above the Landnám tephra, the percentages of Cyperaceae increase rapidly, mostly at the expense of Gramineae. This switching is noted at earlier times, however, and may be a natural process, responding to changes in local hydrology with sedges thriving in wetter conditions. Although it is tempting to point to the rapidity of this particular change, it must be noted that earlier switches have not been analysed in such detail.

Many taxa are present in low numbers before, during, and after the period of occupation suggested by the beetle evidence. Figure 13.7, however, shows that some taxa change in abundance in a way not shown to have occurred beforehand; in other words, this appears to be a botanically significant horizon. *Potentilla* and *Rumex acetosa* type show significant increases in samples at 820 mm and 810 mm respectively, 10 and 20 mm above the ash layer. *Succisa pratensis* occurs for the first time at 700 mm. Filicales spores are present in small numbers in most samples below the Landnám Ash, but are absent in the samples immediately above it. Gramineae are reduced in frequency and *Betula* numbers fall. Pollen concentration figures support the percentage data for these taxa. *Potentilla* and *Rumex* reach their highest concentration values just after the ash layer, and the fall in *Betula* is shown in concentration as well as percentage terms.

### *Discussion of pollen data*

In many north-west European pollen diagrams, the concept of Landnám is associated with tree clearance, and the replacement of arboreal species with open-ground indicators. A similar pattern has been recorded from south Iceland, with the clearance of birch forest shown in many diagrams at around and shortly after *c.*AD 900 (Einarsson 1961, Hallsdóttir 1987). Figure 13.7 shows that *Betula* percentages and concentrations decline at a depth of 820 mm, 10 mm below the Landnám layer, perhaps reflecting changes on the mainland. Papey presents a different problem in that it has evidently been treeless for at least the period of the accumulation of the Goðatættur core. This makes the detection of the first human visitors, or residents, less straightforward. One possible anthropochore has been detected. *Succisa pratensis* was probably introduced with the hay and dunnage of the early settlers (Steindórsson 1962b); its present distribution in Iceland is largely southeastern (Kristínsson 1987). A rich suite of anthropochores would perhaps not be expected at this exposed site, and the pollen flora is even more restricted than those published by Hallsdóttir (1982) from around 400 m a.s.l. in Hrafnkelsdalur, on the mainland, some 75 km to the north-west, where cereal-type pollen was also represented.

Increases in frequencies of *Rumex acetosa* type have been linked to human occupation in high latitudes (Hicks 1985), and the percentages and concentrations of this taxon increase 20 mm above the Vö-900 tephra. Hicks (1988) associated increased quantities of *R. acetosa* with hay meadows. *Potentilla* pollen has been shown to increase proportionally in mixed heathlands as a response to grazing pressure (Moore et al. 1986). The rise in *Potentilla* at Goðatættur, 10 mm above the Landnám Ash, could be associated with introduced sheep, for which there is sound evidence from the insect record. A grazing hypothesis could explain the rapid transition from a Gramineae to a Cyperaceae domination of the pollen assemblage, with Gramineae being eaten more by grazing animals, thus inhibiting flowering (Edwards et al. 1994). Alternatively, as many species of Cyperaceae flower in May and June, whereas many Gramineae flower later in the year (Clapham et al. 1987), cutting a mixed sward for hay in mid-summer may have favoured Cyperaceae in the pollen record.

Filicales frequencies have been shown to fall in pollen records at times of human occupation of North Atlantic islands, including Shetland (Bennett et al. 1992) and Faroe (Jóhansen 1985), perhaps because of trampling effects. On mainland Iceland, however, the reverse is sometimes the case, and increases in fern representation have been attributed to a change in the structure of the vegetation, with a cleared landscape allowing easier spore distribution (Hallsdóttir 1987). On Papey, the frequencies fall to zero around the Landnám tephra, and do not return until after the Ö 1362 tephra horizon, some time after the abandonment of Goðatættur. This could be attributed to domestic animals being present on the island, although the number of spores involved are low (see figures 13.6 and 13.7). The first record of the demise of Filicales is at 840 mm, directly below the Vö-900 tephra.

### *Microscopic charcoal*

No macroscopic charcoal fragments are visible in the core stratigraphy. Microscopic charcoal, mostly in the size range 15–50 μm, have been counted along with the pollen and spores, and data are presented in relative and concentration terms in figures 13.6 and 13.7. Background levels, presumably from long-distance transport, natural fires in the region or small natural fires nearby, are quite high, perhaps indicating slow sediment accumulation. An additional factor is the probability that fires caused by volcanic eruptions on mainland Iceland have contributed to the charcoal record on Papey. Sharp peaks in charcoal frequency occur directly beneath and above the Vö-900 tephra, suggesting a change in the fire regime. One explanation of this change might be human occupancy, with charcoal being derived from fires set either for cooking, warmth or regular smithing activities at the farms. Such is thought to have accounted for much of Iceland's birch woodland (Þórarinsson 1974). While the location of such fires would be difficult to determine palynologically (Patterson et al. 1987, Edwards 1988), the proximity of the archaeological remains and the fossil insect evidence give very strong indications of its origin at least partially in domestic fires less than 25 m away, associated with the farm at Goðatættur.

## CONCLUSIONS

Despite the excavation of several sites on Papey, archaeological research has failed to reveal any evidence of pre-Norse occupation. The frequent use of driftwood in a landscape lacking in trees and the extensive utilization of peat provide problems in the use of radiocarbon dating. The tephrochronology provides a temporal framework for the environmental evidence and allows a number of conclusions to be drawn about the earliest occupation of Papey. The pollen data show the probable introduction of at least one species, *Succisa pratensis*, and changes in frequency of various native taxa. The presence of ectoparasites of sheep and humans, as well as elements of the stored hay fauna, in a relatively restricted part of the succession suggests that occupation may have been restricted to only about 200 years after Landnám. The microscopic charcoal evidence suggests that, while the site at Goðatættur may not have been occupied before *c.*900, occupation elsewhere may have begun slightly earlier, but insufficiently early to imply the presence of pre-Norse papar. The supposed Irish settlers on the North Atlantic islands remain as elusive as ever.

## ACKNOWLEDGEMENTS

This project developed out of discussions with the late Kristján Eldjárn, to whose memory the work is dedicated. It has also benefited from the comments of Sheila Hicks, Jim Innes and Guðmundur Ólafsson. Field sampling was carried out in association with John Hunt, and the insect samples were processed and sorted by Philip Buckland. Research was funded by the Leverhulme Trust and NATO Scientific Affairs Division and carried out under permits from the Icelandic Science Research Council. Access to Papey was kindly made available by the people of Djúpivogur.

## REFERENCES

Amorosi, T., Buckland, P. C., Ólafsson, G., Sadler, J. P. and Skidmore, P. (1992) 'Site status and the palaeoecological record: a discussion of the results from Bessastaðir,

Iceland', in Morris, C. D. and Rackham, D. J. (eds) *Norse and Later Settlement and Subsistence in the North Atlantic* (University of Glasgow) pp. 169–91

Amorosi, T., Buckland, P. C., Sadler, J. P., Magnusson, K. and McGovern, T. H. (1994) 'An Archaeozoological examination of the midden at Nesstofa, Reykjavík, Iceland', in Rowley-Conwy, P. and Luff, R. (eds) *Whither Archaeology? Proceedings of the 1990 Cambridge Meeting, Association of Environmental Archaeologists* (Oxbow, Oxford) pp. 69–80

Askew, R. R. (1971) *Parasitic Insects* (Heinemann, London)

Bennett, K. D., Boreham, S., Sharp, M. J. and Switsur, V. R. (1992). 'Holocene history of environment, vegetation and human settlement on Catta Ness, Lunnasting, Shetland', *Journal of Ecology* 80: 241–73

Buckland, P. C. (1988) 'North Atlantic faunal connections – introductions or endemics? *Entomologica Scandinavica, Supplement* 32: 7–29

Buckland, P. C. (1990) 'Insects, man and the earliest settlement of the Faroe Islands: a case not proven', *Fróðskaparrit* 38–9: 107–13

Buckland, P. C. and Perry, D. W. (1989) 'Ectoparasites of sheep from Stóraborg, Iceland and their palaeoecological significance (piss, parasites and people: a palaeoecological perspective)', *Hikuin* 15: 37–46

Buckland, P. C. and Sadler, J. (1991) 'Farm or shieling: an entomological approach', in Sveinbjarnardóttir, G. Shielings in Iceland – an Archaeological and Historical Survey, *Acta Archaeologia* 61: 93–6

Buckland, P. C., Dugmore, A. J., Perry, D., Savory, D. and Sveinbjarnardóttir, G. (1991) 'Holt in Eyjafjallasveit, Iceland: a palaeoecological study of the impact of Landnám', *Acta Archaeologica* 61: 252–71

Clapham, A. R., Tutin, T. G. and Moore, D. M. (1987) Flora of the British Isles 3rd edn (Cambridge University Press, Cambridge)

Coope, G. R. and OSBORNE, P. J. (1968) 'Report on the Coleopterous fauna of the Roman Well at Barnsley Park, Gloucestershire', *Transactions of the Bristol and Gloucester Archaeological Society* 86: 84–7

Edwards, K. J. (1988) 'The hunter–gatherer/agricultural transition and the pollen record in the British Isles', in Birks, H. H., Birks, H. J. B., Kaland, P. E. and Moe, D. (eds) *The Cultural Landscape – Past, Present and Future* (Cambridge University Press, Cambridge) pp. 255–66

Edwards, K. J., Buckland, P. C., Blackford, J. J., Dugmore, A. J. and Sadler, J. P. (1994) 'The impact of tephra: proximal and distal studies of Icelandic eruptions', *Münchener Geographische Abhandlungen* B12: 79–99

Einarsson, Th. (1961) *Pollenanalytische Untersuchungen zur spät- und postglacialen Klimageschichte Islands*. Sonderveröff. d. Geol. Inst. Univ. Köln 6, 1–52

Eldjárn, K. (1989) 'Papey – fornleifarannsóknir 1967–1981 (edited and with introduction by G. Sveinbjarnardóttir), *Árbók Hins Íslenzka Fornleifafélags* (1988): 35–188

Faegri, K. and Iversen, J. (1989) *Textbook of Pollen Analysis* (4th edn, revised by K. Faegri, P. E. Kaland and K. Krzywinski) (John Wiley, Chichester)

Grimm, E. C. (1991) *TILIA and TILIA.GRAPH* (Illinois State Museum, Springfield)

Hallsdóttir, M. (1982) 'Frjógreining tveggja jarðvegssniða úr Hrafnkelsdal', in Þórarinsdóttir, H., Óskarsson, Ó., Steinþórsson, S. and Einarsson, Þ. (eds) *Eldur er í Norðri* (Sögufélag, Reykjavík) pp. 253–66

Hallsdóttir, M. (1987) 'Pollen analytical studies of human influence on vegetation in relation to the Landnám tephra layer in southwest Iceland', Department of Quaternary Geology, University of Lund, LUNDQUA thesis 18

Hermannsdóttir, M. (1986) 'Merovingertida bosattning på Island', *Viking* 49: 135–45

Hermanns-Auðardóttir, M. (1989) *Islands Tidiga bosattning*. Studia Universitatis Umensis 1, Umeå Universitet

Hicks, S. (1977a) 'Modern pollen rain in Finnish Lapland investigated by analysis of surface moss samples', *New Phytologist* 78: 715–34

Hicks, S. (1977b) 'Modern pollen deposition in Kuusamo, Finland', *Grana* 25: 183–204

Hicks, S. (1985) 'Problems and possibilities in correlating historical/archaeological and pollen analytical evidence in a northern boreal environment: an example from Kuusamo, Finland', *Fennoscandia Archaeologica* 2: 51–84

Hicks, S. (1988) 'The representation of different farming practices in pollen diagrams from northern Finland', in Birks, H. H., Birks, H. J. B., Kaland, P. E. and Moe, D. (eds) *The Cultural Landscape – Past, Present and Future* (Cambridge University Press, Cambridge) pp. 189–207

Hoppe, G. (1982) 'The extent of the last inland ice sheet of Iceland', *Jökull* 32: 3–11

*Íslenzk fornrit 1*. (1968) *Íslendingabók – Landnámabók* (Jakob Benediktsson, Reykjavík)

Jones, G. (1986) *The Norse Atlantic Saga* (Oxford University Press, Oxford)

Jowsey, P. C. (1966) 'An improved peat sampler', *New Phytologist* 65: 245–48

Jóhansen, J. (1971) 'A palaeobotanical study indicating a preViking settlement in Tjörnuvík, Faroe Islands', *Fróðskaparrit* 19: 147–57

Jóhansen, J. (1975) 'Pollen diagrams from the Shetland and Faroe Islands', *New Phytologist* 75: 369–87

Jóhansen, J. (1985) *Studies in the Vegetational History of the Faroe and Shetland Islands* (Føroya Fróðskaparfelag, Tórshavn)

Kristínsson, H. (1987) *The Flowering Plants and Ferns of Iceland* (Örn og Örlygur Reykjavík)

Larsen, G. (1982) 'Gjóskutímatal Jökuldals og nágrennis' in Þórarinsdóttir, H., Óskarsson, Ó., Steinþórsson, S. and Einarsson, Þ. (eds) *Eldur er í Norðri* (Reykjavík, Sögufélag) pp. 51–65

Larsen, G. (1984) Recent volcanic history of the Veidivötn fissure swarm, southern Iceland – an approach to volcanic risk assessment. *Journal of Vulcanological and Geothermal Research* 22: 33–58

Lindroth, C. H., Andersson, H., Bödvarsson, H. and Richter, S. H. (1973) 'Surtsey, Iceland, the development of a new fauna, 1963–1970, Terrestrial invertebrates', *Entomologica Scandinavica Supplement* No. 5

Moore, P. D., Evans, A. T. and Chater, M. (1986). Palynological and stratigraphic evidence for hydrological changes in mires associated with human activity. In Behre, K.-E. (ed.) *Anthropogenic Indicators in Pollen Diagrams*. (A. A. Balkema, Rotterdam) pp. 209–20

Nørlund, P. (1924) 'Buried Norsemen at Herjolfsnes', *Meddelelser om Grønland*, 67

Ólafsson, E. (1991) *Íslenskt Skordýratal*, Fjölrit Náttúrufræðistofnunar 17

Patterson, W. A. III, Edwards, K. J. and Maguire, D. J. (1987) 'Microscopic charcoal as a fossil indicator of fire', *Quaternary Science Reviews* 6: 3–23

Rafnsson, S. (1990) *Byggðaleifar í Hrafnkelsdal og á Brúardölum* (Hið Íslenska Fornleifafélags, Reykjavík)

Sadler, J. (1991) 'Beetles, boats and biogeography: insect invaders of the North Atlantic', *Acta Archaeologica* 61: 199–211

Steindórsson, S. (1962a) 'Um gróður í Papey', *Náttúrufræðingurinn* 33: 214–32

Steindórsson, S. (1962b) 'On the age and immigration of the Icelandic flora', *Vísindafélag Íslendinga*, 35

Stockmarr, J. (1972) 'Tablets with spores used in absolute pollen analysis', *Pollen et Spores* 13: 615–21

Stuiver, M. and Reimer, P. J. (1986) 'A computer program for radiocarbon age calibration', in Stuiver, M. and Kraa, R. S. (eds), *Proceedings of the 12th International $^{14}C$ Conference, Radiocarbon* 28(2B): 1022–30

Sveinbjarnardóttir, G. (1972) 'Papar', *Mímir* 19; 2–20

Sveinbjarnardóttir, G. and Buckland, P. C. (1983) 'An uninvited Guest', *Antiquity* 48: 25–33

Sveinbjarnardóttir, G., Buckland, P. C. Gerrard, A. J., Greig, J. R. A., Perry, D. W., Savory, D. and Snaesdóttir, M. (1981) 'Excavations at Stóraborg: a palaeoecological approach', *Árbók Hins Íslenzka Fornleifafélags* 1980: 118–29

Thórarinsson, S. (1958) 'The Öraefajökull eruption of 1362', *Acta Naturalia Islandica* 2: 2

Tierney, J. J. (ed.) (1967) *Dicuilus. Liber de Mensura Orbis Terrae. Scriptores Latini Hiberniae* 6 (Dublin)

Tyldesley, J. B. (1973) 'Long range transmission of tree pollen to Shetland III. Frequencies over the past hundred years', *New Phytologist* 72: 691–7

Vilhjálmsson, V. Ö. (1991) 'The application of dating methods in Icelandic archaeology', *Acta Archaeologica* 61: 97–107

Woolner, D. (1967) 'Peat charcoal', *Devon and Cornwall Notes & Queries* 30: 118–20

Þórarinsson, Þ. (1974) 'Þjóðin lifði en skógurinn do', *Ársrit Skógræktarfélags Íslands* (1974), 16–29

# PART IV

## North America and the Caribbean

14

# Ecological Responses to Ecosystem Shock in the Island Caribbean: the Aftermath of Columbus, 1492–1992

David Watts

ABSTRACT

If ecosystem shock is defined as a sudden but major break in the existing ecological pattern, then two such shocks, both human-induced, have been experienced in the island Caribbean: one, from 1492 to *c.*1524, during which time virtually all the 2–4 million native peoples were eradicated by Hispanic colonizers, and replaced by 'dogs and beasts' (the contemporary phrase of Las Casas); and the other from 1645, when sugar estates began quickly and extensively to replace seasonal rain forest, a process which continues to this day. The consequences have been profound and ever-increasing ecological and biogeographical instability. This chapter considers the bioenvironmental responses to these shocks, both short-term and long-term, and the causes of such instability. Dissimilarities with the MacArthur and Wilson (1967) model for species immigration to oceanic islands and turnover are noted and discussed.

## INTRODUCTION

The island Caribbean lies along a broad arc some 4000 km in length from the western tip of Cuba to the south easternmost extremity of Trinidad. Within it, there is an immense diversity in island size (some are only 1 km$^2$ or so in extent, while Cuba is over 110 000 km$^2$; figure 14.1), landforms, geology, relief, climate and biological forms. Much of the region is tectonically instable, subject to earthquakes, and to geothermal and volcanic activity – both sometimes on a massive scale – as in 1902, when Mont Pelée engulfed with ash and lava the then capital of Martinique, St Pierre, killing virtually all of its 30 000 inhabitants. Only in the stable block of the Bahamas are signs of recent tectonic instability lacking: elsewhere, the uplifting of mountain

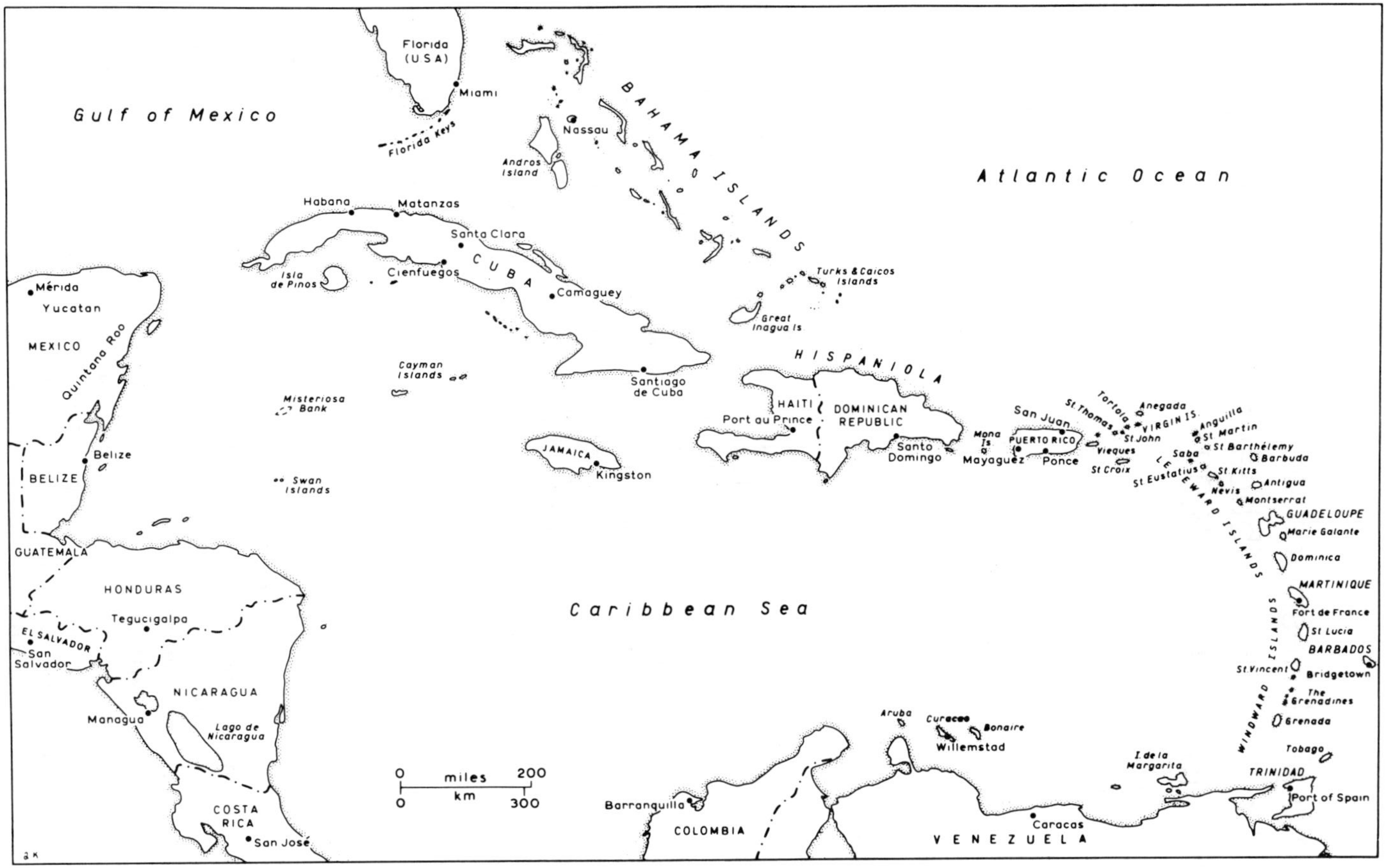


**Figure 14.1** A locational map of the Caribbean.

ranges, downwarping of oceanic trenches and building of volcanic peaks have produced spectacular landscapes, ranging from the 3175 m high Pico Duarte massif in the Dominican Republic to depths of over 8500 m below sea level in the Puerto Rico trench. Indeed, flat, low landscapes are rare in the island Caribbean, except in the Bahamas and parts of Cuba: mountain relief features are much more common, and these were accentuated at the peak of the last, Wisconsin glaciation, when sea levels stood some 120 m lower than today. Present-day climates, and associated natural vegetation, range from tropical rain forest, tropical seasonal forest and mountain forest to savanna (sometimes with palms or pines), dry scrub and desert. Endemicity in both floristic and faunal groups still is high in the larger islands – over 20 per cent among land bird populations in Cuba, Jamaica and Española – and is likely to have been substantially higher before post-Columbian forest clearance set in. Ecologically, all native plant and animal communities are attuned to severe periodic and repetitive environmental disturbance, such as the passage of hurricanes and (particularly in the eastern Caribbean) volcanic eruptions, so that many never have time to reach maturity before a renewed phase of disturbance sets in. Superimposed on this pattern, in historical time, are two sudden and major breaks in the regional faunal and floristic pattern – ecological shocks produced by human activity, the one created by the after-effects of Hispanic colonization and the other by the introduction of commercial sugar cane agriculture, which began some 150 years after the first. Most island Caribbean ecologies still are profoundly affected by the ramifications of these severe events.

## DATA SOURCES

Perhaps more so than in most parts of the world, the elucidation of such events depends less on palynology than on an interpretation of the historical record. Because of its volatile tectonic background, good palynological data are rare in the island Caribbean, and the two good pollen diagrams which exist to date (from Lowland Haiti: Higuera-Gundy 1989, Hodell et al. 1991) in any case provide little of substantive detail relating to the nature of vegetation change in the post-Columbian era, other than the expected long-term diminution of tree pollen, and the enhancement of grass and other pioneer species (NAP pollen) in the record as forest was removed to make way for agriculture. Pollen from successional trees species such as *Cecropia* and *Trema* also became more common, as some vegetation recovery took place following the initial phase of development. The trends derived from such sparse data may be – and indeed probably are – more local than regional in significance, and for further information the historical record needs to be consulted. This varies from island to island, and from period to period, but overall it is quite good. For the early Hispanic period, most of our information concerning both aboriginal conditions, and the contemporary environment and natural history still comes from three primary sources, the works of Peter Martyr, Bartoloméo de Las Casas and Oviedo y Valdes (Martyr 1493–1525; Las Casas 1520–61, 1542; Oviedo y Valdes, 1526, 1532–46). Of these, the works of Las Casas are particularly useful in that he witnessed first hand in both Española and Cuba events at the time, while those of Martyr and Oviedo y Valdes are retrospective, although nonetheless, important. Martyr provided in his *Decades* a good picture of aboriginal life and environment, largely by interviewing those who had been to the New World; and Oviedo y Valdes also interviewed large numbers of old settlers, and described the plants and animals used by aboriginal peoples, in his description of native cultures entitled *Historia General y Natural de las Indias*, which was not begun (1532) until long after the last of the Indians in the region had died (see below). For records of the later French and English settlement in the region, large numbers of documents and several important contemporary books, are available in repositories such as the British Museum, the Public Records Office of London, the Bibliothèque Nationale in Paris, and a host of both public and private libraries in both Europe and the island Caribbean (see Watts 1966 for a more detailed discussion of this). Among the books are those by Edwards (1793), Hall (1764), Hughes (1750), de Laët (1640), Ligon (1657), de Rochefort (1658), Schomburgk (1848) and Sloane (1707). Many herbarium records, and details of biological changes, also were documented, in particular between 1650 and 1750, when botanic gardens in Europe were expanding rapidly, seeking new plant species; and many of these are especially useful in that they were compiled at a time when sugar estates were rapidly replacing native plant communities in many islands. As in all such instances, a rare combination of skills in both taxonomy and the interpretation of historical documents is required to make best use of this type of information, and the island Caribbean is fortunate in having attracted many people interested and competent in such matters, not least in recent years Richard Howard (Howard 1987–8), the now-retired director of the Arnold Arboretum in Cambridge, Massachusetts; and J. D. Maycock in the nineteenth century (Maycock 1830), who carefully recorded and compared the Linnaen and pre-Linnaen taxonomies of many West Indian plants. Methodologically, such taxonomics are exceptionally important, for they allow comparisons with earlier plant data sets, which might not otherwise be possible; looked at over time, these then allow both a comparison with present-day floras (albeit an incomplete one in most instances), as well as a chronological construction of the dates of arrival of many plant species in particular islands, and also some guide as to what was happening to the native flora at the time of initial ecosystem shock. This methodology has been extensively used in this paper.

Added to the evidence from historical documentation, there is now a growing quantity of relevant archaeological material, arising from a renewed interest in the prehistory of this region (see, for example, Drewett et al. 1991); and ecological evidence from long-standing remnant patches of lowland vegetation, at least some of which date back to the pre-Columbian forest (Watts 1978).

## ABORIGINAL BIOENVIRONMENTS

What may be deduced from this array of evidence, concerning the detailed ecological responses to such shocks in the region? First, it should be noted that there is much that is still not known. Even the size of the aboriginal population of the island Caribbean is uncertain, though the most accurate current estimates put it at between 2 and 4 million (Denevan 1992). Most of these appear to have lived in harmony with their environment since at least AD 500, supporting themselves in multi-family houses, under tribal structures headed by chieftains (*caziques*), by means of shifting agriculture termed *conuco*, which was utilized on all lands except the very wet, and which produced a multiplicity of carbohydrate-rich foods such as cassava, sweet potato, xanthosoma, calathea and American yams, along with separate plots for protein-rich peanuts, beans and maize, peppers and the calabash gourd (Sauer 1966). These were supplemented by animal protein sources, including many ground-nesting and tree-top birds, hutias and iguanas on land, and a rich inshore mammal and fish fauna, especially the green turtle, manatee, mullet, shell fish and crab.

Occasionally, population pressure was sufficiently high that some environmental modification took place, most conspicuously on a local scale. Thus in Española, early in the 1500s, both Las Casas and Oviedo y Valdes reported single Indian fields, each containing tens of thousands of *montones*, these being mounds 3–4 m in circumference and *c.*1 m in height, on which manioc (cassava) and sweet potato were grown. There is some slight evidence that the clearance of forest necessary to set up such fields was great enough, along with the cultivation intensity, to have caused soil erosion, although the extent of this is unclear (Denevan 1992). Soil erosion in the pre-Hispanic period has also been recorded by Drewett et al. (1991) along coastlands in Barbados, where population density also was high. The prevalence of aboriginal soil erosion patterns on a broader scale in the island Caribbean, however, is simply not known, and much more work needs to be directed towards this before conclusions may be drawn as to the extent of this form of environmental impact by aboriginal peoples in the region.

## THE EFFECTS OF EARLY HISPANIC SETTLEMENT

The first ecological shock to this long-standing system of food production in native forest lands came with Hispanic settlement, commencing in 1493. Small numerically though this was (less than 10 000 settlers at a maximum during the fifteenth and sixteenth centuries), it was extremely significant in that it resulted in the rapid decimation of native peoples in the region, along with the demise of native agricultural systems, and at the same time began the process of introduction of Old World plants and animals into the New World. All but the smallest groups of native peoples – and *all* of them in the Greater Antilles and the Bahamas – had died through exposure to Old World diseases to which they had no natural resistance, or by other means, by 1524 (Watts 1987). As this happened, *conucos* and other agricultural lands were abandoned or destroyed, to be taken over very quickly by the native forest. The process of rapid tropical-forest advance in post-Columbian America, following the population decline of native peoples, is now understood to have been very widespread, having been documented from the Peten in Guatemala within

Mayan territory (Murdy 1990) to the isthmus in Panama, where some of the densest tropical forests now cover former open land at the time of the Conquest (Sauer 1966, Parsons 1975), and also the Amazon basin (Butzer 1992): similar processes also have been documented for the northeastern U. S. A. (Williams 1989). In the island Caribbean, forest reclamation of former aboriginal planted land was extraordinarily successful, so that in many areas little obvious evidence was left of the activities of the former inhabitants. In Barbados, for example, only 100 years after the removal of the last native Indians by the Spanish, the first English settlers could find no trace of them or of their former agricultural systems – thick, all-pervasive forest then growing down to sea level (Ligon 1657), having covered all the former settlement sites – and the same was true in St Kitts, Montserrat, Nevis and Antigua (Raynal 1788), Martinique and Guadeloupe (de Rochefort 1658, Taylor 1688).

As for the plants and animals introduced by the Spanish into the region (including many Mediterranean weeds), most of the plant species were unsuccessful in establishing themselves in their new environments at this time, not so much because they were unsuited to it (many succeeded later on, when the forest was removed again by later European settlers – see below, p. 275), but as a result of direct competition from species within the advancing native forest, or due to a linked problem, the lack of available space within it. Only the common purslane, a sea-coast weed, and possibly some *Citrus* species, were noted by contemporaries (Las Casas 1520–1561, Acosta 1894) as being distributed widely in Española, at least for a while. In contrast, many of the introduced animals multiplied rapidly, roaming, foraging and breeding in the wild in habitats in which they had no natural predators, and which previously had not known any large hoofed creatures. Pigs from the Spanish meseta were dramatically successful, being left as a potential food resource on virtually every island the Spanish came into contact with, and they adapted superbly to the environmentally very different island Caribbean forest lands; over 30 000, mainly wild, hogs were in Cuba by 1514 (Oviedo y Valdes 1532–46). Cattle also succeeded to a lesser extent, herds of 8000 beasts having been noted in Española by the 1520s (Crosby 1972). The hooves of all these alien domesticates eventually helped to compact soil horizons along their trackways, leading to the channelling of water downslopes in some areas, to create the first recorded gully-erosion episodes in the region (*barrancas* and *arroyos* in Española) before the end of the sixteenth century. In grubbing up root crops for food, the animals also assisted in the destruction of native *conucos*, so accelerating both the population decline of native peoples, and the ensuing forest advance.

## SUGAR-CANE AGRICULTURE AS AN AGENT OF MAJOR BIOGEOGRAPHICAL CHANGE

If the first major ecological shock in the region centred around the demise of native peoples and the introduction of Old World animal domesticates, the second – which both directly and indirectly created the conditions for much more massive and long-term ecological and biogeographical change (following earlier less successful attempts at growing tobacco, cotton, indigo and ginger) –

was the adoption of sugar-cane plantation agriculture, backed by a slave labour force. This was set up by the English and French, initially in two small islands of the eastern Caribbean, Barbados and St Kitts, from the 1640s, and thereafter in time spread both to the Greater Antilles and virtually all the other small islands in the region. Sugar cane, a C-4 tropical grass cultigen, grows well throughout all but the driest parts of the lowland Caribbean, but it dislikes all vegetative competition for solar energy, water and nutrient resources. This means that widespread clear-felling of forest land always was a prerequisite for successful cultivation of this crop, followed by continuous clean-weeding. Since lowland seasonal-forest soils provided the best soil-water and soil-nutrient conditions for cane growth, it was this community which suffered most in the land-clearance processes; and time after time the same process repeated itself in different islands, as planters extended their activities into new lands in the search for substantial economic gain from this one species. In Barbados, cane had almost totally replaced seasonal forest by 1665 (Watts 1966), in St Kitts, Nevis and Montserrat by the 1680s (Sloane 1707), Antigua, Guadeloupe, Martinique and much of lowland Jamaica by the 1750s (Harris 1965, Lasserre 1961, Kimber 1988, Stewart 1808) and later in other territories such as St Dominique, Cuba, Puerto Rico, Trinidad and the Dominican Republic. The story is by now well known, but the ecological consequences bear repeating. We simply do not know how many plant and animal species were lost during clearance, but the number is likely to have been substantial, possibly involving extinctions in the hundreds or thousands, and especially in the small islands such as Barbados, St Kitts, Nevis and Montserrat in which clearance of timber was close to 100 per cent. One tiny clue comes from Barbados: there, writers of the 1650s (Von Ucheritz 1652, Ligon 1657) commented on the marked absence of birdsong, as compared to adjacent islands, and indeed we now know that, other than a few humming birds, all the forest-based birds, including song birds, had by then gone, along with the forest that had supported them.

However, the ecological shock which arose directly from the replacement of lowland forest with cane agriculture did not stop with clearance; further substantial instability arose from the nature of cane cultivation itself, and the relationship of the crop to its main markets. It has been suggested (e.g. Pares 1960) that sugar cane is an easy crop to grow. West Indian experience suggests that this is far from being the case, and indeed that it is very vulnerable to changes in the condition of the physical environment, and also of course to market conditions. Barbados, the first island devoted solely to the raising of sugar cane and, due to space limitations, the main example in this chapter, is a good case in point. By 1665, 80 per cent of all land in this island was under this one crop, and it was giving rise to high prices (with a return of 50 per cent on investment; Ward 1978) in European markets. Subsequently, the amount of land set under cane then started to vary substantially, according to market conditions, or developing environmental or biological problems. Thus, from 1680 to 1700, declining yields created by diminishing soil quality resulted in one-third of the plantation land being temporarily taken out of cultivation and being allowed to revert to fallow; and when this problem had

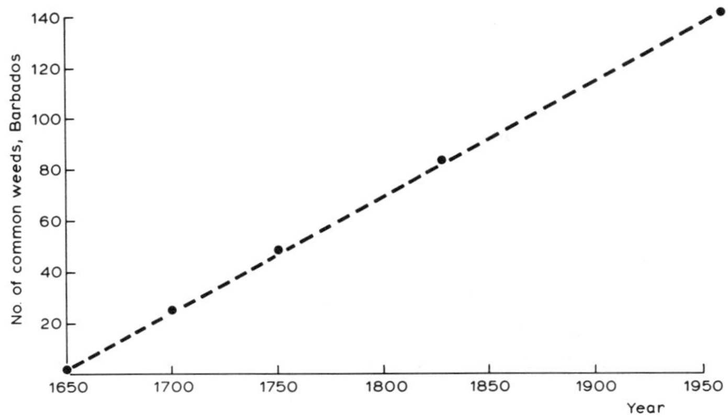


**Figure 14.2** The increase in the number of common weeds in Barbados, 1650–1950.

been resolved through the adoption of new, conservative planting and manuring methods, a market glut during the 1720s and 1730s caused a repetition of fallowing on the same scale (Watts 1987). In contrast, the years between 1740 and 1760 were extremely profitable to planters, giving rise to planted acreages of cane of up to 80 per cent of the whole territory once more. Then, for 20 years after 1760, many settlers were forced to return to fallow, their crops having persistently been devastated by the sugar ant, which ate cane stems from their base, and which was accidentally introduced into the island from Guyana in that year. This pest was only eradicated by the severe hurricane of 1780, which also of itself flattened most of the cane crop. From 1790 to 1800, sugar acreages rose again to maximum levels, especially after the introduction of new, high-yielding, Pacific-island strains of cane (especially Otaheite and Bourbon) in the 1790s. Similar boom–bust cycles were maintained throughout the nineteenth and early twentieth centuries for a variety of similar reasons, and until a few years ago the amount of land under cane stood at 80 per cent once again: in contrast, it is now substantially in decline. Cycles such as these also prevailed in Antigua, except that cane cultivation there did not begin in earnest until 1680, and the crop did not first reach its maximum areal extent until 1720. Furthermore, it may be expected that they are present in other Caribbean islands as well, although detailed evidence for them thus far is lacking.

Such alternative periods of economic boom and bust in small islands appear to have extended and intensified the period of ecological instability initiated by forest clearance, with interesting ecological and biogeographical results. First, after forest decimation, the island became for a while a virtual biological wasteland, dominated as it was by one introduced cultigen, the sugar cane. Then weed species, including both ground plants and shrubs, moved in to colonize any uncultivated ground. But this was not an immediately effective process: it took a few years to get under way, at least on anything other than a very minor scale. Thus Richard Ligon (1657), in his detailed description of Barbados during the late 1640s, at the very beginning of cane cultivation,

Table 14.1 Common weeds in Barbados

| *Habitat* | *Total* | *Origin of species* | | | | | | |
|---|---|---|---|---|---|---|---|---|
| | | *End.* | *WI* | *SA* | *NA* | *OW* | *Asia* | *Afr.* |
| Dry ground | 81 | – | 67 | 1 | 1 | 4 | 6 | 2 |
| Gullies (wetter) | 42 | 1 | 36 | 1 | – | 1 | 2 | 1 |
| Paths and roads | 6 | – | 5 | – | 1 | – | – | – |
| Canefields | 7 | – | 4 | – | – | 2 | 1 | – |
| Windward side | 4 | – | 5 | – | – | – | – | – |
| Total | 140 | 1 | 116 | 2 | 2 | 7 | 9 | 3 |

*Sources*: Gooding et al. 1965; author's own field research.
End., Endemic; WI, West Indies; SA, South America; NA, North America; OW, Old World tropics; Afr., Africa.

noted only four widely distributed weeds, these being: the native shrub guava (*Psidium guajava*); two shrubs from elsewhere in the region, the rodwood (*Eugenia monticola*) and the physic nut (*Jatropha curcas*); and, as in Hispanic territories earlier, the European ground species purslane. However, by the end of the seventeenth century, many more were present, and their numbers have since continued to increase substantially (figure 14.2).

The present *Flora of Barbados* (Gooding et al. 1965) indicates that of the 529 non-cultivated plants currently listed, seven are endemic, 333 are likely to have arrived from elsewhere in the Caribbean, and 137 have come from other parts of the world; 52 additional introduced species have no defined regional source affinities. In other words, 88.7 per cent of the existing wild plant species come from outside the island, a remarkable proportion in any terms. Of those introduced species not originating from the Caribbean (*c*.25 per cent of the total wild flora: a similar figure is also true for Antigua), most (43) come from Asia, the American mainland tropics (29), and Africa and Oceania combined (29). It is also interesting to note that, among weedy species, many have remained common from their arrival to the present day. Currently, some 140 species may be described as common weeds, and their distribution, along with their probable areas of origin, is given in table 14.1. The figures are noteworthy for re-emphasizing how few of the many plant arrivals from beyond the Caribbean region have become common weeds; most of the latter have established themselves from adjacent islands. This perhaps reflects the greater opportunity of repetitive and successful invasions by plants from nearby source areas as compared to elsewhere, a principle which was first put forward, *a priori*, by MacArthur and Wilson (1967) in respect of all oceanic islands, and which appears to be supported in this case. The one endemic is *Phyllanthus andersonii*, a gully shrub.

A temporal pattern of arrival of common weeds into the island (table 14.2, figure 14.2) displays a steady rate of establishment over 315 years, of slightly less than one species every two years. At first sight, and despite similar straight-line graphs for weed species introductions having recently also been produced for south-east Australia (Bulmer, personal communication 1993), this seems ecologically peculiar, and needs some explanation. For such a rate to

**Table 14.2** The cumulative arrival of weeds in Barbados over time

| | *1650* | *1700* | *1750* | *1830* | *1965* |
|---|---|---|---|---|---|
| All | | 61 | – | – | 328 |
| Common weeds | 4 | 25 | 48 | 83 | 140 |

be maintained, not only must successful colonization be effected, but opportunities for species, once colonized, to disseminate widely into a range of suitable habitats must also be presented if an immigrant is then to become 'common'. In the case of Barbados, it is suggested that these opportunities were created indirectly by the successive boom–bust phases of the sugar-cane economy, and that the required suitable habitats were derived from the extension of fallow land in periods of production downturns.

To take this argument further, a set of introduced species might be expected, through rapid seeding, to spread relatively easily on to fallow land, only to have their spatial distributions restricted again during the ensuing production upturn, when cane would replace the fallow. One might postulate that, as in the case of many oceanic island species (Williamson 1981), such a relative restriction (relative only in the sense that the species remained common, though not as widely dispersed as formerly) might lead to some loss of dispersability among them. Accordingly, during the next 'bust' phase, a set of new weed species arrivals would be the superior competitors for fallow space, and so become widely established themselves, in time giving rise to a repetition of the whole sequence. In terms of population dynamics, each new set of species would undergo rapid growth and expansion, followed by contraction, in a highly volatile way. Overall, the steady rate of arrival, which apparently has been the case for common Barbadian weeds, does not conform in any way to the general ideas respecting the *rate* of colonization of oceanic islands put forward by MacArthur and Wilson (1967), all of which presupposes a reduction in the immigration rate over time; and these ideas accordingly need to be reviewed, certainly for intensely cultivated islands. Nor are there any signs that, under the prevailing circumstances, species population equilibrium is close to being attained in Barbados: for this to occur, an increasing turnover of species would need to be recorded, and there are no serious indications of this. The absence of any major turnover pattern in the region has been occasionally noted before in respect of other organizms, as in the case of reef fish in the western Caribbean (Smith 1979).

Under such volatility, however, one might expect that there should be some slight turnover of species, and in the case of a few formerly common weeds, this does indeed seem to be the case. From the 1680–1700 collections, four of the species then identified as common have largely or completely disappeared: two shrubs, the hopshrub (*Dodonea viscosa*) and the bastard senna (*Cassia emarginata*), the first of which is likely to have been indigenous, were last recorded respectively in 1848 and 1830 (Maycock 1830, Schomburgk 1848); the herb spiritweed (*Aegiphila alata*) also has not been noted since 1830, and the Spanish gooseberry (*Pereskia aculeata*) is now found only in cultivation. The common physic nut (*Jatropha curcas*), named by Ligon in 1657, is not common now; and neither is the dartwood (*Anisomeris fasciculata*), which had disseminated itself widely by 1750 (Hughes 1750). Well over 20 weeds have disappeared since J. D. Maycock's time (1830). To replace these, others have continued to arrive. Thus in 1960, *Trichodesma zeylanicum*, an Asian herb, was accidentally introduced, subsequently to spread widely over the northern part of the island; and even more recently, *Cyanotis cucullata* has also arrived from the Old World and spread, in the process being recorded for the first time in the Western Hemisphere.

## CONCLUSIONS

I have not had space in this chapter to discuss in detail some of the additional, clear signs of ecological shock which have developed since, and because of the introduction of intensive, sugar-cane plantation agriculture, such as the accelerated rate of faunal extinctions arising both from prolonged habitat loss, and the sometimes planned introduction of new faunal predators. To take one example only of the latter situation, the mongoose, introduced by planters in 1872 to rid canefields of snakes, also caused the extinction of other benign creatures including the rice rats of Jamaica and St Vincent, the Jamaican wood rail, the Martinique wren, and the bank-nesting burrowing owls of Antigua and Guadeloupe (Day 1981). With regard to the flora, one may say that the current persistently disturbed environmental and biogeographical conditions induced by such ecological shock in Barbados, and probably all other present and former sugar-producing islands as well, have led to a present-day wild flora in lowland districts of largely introduced species, most of which have been derived opportunistically from the Caribbean rather than further afield. At least for Barbados, among these are a number of common weeds, which have arrived at a remarkably steady rate over time, and apparently continue to do so. It is argued that the ecological success of these weeds reflects, in large measure, the past and continuing boom–bust phases of the sugar-cane economy, which have provided the sequential extensions and contractions of fallow land on which such species could colonize, and then become established. Both the temporal patterns of arrival of common weed species, and the absence of any signs of approaching species equilibrium, do not conform to the model for oceanic islands put forward by MacArthur and Wilson (1967). It is likely that these patterns, linked as they are into a two-tier pattern of ecological shock, are replicated in all other present and former sugar-producing islands in the region, but a good deal of further investigation is required before this can be absolutely confirmed.

## REFERENCES

Acosta, J. de (1894) *Historia Natural y Moral de las Indias* (R. Inglés, Madrid)

Butzer, K. W. (1992) 'The Americas before and after 1492: an introduction to current geographical research', *Annals, Association of American Geographers* 82: 345–68

Crosby, A. W. Jr. (1972) *The Columbian Exchange: Biological and Cultural Consequences of 1492* (Greenwood Press, Westpoint, Connecticut)

Day, D. (1981) *The Doomsday Book of Animals* (Ebury Press, London)

Denevan, V. M. (ed.) (1992) *The Native Population of the Americas in 1492*, 2nd edn, (University of Wisconsin Press, Madison, Wisconsin)

Drewett, P. L., Harris, M. H. and Wing, E. (1991) *Prehistoric Barbados* (Archetype, Denbigh)

Edwards, B. (1793) *The History, Civil and Commercial, of the British Colonies in the West Indies* (London)

Gooding, E. G. B., Loveless, A. R. and Proctor, G. R. (1965) *Flora of Barbados* (HMSO, London)

Hall, R. (1764) *Acts Passed in the Island of Barbados* (London)

Harris, D. R. (1965) 'Plants, animals and man in the outer Leeward Islands, West Indies', *University of California Publications in Geography* 18: 1–184

Higuera-Gundy, A. (1989) 'Recent vegetation changes in Southern Haiti', (in Woods, C. A. (ed.) *Biogeography of the West Indies, Past, Present and Future,* (University of Florida Press, Gainesville) pp. 191–200

Hodell, D. A., Curtis, J. H., Jones, G. A., Higuera-Gundy, A., Brenner, M., Binford, M. W. and Dorsey, K. T. (1991) 'Reconstruction of Caribbean climate change over the past 10,500 years', *Nature* 352: 790–2

Howard, R. A. (1987–8) *Flora of the Lesser Antilles*, 6 vols (Arnold Arboretum, Harvard University, Boston)

Hughes, G. (1750) *The Natural History of the Islands of Barbados* (London)

Kimber, C. T. (1988) *Martinique Revisited: the Changing Plant Geographies of a West Indian Island* (Texas A. & M. University Press, College Station, Texas)

de Laët, J. (1640) *L'histoire du Nouveau Monde, on Description des Indes Occidentales* (Leyden)

Las Casas, B. de (1520–61) *Historia de las Indias* (Madrid)

Las Casas, B. de (1542) *Brevissima Relacion de la Destruccion de Las Indias* (Madrid)

Lasserre, G. (1961) *La Guadeloupe* (University of Bordeaux Press, Bordeaux)

Ligon, R. (1657) *A True and Exact History of the Island of Barbados* (London)

MacArthur, R. H. and Wilson, E. O. (1967) *The Theory of Island Biogeography* (Princeton University Press, Princeton)

Martyr, P. (1493–1525) *De Orbe Nove* (Madrid)

Maycock, J. D. (1830) *Flora Barbadensis* (London)

Murdy, C. N. (1990) 'Prehispanic agriculture and its effects in the valley of Guatemala', *Forest and Conservation History* 34: 179–90

Oviedo y Valdes, G. F. de (1526) *Sumario de la Natural Historia de las Indias* (Madrid)

Oviedo y Valdes, G. F. de (1532–46) *Historia General y Natural de las Indias* (Madrid)

Pares, R. (1960) Merchants and planters. *Economic History Review, Supplement* 4: 1–91

Parsons, J. J. (1975) 'The changing nature of New World tropical forests since European colonisation', in *The Use of Ecological Guidelines for Development in the American Humid Tropics*, (IUCN, Morges) pp. 28–38

Raynal, A. (1788) *A Philosophical and Political History of ... the East and West Indies* (O. Justamond, London)

de Rochefort, C. (1658) *Histoire Naturelle et morale des Isles Antilles de l'Amérique* (Paris)

Sauer, C. O. (1966) *The Early Spanish Main* (University of California Press, Berkeley)

Schomburgk, R. H. (1848) *The History of Barbados* (London)

Sloane, H. (1707) *A Voyage to the Islands of Madeira, Barbados, Nieves, St. Christophers and Jamaica* (London)

Smith, G. B. (1979) 'Relationship of eastern Gulf of Mexico reef-fish communities to the species-equilibrium theory of insular biogeography', *Journal Biogeography* 6: 49–61

Stewart, J. (1808) *An Account of Jamaica and its Inhabitants* (London)

Taylor, J. (1688) *Multum in Parvo, or Taylor's Historie of his Life and Travells in America and Other Parts* (Jamaica)

Von Uchteritz, H. (1652) *A German Indentured Servant in Barbados* (trans. A Gunkel and J. S. Handler) (Barbados Museum, Barbados)

Ward, J. R. (1978) 'The profitability of sugar planting in the British West Indies' *Economic History Review*, 2nd Series 31, 197–213

Watts, D. (1966) 'Man's influence on the vegetation of Barbados, 1627–1800', *University of Hull Occasional Papers in Geography* 4: 1–96

Watts, D. (1978) 'Biogeographical variations in the mahogany (*Swietenia mahogani*) woodlots of Barbados', *Journal of Biogeography* 347–63

Watts, D. (1987) *The West Indies: Patterns of Development, Culture and Environmental Change Since 1492* (Cambridge University Press, Cambridge)

Williams, M. (1989) *Americans and their Forests: a Historical Geography* (Cambridge University Press, Cambridge)

Williams, M. (1981) *Island Populations* (Oxford University Press, Oxford)

15

# Human Impact on the Wetlands of Belize, Central America

Peter A. Furley, David M. Munro, Janice P. Darch and Roland R. Randall

ABSTRACT

The ancient Maya civilization which dominated much of southern Mexico, Guatemala and Belize until the arrival of the Europeans in the sixteenth century, developed a form of intensive wetland agriculture that involved the creation of a drained field system and an intensive trading network traversing the swampy coastal zones. Soil, pollen and contemporary vegetation studies in the swamps and river floodplains reveal how the fields were constructed and what types of crops were grown, and provide the basis of estimates of population carrying capacity. Following the collapse of the Maya culture the wetland management fell into disuse. Around 300 years later, British loggers extracted a handful of valuable tree species from hardwood forest, and attention was once again directed towards the agricultural development of the coastal and interior wetlands. During the 1920s, a major scheme was devised to open up and drain the Sibun and Manatee areas to the south of Belize City. Maps and manuscripts of the time indicate how this ambitious scheme centred on the creation of a network of canals linking rivers and lagoons. More recently, since the early 1980s, increased pressure on the coast and barrier reef has prompted studies to gauge the nature, extent and resources of mangrove and associated wetlands. Despite current research contributing to a coastal management plan, the handling of the Belizean wetlands has so far never matched the sophistication and sustained resource use of the indigenous Maya.

## INTRODUCTION

The atlantic wetlands of Central America have at times been part of a vigorous arena of human activity and, at other times, have been by-passed entirely during the historical evolution of the region. Nowhere is this more evident than in Belize, a country which occupies around 22 000 km$^2$ and which has witnessed at least three major and quite different periods of occupation and settlement.

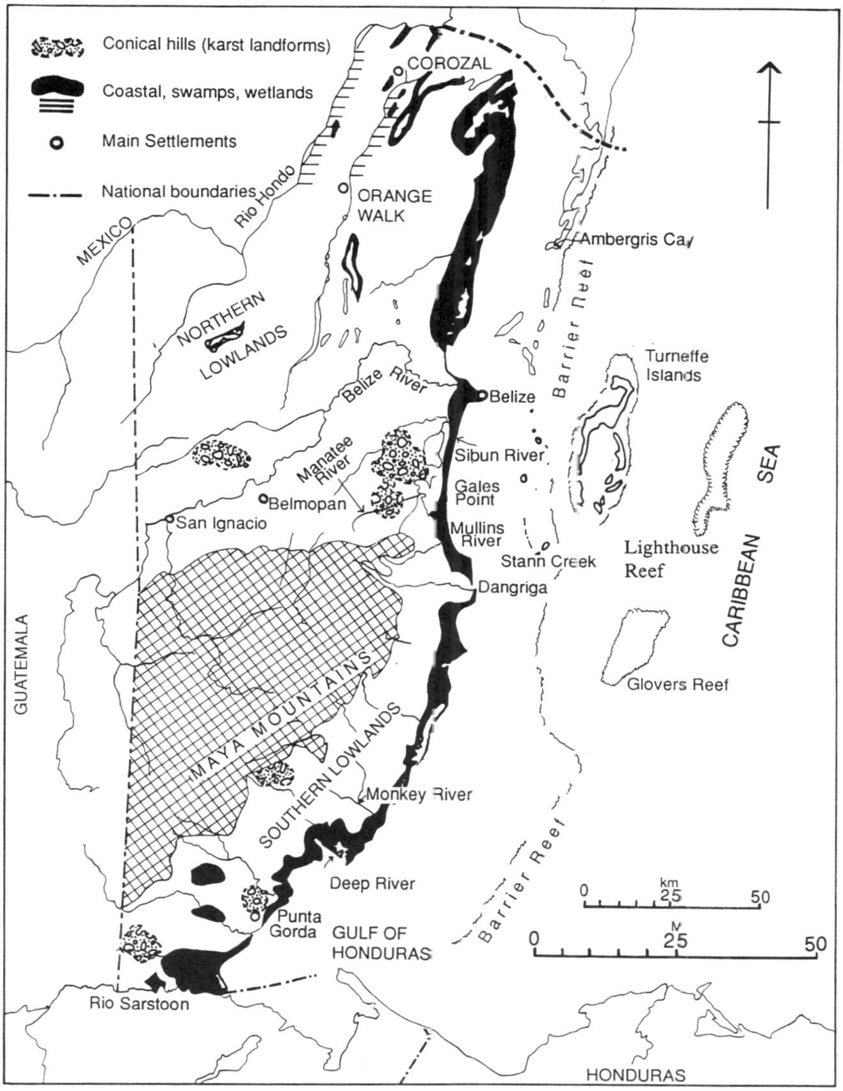


Figure 15.1 Natural regions and the distribution of wetlands in Belize.

From the earliest records several thousand years ago, the original inhabitants, the Maya, progressively expanded and at their peak, in the ninth and tenth centuries, settled virtually the whole of what is now Belize. During this period, their utilization and management of the wetlands reached high levels of sophistication unrivalled by subsequent events. After the collapse of the Maya, the country became curiously depopulated and few Indians remained to confront the second wave of settlers, this time from Europe. The Spanish hardly penetrated into Belize despite their prodigious feats of conquest throughout Central and South America. Both Dutch and British privateers and later

British logwood cutters skirted the coast and penetrated a few of the rivers, but the wetlands, particularly the swampy coastline, inhibited settlement and served more as a refuge for piracy. Gradually more permanent settlement increased, though still essentially coastal, even at the height of the British colonial period in the late nineteenth to early twentieth centuries. The inland swamps and wetland forests were hardly explored and the only occupation consisted of scattered coastal villages which slowly grew into small towns – with the exception of Belize City which became the capital and by far the most populous settlement (figure 15.1). Surrounded by mangroves or wet grassy swamplands and scrub woodland, these settlements remained static and the first significant road inland was only built in the mid-1950s. A third phase of occupation and settlement may be distinguished from about this period, when the first self-government was introduced in 1964, the new capital – Belmopan – being built in 1970, away from the hurricanes and coastal flooding, and when the country became fully independent in 1981. In the last few decades, the true nature of the archaeological richness of Belize has become apparent. There is now a revived interest in the wetlands, but one which is attempting to recognize the environmental and biological as well as the historical, economic and conservationist value of this under-valued resource.

The aim of this chapter is to analyse the ways in which the wetlands have been utilized or managed over the course of these three major periods of historical development.

## USE AND MANAGEMENT OF THE WETLANDS UNDER THE MAYA

The ancient Maya civilization flourished in the lowlands of Central America for over a thousand years, with evidence from palynological records of deforestation and early cultivation going back 4500 years BP (Hammond 1986). The Classic Maya civilization began to collapse from the latter part of the eighth century AD (Culbert 1973), and a century later the majority of cities in the southern lowlands had been abandoned. The causes of the break up of what might be termed a loosely knit federation of city–states, have been a source of enquiry and speculation since the rediscovery of the Mayan ceremonial sites by Stephens and Catherwood (1841). Environmental degradation and ecological change provide one of the substantive hypotheses (Turner and Harrison 1978).

Until around two decades ago, it was commonly believed that the wetlands of Belize had remained an obstacle to occupation and development, as much to the Maya as to present-day inhabitants. It is now known from a number of archaeological sources that this view is incorrect, and that many of the wetlands were intensively cultivated or utilized for a variety of purposes and criss-crossed by routes. Consequently, the contemporary wetland landscapes of Belize bear little relation to their intensive use during the peak of Mayan activity, or to the types of vegetation that preceded human impact.

The Maya impressed their personality on the landscape in no uncertain manner. On the drier land, slash and burn agriculture (referred to as *milpa* farming throughout the region) is believed to have progressively consumed the forest barriers between rival cities (Hammond 1978). It is extremely unlikely that such a land-use system could have provided food surpluses capable of supplying the estimated densities of population. The human carrying capacity of the adjacent Guatemalan Petén, based on a *milpa* system, is estimated by Cowgill and Hutchinson (1963) to be between 40 and 70 people/km$^2$ – figures supported by a number of other authors. Yet estimates for the peak Classic period population exceed 200 people/km$^2$, with local densities of over 1000 (Turner 1976, Santley et al. 1986, Rice and Culbert 1990). Belize may conceivably have had a Mayan population of a million or more, compared with a total population of around 200 000 today, and it is argued that this density of settlement could only have been achieved by intensive utilization of favourable areas. These include the hydrologically buffered, shallow wetlands and karstic riverine plains (White 1988) as well as the combined systems of terraced hillslopes and wet depressions (Turner 1983).

### *Utilization of the coastal lowlands*

The littoral zone, including lagoons, semi-permanent wetlands and wet savannas together with the offshore cays and barrier reef, played an important part in Mayan history, although it has only been appreciated and investigated in any detail recently (Vail 1988). The coastal wetlands were by and large unsuited to agriculture, but were criss-crossed by routes running inland from 'ports' (Andrews 1987), and were used for salt-making (both by evaporation and boiling), trade in minerals such as obsidian, jade and chert, lime manufacture, fishing and shellfish exploitation, especially conch, and for the hunting of larger marine animals such as manatee and turtle, especially from Moho Cay (Graham 1983). Many localities were also residential as illustrated for example, in Guderjan's (1988) work on Ambergris Cay.

Coastal settlements flourished, mostly in the northern and central areas of what is now Belize (figure 15.2) and were linked by well developed trade routes. The settlements appear to have been relatively small and served inland centres such as Colha, Cuello, Nohmul and Altun Ha, with access across the coastal wetland mosaic of vegetation formations (Furley and Ratter 1992). Lack of inland access has been quoted as a possible reason for the success of a site such as Cerros, close to the present day town of Corozal, as against Sarteneja or Shipstern. Nevertheless items from as far away as Teotihuacan in central Mexico indicate a long-distance as well as local trade. Altun Ha and Santa Rita both possess artefacts indicating long-distance trade. The presence of very large Mayan cities inland at Caracol in the western Maya Mountains of Belize, or at Tikal in the northern Guatemalan Petén province, suggest that the pattern of commercial exchange must have overcome the obstacles of access through the littoral and inland wetlands.

### *The drained field systems*

On the seasonally and semi-permanently wet lands, such as the shallow karstic

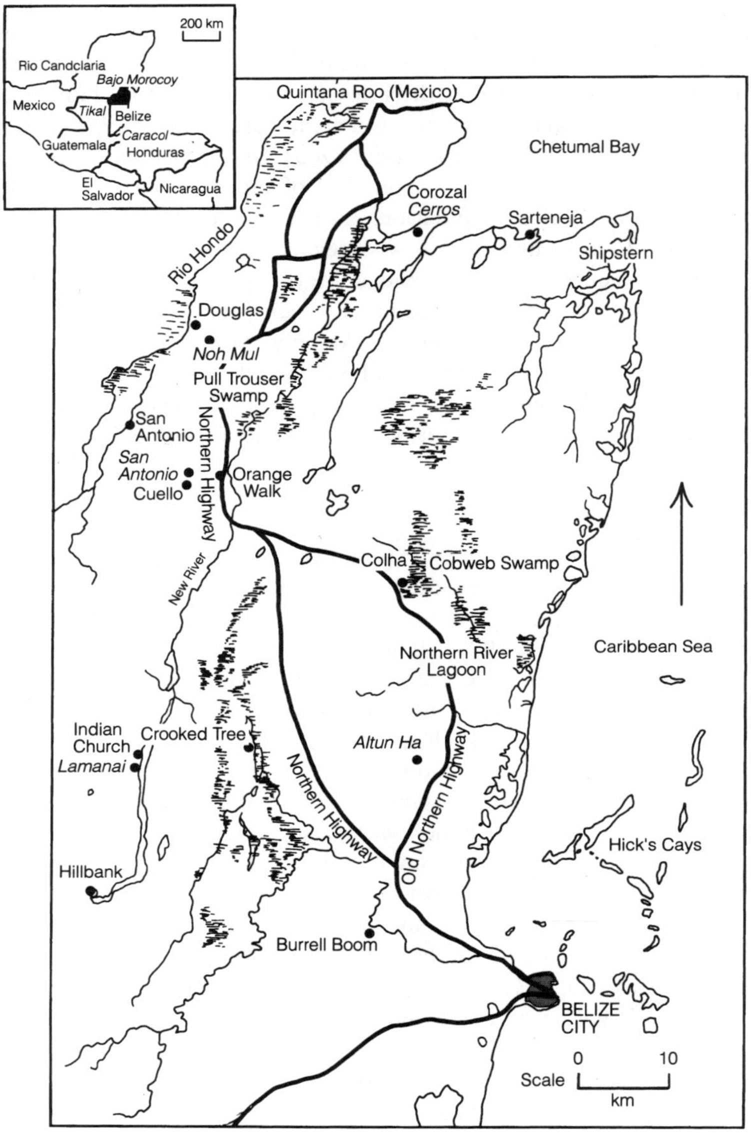


Figure 15.2 Mayan sites in northern Belize.

depressions (*bajos*) and on the non-saline swampy lowlands, drainage ditches were excavated and the silty organic fill spread as a mulch across adjacent fields forming raised beds (*chinampas*) and 'floating gardens'. The process is well described in Gliessman (1990) and Darch (1983) and dates from the late Pre-Classic (around AD 150), although there is a good deal of argument as to the extent and sophistication of the system. In some cases in Belize, these field areas were also elevated above the water table by the accumulation of erodate, or deposits derived from run-off and mass movement from nearby higher land (as at Cobweb Swamp; Jacob 1992), or by flood deposits (as at San Antonio, Olson 1977). Each site seems to have established individual responses to the particular hydrological patterns and environmental peculiarities of the area. Therefore, the picture painted by Shafer and Hester (1983) or Jacob (1992) at Cobweb Swamp differs in detail from those at Pulltrouser Swamp (Turner and Harrison 1983), Albion Island (Antoine et al. 1982, Pohl 1990), Nohmul (Pyburn 1989) or at San Antonio – Rio Hondo (Bloom et al. 1983).

It is now recognized that the Maya developed several agricultural strategies, some of which were highly intensive and some dependent on the more extensive slash and burn system. The wetland fields, first observed from the rectilinear patterns on aerial photographs in Campeche, Mexico in the 1960s, were shown to be similar to the *chinampas* of Central Mexico, with shaped fields 5–15 m wide separated by canals. Later, radar was utilized to detect the distribution of canals and ditches throughout the Maya lowlands (Adams et al. 1981). Much of the wetland agriculture in Belize appears to have been in the form of discrete fields with dimensions of 5–10 m by 20–30 m (Jacob 1992), separated by a network of drainage channels varying from narrow ditches to broad canals and from rectilinear to sinuous in shape. As well as modifying the hydrology by channelling the wetlands, the Maya also raised the height of fields or parts of fields, and used mulching techniques to improve fertility. For example, waterlilies (Nympheaceae) and other marsh plants seem to have been harvested or carried with silty organic muds in the canals and spread over the fields (Wiseman 1983). Siemens (1983) estimated that raised fields in northern Belize covered around 3400 ha. They have been described around Cerros (Scarborough 1983), Nohmul (Hammond et al. 1987), Lamanai near Indian Church (Lambert et al. 1984) and Bajo Morocoy across the border in Mexico (Gliessman et al. 1983) as well as the fields at Pulltrouser Swamp and Albion Island mentioned earlier.

### *The Nohmul drained field site*

The fields at Nohmul have been described as being amongst the most reticulate in form of all the Belizean sites (Hammond 1986, Jacob 1992). The site has been written up by Pyburn (1989) but the example is discussed in more detail here because it formed the object of an investigation into the nature and management of raised fields from soil profile evidence (Darch and Randall 1989) and palynological records (Jackson and Paige 1989).

A survey was undertaken in a complex of drained fields, close to the ceremonial centre at Nohmul and the present-day Douglas Village. The swamp lies about a mile from the Rio Hondo, but is totally isolated from the river and receives drainage water from the nearby limestone uplands. Although used currently to graze cattle during the day, the area was, at one time, intensively cultivated (figure 15.3). There appear to be three types of drained field: the first consists of plots totally surrounded by water for most of the year and located close to the centre of the topographic depression (numbers 5, 7, 8, 9,

**Table 15.1** Results of soil analyses at Nohmul (Douglas Village)

| Profile no. | Sample no. | Organic C (%) | pH | Ex. Ca (cmol/kg) | Ex. Mg (cmol/kg) | Ex. K (cmol/kg) | Ex. Na (cmol/kg) | TEB (cmol/kg) | Ex. H (cmol/kg) |
|---|---|---|---|---|---|---|---|---|---|
| 1 | 1 | 13.9 | 7.4 | 108 | 17 | 0.18 | 0.88 | 127 | 0.28 |
| | 2 | 5.7 | 7.7 | 94 | 16 | 0.31 | 3.13 | 114 | 0.68 |
| | 3 | 4.4 | 7.5 | 92 | 24 | 0.21 | 4.71 | 122 | 0.70 |
| | 4 | 10.1 | 7.8 | 119 | 15 | 0.12 | 3.92 | 139 | 0.52 |
| | 5 | 6.0 | 7.9 | 120 | 2 | 0.14 | 1.90 | 125 | 0.48 |
| | 6 | na[a] | na | 138 | 7 | 0.33 | 2.35 | 148 | 0.52 |
| | 7 | 9.0 | 7.7 | 128 | 6 | 0.31 | 2.55 | 138 | 0.40 |
| 2 | 8 | 7.6 | 7.6 | 98 | 25 | 0.25 | 1.76 | 126 | 0.70 |
| | 9 | 3.5 | 7.8 | 79 | 19 | 0.33 | 1.57 | 102 | 1.40 |
| | 10 | 9.5 | 7.9 | 119 | 8 | 0.31 | 1.17 | 129 | 0.48 |
| | 11 | 4.6 | 7.8 | 123 | 6 | 0.44 | 1.08 | 131 | 0.40 |
| | 12 | na | na | 118 | 6 | 0.31 | 1.17 | 125 | 0.40 |
| 3 | 13 | 14.5 | 7.5 | 87 | 26 | 0.18 | 3.30 | 117 | 0.64 |
| | 14 | 4.6 | 7.7 | 87 | 26 | 0.31 | 3.14 | 117 | 0.80 |
| | 15 | 4.4 | 7.9 | 92 | 8 | 0.25 | 1.14 | 134 | 0.70 |
| | 16 | 1.9 | 7.9 | 121 | 4 | 0.25 | 1.37 | 128 | 0.60 |
| 4 | 17 | 29.2 | 7.3 | 85 | 36 | 0.33 | 6.28 | 130 | 2.20 |
| | 18 | 4.4 | 7.6 | 119 | 24 | 0.22 | 3.30 | 148 | 1.80 |
| | 19 | 23.5 | 7.7 | 120 | 9 | 0.25 | 1.17 | 131 | 0.62 |
| | 20 | 10.4 | 7.7 | 127 | 1 | 0.31 | 1.08 | 130 | 0.70 |
| | 21 | 1.1 | 7.8 | 127 | 3 | 0.44 | 0.64 | 132 | 0.90 |
| 5 | 22 | 14.7 | 7.4 | 90 | 18 | 0.31 | 1.96 | 113 | 1.84 |
| | 23 | 2.7 | 7.7 | 128 | 6 | 0.31 | 0.93 | 138 | 2.00 |
| | 24 | 4.1 | 7.7 | 112 | 1 | 0.36 | 0.72 | 114 | 0.90 |
| | 25 | 1.1 | 7.6 | 92 | 5 | 0.72 | 0.54 | 100 | 1.22 |
| 6 | 26 | 20.7 | 7.3 | 116 | 7 | 0.31 | 0.58 | 121 | 1.04 |
| | 27 | 2.7 | 7.5 | 100 | 24 | 0.25 | 0.60 | 127 | 1.60 |
| | 28 | 0.5 | 7.8 | 138 | 1 | 0.18 | 0.76 | 140 | 0.80 |
| | 29 | 1.1 | 7.8 | 144 | 1 | 0.35 | 0.60 | 147 | 0.75 |
| | 30 | 0.8 | 7.3 | 139 | 1 | 0.33 | 0.67 | 143 | 1.02 |
| | 31 | 0.3 | 7.3 | 141 | 5 | 0.31 | 0.98 | 148 | 0.96 |

[a] na = Too organic to analyse.

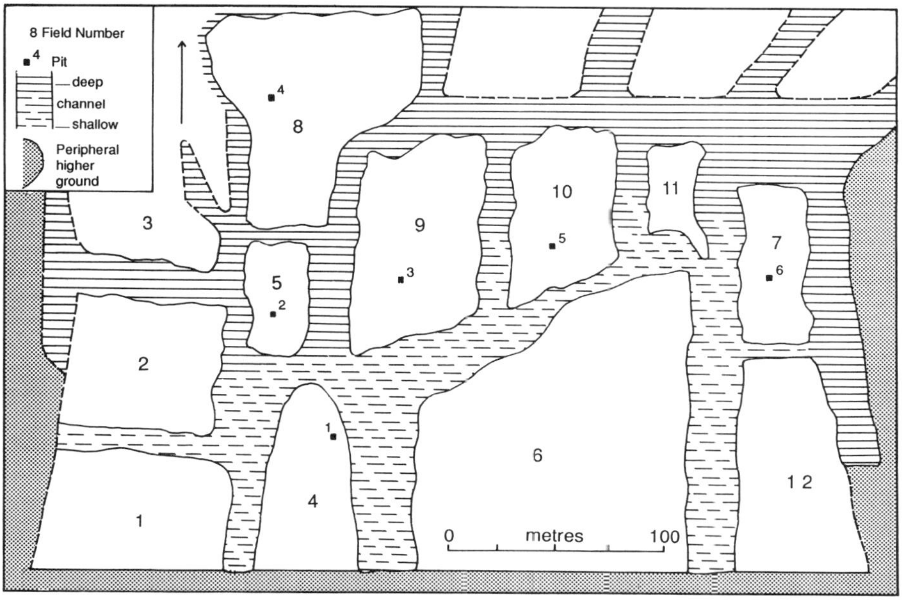


**Figure 15.3** The plan of the raised field system and associated channels at Nohmul, near Douglas Village (drainage towards the north).

10 and 11 in figure 15.3); the second type represents plots where boundary channels are cut into the ground surface for drainage (numbers 1 and 6); a third type presents a similar channel pattern but with the downslope sides of the fields showing evidence of field-raising to extend their surface areas in to the swamp (number 4).

Six profiles were excavated in the drained fields (at sites 4, 5, 8, 9 and 10). The profiles were extended to reach what seemed to be the original surface of swampy woodland, identified by black-coloured buried horizons at about 60 cm. At the time, this was the first encounter with buried soils on such a large drained field complex. Gleying indicated that the water table reached to within 20 cm of the present-day surface in the wet season (falling to 2 m or more in the dry season). Thick, black, highly organic surfaces result from the slow rates of organic decomposition, and the extreme wetness also inhibits the perturbing activities of soil fauna. Simplified profile descriptions of the six sites are given in figure 15.4.

Preliminary soil analyses were carried out to assess whether or not the dark layer, observed at around half a metre depth, was in fact a buried A horizon. Increased levels of organic carbon, pH and exchangeable potassium (table 15.1), corroborated the field evidence of dark colour, comminuted ligniceous fragments and the presence of shells of freshwater snails. Nevertheless, the horizon was not as humic as appeared from visual inspection and the highest carbon figures sometimes lay just above the organic horizon as distinguished in the field. One reason for this may lie in the method of raising field levels.

**Table 15.2** Total number of pollen grains and spores for each family identified within each horizon, expressed as a percentage of the pollen sum

| | | Percentage of pollen sum | | | | | |
|---|---|---|---|---|---|---|---|
| *Horizon* | *Pollen sum* | *Pinus* | *Quercus* | *Myrica* | *Gramineae* | *Ferns* | *Melastromataceae* |
| Surface | 48 | 18.75 | 2.00 | 12.50 | 22.90 | 43.75 | – |
| A/O | 78 | 1.28 | – | – | 12.80 | 85.89 | – |
| B | 16 | 12.50 | – | – | 75.00 | 12.50 | – |
| Buried | 17 | 38.90 | 19.48 | 10.38 | 1.29 | 22.00 | – |
| Pulltrouser Swamp | 116 | 2.58 | – | 1.72 | 3.44 | 50.86 | 41.37 |

Profile / Pit 1

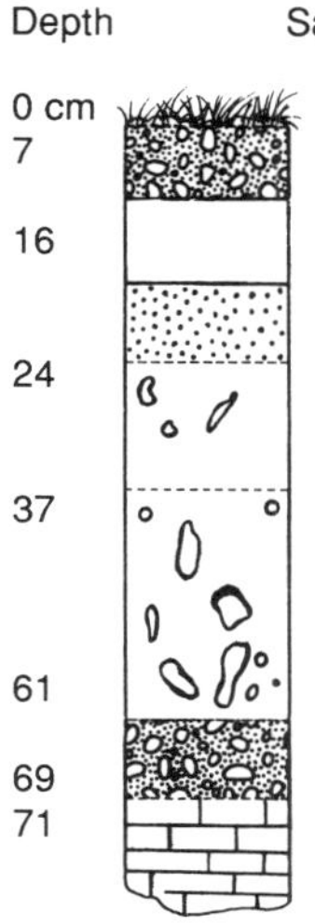


| Sample No. | Horizon | Description |
|---|---|---|
| 1 | OA | Highly organic, with many roots. |
| 2 | (B) | Argillaceous muds; organic. |
| 3 | II Ag | Buried milpa; gleyed. |
| 4 | II $B_1g$ | Slightly mottled; gleyed; massive structure. |
| 5 | II $B_2g$ | More mottled "fill" ; mollusc shells; gleyed; massive structure. |
| 6 | III A | Buried A-horizon highly organic; massive structure. |
| 7 | AC | Top of C-horizon;chalky sascab; massive with chalk fragments. |

Profile / Pit 2

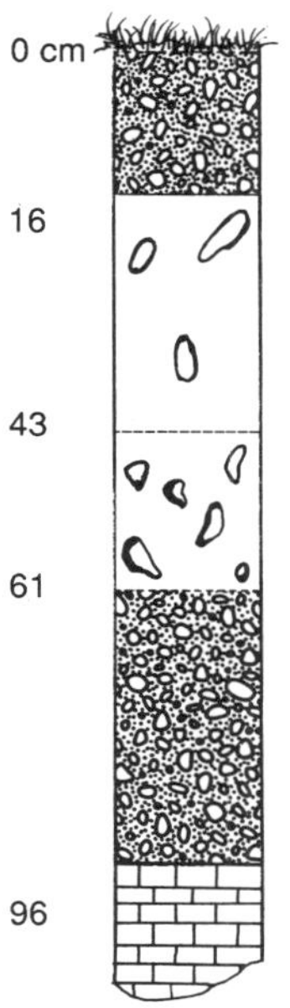


| Sample No. | Horizon | Description |
|---|---|---|
| 8 | 0A | Contemporary surface; organic. |
| 9 | $B_1g$ | Slightly mottled "fill" ; gleyed; massive structure. |
| 10 | $B_2g$ | Fill with molluscs; gleyed; massive structure. |
| 11 | II A | Buried A-horizon of black soil; richly organic with fine roots; massive structure. |
| 12 | II C | Sascab; mollusc-rich. |

**Figure 15.4** Soil profiles in the drained fields at Nohmul.

Profile / Pit 3

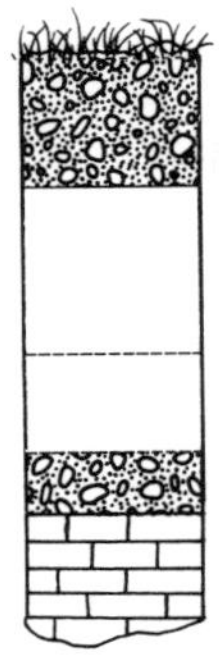

| Depth (cm) | Sample | Horizon | Description |
|---|---|---|---|
| 0 cm | | | |
| | 13 | 0 A | Highly organic with root mat. |
| 13 | | | |
| | 14 | $B_1g$ | Gleyed; massive structure. |
| 30 | | | |
| | 15 | $B_2g$ | Fill; massive structure. |
| 40 | | | |
| 46 | 16 | II A | Buried A-horizon; rich in molluscs; highly organic; massive with fine roots. |
| | | II C | Sascab. |

Profile / Pit 4

| Depth (cm) | Sample | Horizon | Description |
|---|---|---|---|
| O cm | | | |
| 6 | 17 | 0 A | Highly organic with dense root mat. |
| | 18 | B g | Gleyed; massive structure. |
| 25 | | | |
| | 19 | C g | Sascab; mottled; massive structure. |
| 58 | | | |
| | 20 | II A | Buried, dark-coloured; highly organic; some mixing with mottled fill. |
| 67 | 21 | II C | Sascab. |

Figure 15.4 *(continued)*

Most of the organic properties seem to have been severely depleted from the presumed original surface condition, compared with contemporary surfaces which might act as analogues. It would require more elaborate tests of the dark coloured staining to establish the nature of the residual organic matter. However, the evidence for the existence of buried soil profiles is convincing and is, furthermore, found in all of the sites investigated where field-raising

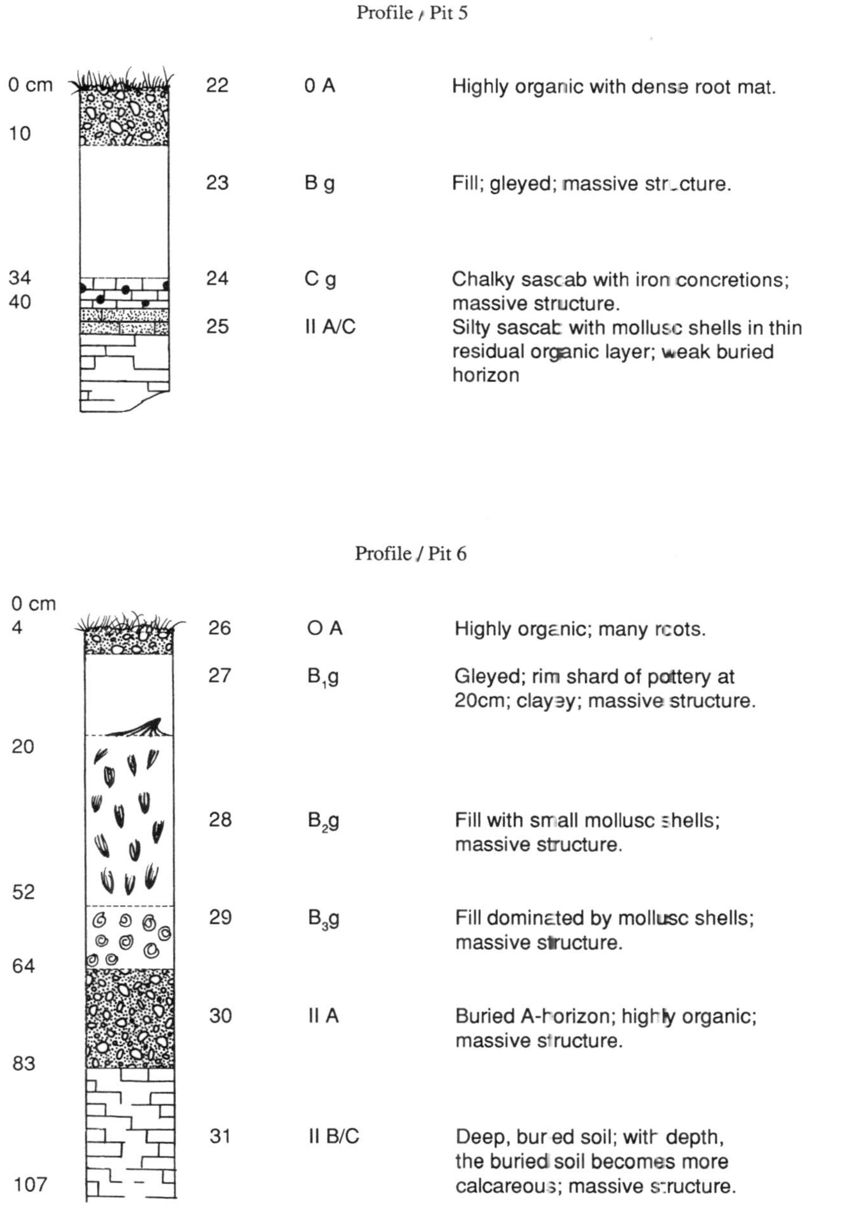


Figure 15.4 *(continued)*

has occurred, indicating an advanced form of management throughout the field system.

Reconnaissance pollen investigations were carried out at the same locations. Jackson and Paige (1989) examined three profiles down to and including the buried surface (figure 15.5). Since the contemporary land surface has been

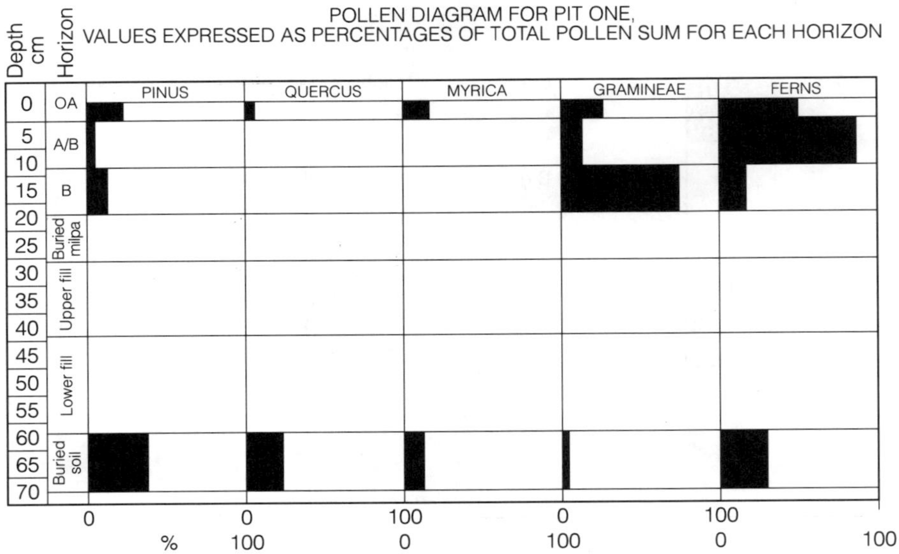


**Figure 15.5** Pollen counts at Nohmul.

disturbed by grazing, comparisons were made with Pulltrouser Swamp where there has been less disturbance (table 15.2). Interpretation of the pollen is constrained by the relatively small number of grains encountered in the samples, but it was considered significant that there was a virtual absence of pollen in the upper 'fill' material made up of weathered limestone, or *sascab* (Darch 1981, Darch and Furley 1983). The high proportion of pine in the buried zone may reflect the dispersal ability of the species (*Pinus caribea*) rather than its predominance in the wetland at the time. Currently pine, favouring coarse, acidic soils, is found some 20 km from the site. *Quercus* spp. are also more abundant in the buried zone, and this is a species more characteristic of marginal forest (Bartlett 1936). On the other hand, some species, such as *Myrica*, appear to be distributed equally abundantly in the contemporary and ancient surfaces. The present-day surface is characterized by grasses and ferns, reflecting the open habitat. Grasses and ferns are totally absent from the buried soil surface, which is dominated by woody genera. While the results of this reconnaissance investigation need to be treated with caution, in view of the low pollen count and the likely over-representation of a few persistent species, there is nevertheless confirmation of two distinct organic layers separated by a nearly abiotic zone, and this can be used to support the idea of a buried surface and superimposed *sascab* fill.

The site shows a number of features characteristic of Maya wetland management. Water manipulation consisted of channelling and adjusting the land surface to the water table and to seasonal water variation. Where necessary, soil drainage was improved at suitable locations by raising the field surface and fertility was enhanced by the addition of silty and organic mulch. From Nohmul and from other *chinampa* sites in Central America, it is evident that the land system was intensive and productive. Since it persisted for 600–800 years or more, it can be argued to have been considerably more sustainable than the dryland *milpa* system. It is likely that the drained field system has been one of the most advanced indigenous forms of land management and it continues to offer a guideline for current practice.

The term 'drained field system' is probably a more accurate reflection of the situation in the Belize wetlands than 'raised field' since, as Darch (1983) observes, the former carries no preconceived definition of field morphology. Despite this, the evidence at Nohmul seems to confirm the presence of a buried surface and further suggests that around half a metre of material has accumulated or been artificially added to the original level, making it a genuine 'raised field'.

In sum, the Maya are likely to have occupied the wetlands in Belize for two or three thousand years, although most settlements were small and never developed to the extent of nearby inland cities (Caracol or Altun Ha) or coastal ports such as Tulum on the Mexican Yucatán peninsula (Quintana Roo). At the height of Maya civilization, there were numerous small coastal settlements and established trade routes through the wetlands to the substantial inland city–states. The wetlands were manipulated in places for agriculture, involving a complex and intensive management system the precise details of which varied greatly according to location, environmental constraints (particularly depth and fluctuation of ground water) and to the nature of coastal and seasonal flooding.

## COLONIZATION AND BRITISH SETTLEMENT

### *Early occupation*

Europeans first caught sight of the coast of Belize in July 1502 when Columbus sailed westwards from Santo Domingo on the fourth and last of his great voyages of discovery. Although subsequently explored by Spaniards such as Cortez, who passed through part of Belize in 1524, no attempt was made to settle the Caribbean coast of Central America until 1531 when an outpost was established at Chetumal on the Mexican side of the present Belize/Mexico frontier (Chamberlain 1966). For the most part the hot, humid and insect-ridden coastal wetlands of Belize were dismissed by the Spanish as being unsuitable for settlement and no observation was made of any coastal occupation by the Maya. A century later, however, its river mouths, lagoons and offshore cayes were to provide ideal cover for British and Dutch buccaneers intent on raiding the homeward-bound Spanish bullion fleets (Haring 1910).

During the mid-seventeenth century the transition from pirate hideout to British settlement took place following the discovery of logwood (*Haematoxylon campechianum*), a tropical wood introduced to Europe by the Spanish as a fixing dye and exported by them in large quantities from the Yucatán peninsula (Wilson 1936). When the British government reached an agreement with Spain in 1667 to suppress piracy in return for freedom of trade, the privateers immediately turned their attentions to the exploitation of logwood, extracting it from accessible land near the coastal lagoons and rivers. By 1670, the population of the Bay Settlement had grown to 700, most of these being logwood cutters (Winzerling 1946). In that year Britain concluded a further treaty with Spain (the Godolphin Treaty), confirming British sovereignty over those parts of the Americas settled by British subjects. Although not formally established as a colony, the Bay Settlement was considered to be one such area, its stature being assured by British possession of Jamaica which Britain had captured and governed since 1655.

The presence of a British community in the heartland of the Spanish Americas was a source of constant irritation to Spain. Despite the 1667 and 1670 agreements, the Spanish considered the Baymen to be 'logwood-stealing foreign trespassers' (Setzekorn 1981). Throughout the eighteenth century British settlers were continually harassed by the Spanish, despite further agreements reached in 1783 and 1786 granting them the right to extract mahogany (*Swietenia macrophylla*) as well as logwood from the forests lying between the River Hondo in the north and the Sibun River in the south. This continued harassment eventually led the Governor of Jamaica to appoint, in 1784, a Superintendent for the purpose of overseeing British interests in the Bay Settlement, an appointment that was to be the first step towards the eventual establishment of a crown colony with a governor.

In September 1798 the Bay Settlement won its final and lasting victory against Spanish authority when a force of 32 Spanish ships was defeated during an offshore confrontation that came to be known as the Battle of St George's Cay (Caiger 1951). Still celebrated as a public holiday, this engagement was the last Spanish attempt to dislodge the British from Belize by force. When Spain's hold over its Central American colonies came to an end in the early 1820s, there seemed little threat to the commercial activities of the Bay Settlement, but with the collapse of the Central American Federation in 1838 there emerged independent republics posing a new threat to British interests. Chief amongst these was Guatemala, which not only questioned the frontiers of Belize but also the legitimacy of British sovereignty. In the face of possible conflict with its neighbour, more than 200 years of ill-defined status was finally brought to an end when Belize was declared a British colony on 12 May 1862.

Although timber extraction throughout the nineteenth and early twentieth centuries continued to be predominantly from the northern forests, elimination of the Spanish threat and the establishment of closer links with Britain encouraged woodcutters to explore the southern coastline in search of mahogany which had, by 1800, overtaken logwood as the leading export timber. Enjoying a period of commercial expansion reflected in a dramatic rise in mahogany exports (table 15.3), loggers worked their way southwards, reaching the Deep River by 1800, the Moho River by 1814 and the Sarstoon River, which forms the present-day international boundary, by 1824.

The environmental impact of both the logwood and mahogany cutters was tempered by the fact that the volume of timber extracted was small and tended to skirt wetland areas. In addition, felling operations were initially restricted to mahogany works which mostly lay within 5 miles of rivers and

Table 15.3 The export of mahogany logs, 1802–46

| *Year* | *Cu. ft* |
|---|---|
| 1802 | 375 000 |
| 1805 | 750 000 |
| 1820 | 500 000 |
| 1824 | 1 083 000 |
| 1837 | 1 420 000 |
| 1840 | 750 000 |
| 1845 | 1 667 000 |
| 1846 | 2 300 000 |

*Source*: Duncan (1968).

lagoons capable of transporting logs to the coast. It was only with the introduction of mechanized transport and the much later construction of forest roads (late 1940s), that the forests became accessible, and luxury hardwoods were then extracted almost to the point of extinction.

Thus the early colonial history of the coastal wetlands was focused on continual skirmishes between the colonial powers, with little penetration inland and limited opportunity for the establishment of thriving permanent settlements, other than that which developed at Belize City. There was virtually no contact with the remaining Indians and practically no knowledge of their earlier civilization and management skills. Despite nearly 300 years of destructive exploitation, the forests survived because the intensity of extraction was light. Regrowth was vigorous and the Belizean forests now constitute one of Central America's last great areas of wilderness.

### *Plantation agriculture in the coastal lowlands*

The emergence of plantation farming began to have a greater impact on the coastal lowlands towards the end of the nineteenth century. By the time British Honduras was designated a Crown colony separate from Jamaica in 1884, the population had climbed to about 30 000, of whom about 1000 were woodcutters. The majority of the people lived in Belize City which was, and still is, the largest settlement. However, not a single road then linked it with any other part of the country. There was next to no tradition of pioneer agricultural colonization and no plantation economy of the kind that had developed in the British West Indies, where the cultivation of coffee, tobacco, sugar cane, rice, cotton and tropical fruit was encouraged. Largely ignored as an area to be developed, Belize was simply looked upon by the British as a source of a limited range of exploitable forest products. The Creoles and Caribs who were employed by them as woodcutters similarly had little interest in agriculture, preferring seasonal work in the forest to the daily year-round toil of farm labour. Setzekorn (1981) even suggests that as far as they were concerned, to be a farmer was to be socially inferior. The Belizean disdain for agriculture resulted in the shunning of local produce and the consequent importing of basic foodstuffs at considerable cost, a tradition that persisted throughout the colonial period.

Ironically, the point at which Belize became a British colony marked a watershed in the history of forest exploitation, with most of the accessible forest in the northern lowlands depleted of quality timber. As early as 1839, the Colonial Superintendent had warned of the economic disaster that lay ahead, recommending the stimulation of agriculture 'which, owing to the exhaustion of Mahogany, is the only industry on which the colony can depend' (Burdon 1935). Little attention was paid to his advice however, since landowners were unwilling to lose their skilled forestry labourers to the land.

Despite a general reluctance to engage in agriculture, farming did eventually emerge in the coastal lowlands as a significant economic activity during the second half of the century. In the 1860s, the migration of sugar-cane farmers into northern Belize from the Yucatán provided a small but much-needed stimulus to agricultural development and attempts were made to export sugar and rum. In the process, over 1200 ha was cleared for sugar-cane production in the northern Districts of Corozal and Orange Walk, a scale of clearance quite different from anything in the preceding colonial period. Over a decade later, plantation crops were further promoted when a new Colonial Governor changed the mail subsidy from the Jamaica to the New Orleans route, thus opening up the possibility of exporting bananas to the United States (Dobson 1973). As this was a tree crop that could be readily grown on smallholdings as well as larger estates, the immediate result was a heavy investment in banana production and by 1891 over half a million stems were being exported, chiefly from Stann Creek District in southern Belize. Now exposed to agricultural export opportunities, landowners began to experiment with a number of different crops including coffee, cocoa and cotton (Morris 1883).

Although largely in private hands, plantations and smallholdings were supported by the government, which recognized the urgent need to diversify the economy and in particular promote agriculture. With this in mind, an Agricultural Board was set up in 1879 for the purpose of giving advice and offering bonus schemes for successful exports. Much later, in 1925, agricultural trade was given a further boost when the growing of good quality citrus gained a foothold in Stann Creek valley alongside banana plantations. Since then, citrus and bananas together with sugar cane have been the prominent agricultural exports from the coastal lowlands. By 1964, the year in which Belize gained self-government, 6800 ha of land was devoted to growing sugar cane in northern Belize, 3265 ha to citrus, primarily in the Stann Creek valley, and a total of 18266 bunches of plantains and bananas was exported from the plantations and smallholdings of Stann Creek and Toledo Districts in southern Belize.

Since the initial forest clearance for sugar cane in the 1860s, the extent of coastal lowland cleared for plantation farming remained small compared with the total area of the country (22 954 km$^2$). However, these clearances were to act as focal points for the more rapid expansion of agriculture that was to take place after the Second World War, and particularly in the 1980s. It is only within the past two decades or so that development has finally reached the more permanently wet lowlands.

*The canal-building period*

During the early years of the twentieth century, a programme of canal construction was initiated in an attempt to penetrate the wetlands and coastal lowlands and open up land for settlement and farming. One of the major drawbacks to development was the isolation of the colony's principal city from the rest of the country. Although a small railway was built in 1908 to link the Stann Creek valley with a new port facility to the south of the town of Stann Creek (now called Dangriga), Belize City was still the largest settlement and chief outlet for exports. In the early 1900s, agricultural produce was brought to the city market in fragile pit-pans and doreys which took to the sea and followed the Caribbean coast northwards from settlements such as Mullins River, Stann Creek and Gales Point. This was a long and sometimes hazardous journey to take on a heavily laden 14 foot river dorey, particularly when negotiating the sand bars at the mouths of the Sibun and Manatee rivers, but it was a journey taken once or twice each month by planters carrying rice, plantains and general produce.

The need for a safe inland route initially prompted the construction of a shallow waterway between the Sibun River and Northern Lagoon, a task that was undertaken between 1906 and 1917 without the aid of mechanical dredgers. This, however, was not the first canal to be excavated by colonists in Belize. As early as 1810 a 22 foot wide channel had been cut by soldiers on the south side of Belize City in order to drain mangrove swamp. Extended to the north side of the city in 1817, this drainage scheme was further developed in 1882, when a 170 yard (155 m) drainage canal was dug through mangrove from the west end of Yarborough Lagoon to the Caribbean Sea just south of Belize City. The continued westward development of Belize City in the early 1900s made it necessary to extend land drainage and reclamation even further, and a series of canals was built after land was granted by the government to First World War veterans (Leslie 1970).

Following the completion of the canal linking the Northern Lagoon to the Sibun River, a plan was drawn up in 1917 for a canal that would link Haulover Creek to the west of Belize City with the Sibun River, following a southwesterly route that passed through Fabers Lagoon and Jones Lagoon (Belize Archives 3371–17). In support of this project it was argued that, in addition to providing a safe waterway from the Sibun and Manatee areas to the south of Belize City, such a scheme might prevent flooding on the banks of the Belize River and help drain the wetlands on either side of the Sibun River. The drainage and reclamation of the Sibun lands, it was thought, might even encourage the production of garden produce suitable for sale in nearby Belize City.

After nearly 10 years of argument and delay, the ultimate decision to go ahead with the canal scheme rested not so much on the need for a safe waterway but on pressure to open up new land for citrus and banana crops. With an approach already made to the Empire Marketing board for a subsidy, the growing of citrus fruit promised to become a major source of revenue. Even more urgent, however, was the need to find new land on which to grow

**Table 15.4** Exports of bananas, 1913–40

| *Year* | *Bunches* | |
|---|---|---|
| 1913 | 617 537 | |
| 1915 | 486 420 | First World War |
| 1917 | 887 481 | |
| 1919 | 604 226 | |
| 1921 | 464 898 | |
| 1923 | 332 500 | |
| 1925 | 269 795 | |
| 1927 | 112 640 | |
| 1929 | 141 840 | |
| 1931 | 87 036 | |
| 1933 | 141 662 | End of Panama disease |
| 1935 | 356 056 | |
| 1937 | 938 985 | |
| 1939 | 552 563 | |

*Source*: Reports on the Customs, Trade and Shipping of British Honduras 1915–31.

bananas, which were a prominent feature of the landscape in the Stann Creek, Mullins River and Monkey River valleys, but which had been seriously devastated since the early 1920s by Panama disease. The scale of the problem created by this root disease was eventually pointed out to the Colonial Secretary in December 1922, when it was reported that 400 acres, accounting for one-third of the total production from the Riversdale area, had been affected (Belize Archives SP 1056). Banana production fell rapidly and exports in the decade between 1917 and 1927 dropped by an alarming 87 per cent (table 15.4), raising fears that the United Fruit Co. would not find it worthwhile keeping British Honduras on its Central America route. The withdrawal of this service, the colony's chief link with markets in North America, would certainly have brought to an end future prospects of developing an agricultural export trade (Bowman 1979).

Believing that the soils of the Sibun area were more resistant to Panama disease than those of the lands from Mullins River to the south, the building of the canal was seen as a matter of urgency. With exports of bananas at an all-time low, a dispatch was sent from Government House in Belize to the Secretary of State for the Colonies on 10 January 1927, pointing out that the opening up of the Sibun area would 'take the fullest advantage of the opportunity provided for replacing their vanishing supply of bananas', adding that 'should the canal materialise there will be no fear of our losing our only regular communication with the outer world; and there will be ample room for both bananas and grapefruit' (Belize Archives Despatch 3, 1927). Only then was the scheme given the final support of the British government.

By the end of September 1927 a survey had been carried out along the proposed route of the new canal by James A. Waight, who was later to become Surveyor General for Belize (see Waight 1927). Passing through open savanna, broken ridge (a mixture of scattered trees and savanna) and extensive areas of red mangrove, the canal was to follow a line that by-passed Fabers Lagoon, which was not reckoned to be navigable or dredgeable. In 1928, a contract

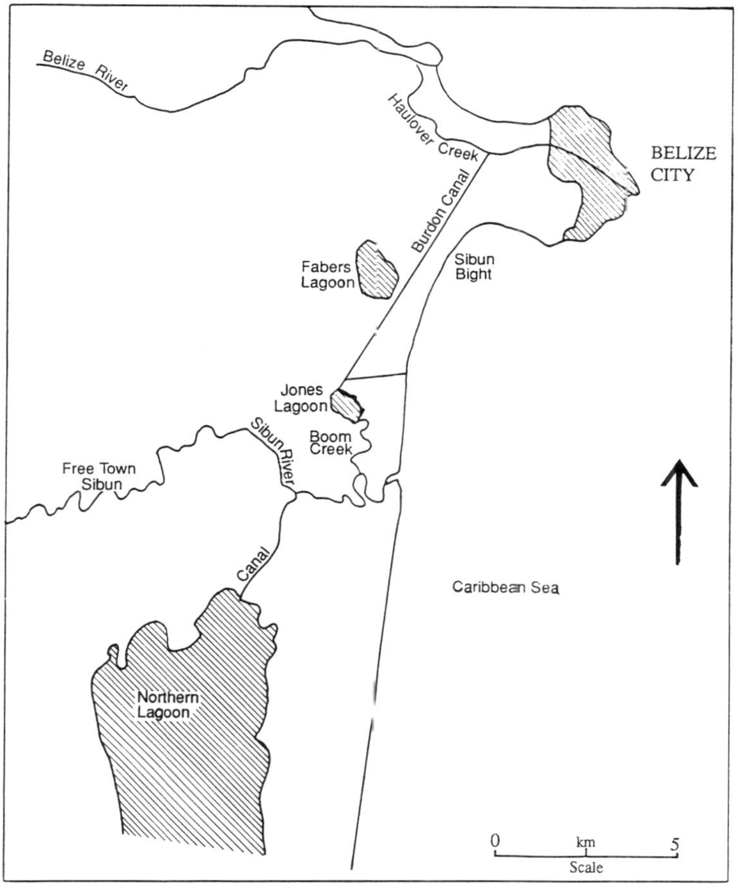


Figure 15.6 The wetlands of central Belize and the Burdon Canal.

was finally offered to a mahogany exporter, Cuthbert Fraser, who had gained considerable engineering experience in making roads into the forest for the extraction of timber but, more significantly, had been engaged in dredging rivers down which logs were floated to the coast.

Employing a Priestman Dredge, Fraser cut a line 23 621 feet (7200 m) in length from Haulover Creek to the northern tip of Jones Lagoon and then proceeded to create a channel linking the southern end of Jones Lagoon to a point on Boom Creek a further 3960 feet (1207 m) to the south (figure 15.6). With a total length of 5.23 miles (8.37 km), a width of 25 feet (7.7 m) and a depth of 5 feet (1.5 m), the canal was designed to be large enough to carry not just small river craft but also large fruit barges and log rafts. Completed on 15 April 1929, the Sibun Canal was later renamed the Burdon Canal in honour of Sir John Burdon, Governor of Belize (1925–32).

In its attempt to open up the Sibun area for farming, the Burdon Canal scheme proved to be a singular failure, principally because the root disease that had created such a rapid decline in banana production was brought under control in the early 1930s and the export of bananas from southern Belize was quickly restored to pre-1920 levels. Similarly, the scheme failed to attract citrus growers, who concentrated their planting in the Stann Creek valley rather than attempt to reclaim the wetlands in the Sibun area. Those who did benefit from the building of the canal included timber exporters and the smallholder farmers of the Sibun lands, who now had a direct and safe waterway linking them to the market at Belize City. For a few years the population of the Sibun area increased, growing from 548 in 1921 to 602 in 1931 (Census of British Honduras 1931), but the settlement and reclamation of the wetlands by smallholder farmers was short-lived. By 1946 the population had fallen by 25 per cent to a total of only 449 individuals (Census of British Honduras 1946). With the subsequent building of a road linking the Sibun River to Belize City, the financial return on the building of the Burdon Canal was, at the end of the day, minimal.

Whereas the Maya concentrated in the interior and used the coastline and rivers for trade, the European colonists arriving by sea stayed close to the coast and only gradually penetrated inland by rivers and lagoons. The wetlands were perceived first by the Spanish and then by the British as an obstacle to development and were therefore not seriously utilized during the colonial period, even though plantation agriculture began to emerge, particularly in the coastal lowlands of southern Belize. Nevertheless, the very fact that colonization began at the coast led to establishment of coastal towns and ports – a trading foothold in an inhospitable and largely disregarded landscape.

## POST-COLONIAL DEVELOPMENT AND THE COASTAL WETLANDS

The emergence of Belize as an independent nation in 1981 marked a watershed in both economic development and conservation, heralding a decade of change that was to focus attention on environmental issues, the most notable of which has been the human impact on coastal wetlands. In 1984 accelerated growth was promoted by the incoming United Democratic Party (UDP), which encouraged the investment of foreign capital and the forging of stronger links with North America. A period of economic expansion followed in which population growth, urban development and an increase in rural industry and tourism all contributed to the destruction and disturbance of the wetlands.

Although it is a recent development in Belize, human impact on tropical wetlands has been steadily growing in Latin America and the Caribbean. At a symposium on the 'Biology and Management of Mangroves', held in Florida in 1975, attention was drawn to the destruction of mangrove in El Salvador, which is the most densely populated country on the mainland of the Western Hemisphere (Dougherty 1975). Between 1900 and 1975, about 50 per cent of that country's mangrove was cleared for agriculture, fuel wood, salt-making and housing development, activities the environmental impact of which has been extensive in both economic and biological terms. Similarly, it has been calculated that by 1990, 50 per cent of mangrove in Costa Rica and 75 per cent in Puerto Rico had been destroyed (Bossi and Cintron 1990). In each of these countries, forest clearance has not only damaged offshore fishing but also caused accelerated coastal erosion and created problems associated with sedimentation in adjacent estuarine ecosystems.

In Belize, massive destruction of coastal mangrove has been avoided as a result of relatively low levels of human occupation. Exports of mangrove products, for example, have been limited and the coastal population has, until recently, never been great enough to be a threat. Consequently, outside Mexico, Belize has the third largest reserve of mangrove in Central America, at a time when the environmental, protective and biological roles of mangrove are just being properly recognized. While development pressures have been highly localized and less than in any other Central American country, tourism, agricultural, industrial and domestic housing developments have all posed an increasing threat to coastal wetlands since the onset of accelerated economic growth in the mid-1980s.

### *The impact of population growth and development*

Since the attempt to open up the Sibun and Manatee areas in the late 1920s, the population of Belize has quadrupled. Despite the movement of the capital in 1970 to a new site at Belmopan in the interior, much of this growth has taken place on the coastal lowlands. Initially, growth was concentrated in urban centres such as Belize City, Corozal, Dangriga and Punta Gorda, and for 50 years urban population was greater than rural population, a remarkable proportion for a developing country. Since the 1950s, when agriculture began to overtake forestry as the leading economic activity, the population of rural areas has risen at a faster rate (table 15.5). In the 1990s, following nearly 15 years of refugee immigration from El Salvador and Guatemala, the rural population is once again greater than the urban population.

The pattern of population growth has varied from decade to decade, and has been closely linked to the fluctuating fortunes of agricultural exports. Over the period 1970–80, rural population growth was greatest in the sugar-cane-growing Corozal District (47.2 per cent), whereas in the period 1980–9 growth was highest in the citrus- and banana-growing Stann Creek and Toledo Districts (52.7 per cent and 51.6 per cent respectively).

**Table 15.5** Urban and rural population growth, 1921–91

| *Year* | *Urban* | *Percentage change* | *Rural* | *Percentage change* | *Total* |
|---|---|---|---|---|---|
| 1921 | 21 500 | | 23 817 | | 45 317 |
| 1931 | 26 417 | 22.9 | 24 930 | 4.7 | 51 347 |
| 1946 | 33 072 | 25.2 | 26 148 | 4.9 | 59 220 |
| 1960 | 48 768 | 47.5 | 41 737 | 59.6 | 90 505 |
| 1970 | 62 970 | 29.1 | 56 964 | 36.5 | 119 934 |
| 1980 | 72 717 | 15.5 | 72 636 | 27.5 | 145 353 |
| 1991 | 91 903 | 26.2 | 98 989 | 36.3 | 190 792 |

*Source*: Extracted from Census reports.

**Table 15.6** Clearance of mangrove in the Belize City area, 1939–91

| *Year* | *Percentage of forest remaining* | *Percentage change* | *Percentage of annual clearance* |
|---|---|---|---|
| 1939 | 90–93 | | |
| 1961 | 84 | 9 | 0.3 |
| 1968 | 82 | 2 | 0.3 |
| 1975 | 76 | 6 | 0.9 |
| 1988 | 63 | 13 | 1.0 |
| 1991 | 52 | 11 | 3.6 |

*Source*: McShane (1991).

Although there is some indigenous use of wetlands, population growth in rural areas is mostly related to the clearance of land for new settlements, for agriculture or for aquaculture – notably shrimp farming. Of the 49 agricultural development concessions granted between 1985 and 1991, 27 related to the expansion of citrus and banana plantations in the southern coastal plains. Although only a fraction of the 14 600 ha of forest currently being cleared for these crops is likely to be wetland, the scale of plantation development has heightened the importance of coastal and riverine communities. Mangrove may well play a vital part not only in trapping sediments created by increased erosion, but also in filtering out residual nutrients and pesticides that find their way into river channels after being applied to plantation crops.

Since the mid-1980s, the most direct assault on the wetlands has resulted from urban expansion, principally to the west of Belize City. McShane (1991), using a time series of aerial photographs, estimated that 48 per cent of the mangrove forest on and adjacent to the peninsula had been removed since the beginning of the century, and the greater part (32 per cent) in the 30 years between 1961 and 1991. Mangrove destruction accelerated with an annual rate of clearance rising from 0.3 per cent in 1961 to 3.6 per cent in 1991 (table 15.6).

While coastal towns such as Punta Gorda, Dangriga and Sarteneja have all spread slowly into wetland areas in order to accommodate larger populations, the large scale clearance of mangrove around Belize City has taken place rapidly for both housing and industry. Of the 75 development concessions granted in Belize District between 1985 and 1991, 34 were related to factory construction on the outskirts of the city for the production of beer, clothing, fertiliser and forest products.

Seen as an impediment to permanent settlement throughout the greater part of the colonial period, the wetlands of Belize have recently experienced the impact of agricultural, aquacultural, industrial and domestic developments. Although highly localized, these activities have entailed the dramatic clearance of large areas of vegetation that was formerly utilized at very low levels and is now replaced by single forms of land use.

### *The impact of tourism*

Since the early 1950s, tourism has grown steadily throughout South and Middle America, and has progressively disturbed and latterly caused extensive

Table 15.7 International tourist arrivals in Belize, 1975–90

| *Year* | *Total number* | *Year* | *Total number* |
|---|---|---|---|
| 1975 | 57 891 | 1983 | 64 229 |
| 1976 | 61 473 | 1984 | 88 430 |
| 1977 | 55 072 | 1985 | 93 440 |
| 1978 | 60 587 | 1986 | 93 753 |
| 1979 | 61 714 | 1987 | 99 266 |
| 1980 | na | 1988 | 142 006 |
| 1981 | 64 219 | 1989 | 172 829 |
| 1982 | 67 219 | 1990 | 216 395 |

*Source*: Belize Tourist Board.

clearance of the coastal wetlands. The total number of international tourist arrivals has climbed from nearly one million in 1950 to just under 20 million in 1985 (World Tourism Organisation 1985, quoted in McMinn 1992). Never a major tourist destination, Belize only began to attract large numbers of visitors in the mid-1980s when ecotourism was promoted and the country was advertised internationally as the 'Adventure Coast'. Between 1984 and 1990, there was a sudden increase of 145 per cent in visitor arrivals (table 15.7). In terms of perentage growth, this far exceeded expansion of tourism in any other regions of the Americas (tourist arrivals in Latin America and the Caribbean averaged 33 per cent growth over 1985–9; McMinn 1992).

Largely concentrated on the coast and a handful of offshore cays, the development of hotels has inevitably involved the clearance and infilling of wetland, not simply to create building sites but also to open up a hinterland relatively free from insect life. Of the 59 hotels for which concessions were granted between 1986 and 1991, 49 were located on the coastal areas of Belize and Stann Creek Districts (Ministry of Economic Development 1991). While this type of development is highly localized, the environmental impact of tourism extends to overfishing, dumping of rubbish, uncontrolled camping and the disturbance of critical habitats by an ever-increasing number of tour groups and individuals who now come to visit the once ignored coast, barrier reef and coral cayes.

### *Conservation and environmental awareness*

Just as it lagged behind its neighbours in developing a tourist industry, so Belize came late to conservation and the designation of protected areas. It now has an enviable 80–90 per cent of the country in some form of protection and one of the highest proportions of protected area to population in the world. A feature of the 1980s, environmental awareness, arrived hand in hand with foreign investment – a decade or more after the establishment of some 250 parks and reserves in other parts of the neotropics. As early as 1960, a team of New York consultants had suggested the development of a series of national parks as a first step towards the promotion of a viable tourist industry (Stanton Robbins 1960), but that step was not taken for another twenty one years. In November 1981, only weeks after gaining independence, a National Parks System Act and a Wildlife Protection Act were passed, creating the essential legal framework for environmental conservation. By 1989 tourism and conservation were inextricably linked with the formation of the Ministry of Tourism and Environment.

In the decade from 1982 to 1991, following the passing of protected area legislation, over a dozen parks, reserves and wildlife sanctuaries were created, extending to some 214 000 ha, nearly 10 per cent of the total area of the country. Of these, the most significant in terms of protecting wetlands are the 1810 ha Crooked Tree Wildlife Sanctuary (1984), the privately run 9022 ha Shipstern Wildlife Reserve (1987), both in northern Belize, and the Hol Chan Marine Reserve (1987), part of which includes mangrove and swamp on the edge of Ambergris Cay. The establishment in 1991 of the Manatee Special Development Area, with over 47 348 ha of coastal lagoons between Belize City and Dangriga, was significant in that it marked the beginning of a move towards the creation of larger physical planning units within which zoned development would take account of the existence of critical habitats and other environmental interests.

Although the National Parks System Act is administered by the Forestry Department, created in the 1920s, the task of managing many of the parks and reserves falls to the Belize Audubon Society, which was formed in 1969. Originally with a membership of enthusiastic ornithologists, the Audubon Society was given charge of a number of bird nesting sites in the 1970s, most of which were located in the coastal wetlands. Along with other conservation NGOs, including the Belize Center for Environmental Studies, the Belize Zoo and the Programme for Belize, the Audubon Society has encouraged public awareness of environmental issues through the media, information leaflets and publications, schools programmes, public lectures and debates. In particular, it has, since 1985, given prominence to the perceived threat to mangrove forest.

The designation of protected areas, the formation of a Coastal Zone Management Unit under the Ministry of Agriculture and Fisheries and the establishment in 1989 of regulations to protect mangrove, are all measures that reflect a renewed interest in the wetlands of Belize. Emphasis has been placed on the role of coastal wetlands as a nutrient and sediment trap and as an environment of considerable biological interest not only to scientists but also to a growing number of tourists. In particular, there is now a greater appreciation of the importance of mangrove in stabilizing the coastline, buffering the effects of tropical storms and hurricanes, acting as a nursery for commercially important species of fish and crustaceans, and protecting the offshore coral reef and sea grass beds from the harmful effects of sediments brought down to the sea by rivers.

## CONCLUSIONS

Both the coastal and the inland wetlands of Belize were utilized, controlled and cared for by the Maya. Ignored for the greater part of the colonial period, recent years have witnessed a renewed interest, particularly in the coastal wetlands. Although threatened by developers, the value of wetlands as an economic, biological and environmental resource is being increasingly appreciated and reassessed. The future zonation of the coastal wetlands will go a long way towards ensuring an appropriate balance between complete clearance for a variety of purposes, low levels of utilization and designation as protected areas. Despite this new-found interest, if it were asked who have been the most effective managers of the wetlands throughout history, the answer would undoubtedly be the original indigenous inhabitants of the region, the Maya.

REFERENCES


Adams, R. E. W., Brown, W. E. J. and Culbert, T. P. (1981) 'Radar mapping, archeology and ancient Maya land use', *Science* 213: 1457–1464

Andrews, A. P. (1987) 'A typology of Maya trading ports', Paper presented to symposium on Coastal Trade and Exchange, 52nd Annual Meeting of the Society of American Archeologists, Toronto (quoted in Vail 1988)

Antoine, P. P., Skarie, R. L. and Bloom, P. R. (1982) 'The origin of raised fields near San Antonio, Belize: an alternative hypothesis', in Flannery, K. V. (ed.) *Maya subsistence, Studies in Memory of Dennis E. Puleston*, (Academic Press, New York) pp. 227–36

Bartlett, H. H. (1936) *Botany of the Maya Area*, Miscellaneous Papers (Carnegie Institute, Washington, D.C.)

Bloom, P. R., Pohl., M., Buttleman, C., Wiseman, F., Covich, A., Moksicek, C., Bell, J. and Stein, J. (1983) 'Prehistoric Maya wetland agriculture and the alluvial soils near San Antonio, Rio Hondo, Belize', *Nature* 301: 417–19

Bossi, R. and Cintron, G. (1990) *Mangroves of the Wider Caribbean: Towards Sustainable Management* (Caribbean Conservation Association, Panos Institute and UNEP)

Bowman, H. T. A. (1979) *Emerald Valley and Twinkling Town: my Autobiography* (privately published, Stann Creek, Belize)

Burdon, Sir J. A. (1935) *Archives of British Honduras* 3 vols (Sifton Praed, London)

Caiger, S. L. (1951) *British Honduras Past and Present* (George Allen and Unwin, London)

Census of British Honduras (931, 1946) *West Indies Population Census* (Department of Statistics, Kingston, Jamaica)

Chamberlain, R. S. (1966) *The Conquest and Colonisation of Yucatán 1517–1550* (Carnegie Institute, Washington)

Cowgill, U. M. and Hutchinson, G. E. (1963) 'Ecological and geochemical archeology in the southern Maya lowlands', *Southwest Journal of Anthropology* 19: 242–60

Culbert, T. P. (ed.) (1973) *The Classic Maya Collapse* (University of New Mexico Press, Alburqueque)

Darch, J. P. (1981) 'The characteristics and origins of sascab in northern Belize, Central America', *Zeitschrift für Geomorphologie*, N. P. 25: 400–19

Darch, J. P. (1983) *Drained Field Agriculture in Central and South America*, International Series 189 (BAR, London)

Darch, J. P and Furley, P. A. (1983) 'Observations on the nature of sascab and associated soils in Cayo and Orange Walk Districts, Belize and in El Petén, Guatemala', in Furley, P. A and Robinson, G. M. (eds) *Resources and Development in*

*Belize*, (Department of Geography, Occasional Publications NS 5, University of Edinburgh) pp. 179–222

Darch, J. P. and Randall, R. R. (1989) 'Aspects of the biogeography of Mayan drained field agriculture at Douglas Village, Orange Walk District', in Furley, P. A. (ed.) *Advances in Environmental and Biogeographical Research in Belize, Biogeographical Monographs* 3: 47–52

Dobson, N. (1973) *A history of Belize* (Longman Caribbean, London)

Dougherty, H. E. (1975) 'Human impact on the forests of El Salvador', Proceedings of the International Symposium on Biology and Management of Mangroves, Institute of Food and Agricultural Science (University of Florida, Gainesville, Florida)

Duncan, K. D. (1968) Aspects of the hardwood forest industry in British Honduras, in Furley, P. A. (ed.) *Edinburgh University Expedition to British Honduras – Yucatán 1966: General Report* (University of Edinburgh) pp. 52–62

Furley, P. A. and Ratter, J. A. (1992) *Mangrove Distribution, Vulnerability and Management in Central America*, Research Contract R4237, Oxford Forestry Institute–Overseas Development Administration (Edinburgh)

Gliessman, S. R., Turner, B. L. and Rosado May, F. J. (1983) 'Ancient raised field agriculture in the lowlands of south eastern Mexico', in Darch, J. P. (ed.) *Drained Field Agriculture in Central and South America*, International Series 189 (BAR, Oxford)

Gliessman, S. R. (1990) *Agroecology: Researching the Basis for Sustainable Development* (Springer-Verlag, New York)

Graham, E. (1983) The Highlands of the Lowlands: environment and archaeology in Stann Creek District, Belize, Central America, unpublished Ph.D. thesis, University of Cambridge

Guderjan, T. H. (1988) 'Maya maritime trade at San Juan, Ambergris Cay, Belize' (unpublished Ph.D. thesis, Southern Methodist University, Dallas, Texas)

Hammond, N. (1978) 'The myth of the *milpa*: agricultural expansion in the Maya lowlands', in Harrison, P. D. and Turner, B. L. (eds) *Pre-Hispanic Maya Agriculture* (University of New Mexico Press Alburqueque) pp. 23–34

Hammond, N. (1986) 'New light on the most ancient Maya', *Man* 21: 399–413

Hammond, N., Donaghey, S., Gleason, G., Staneko, J. C., van Tuerenhout, D. and Kosakowsky, L. J. (1987) 'Excavations at Nohmul, Belize 1985', *Journal of Field Archaeology* 14: 257–81

Haring, C. H. (1910) *The Buccaneers in the West Indies in the 17th Century* (London)

Jackson, J. and Paige, C. (1989) 'Analysis of pollen from the lowlands of northern Belize', in Furley, P. A. (ed.) *Advances in Environmental and Biogeographical Research in Belize, biogeographical Monographs* 3: 53–66

Jacob, J. S. (1992) 'The agroecological evolution of Cobweb Swamp, Belize', unpublished Ph.D. thesis, Texas A&M University

Lambert, J. D. H., Siemens, A. H. and Arnason, J. T. (1984) 'Ancient Maya drained field agriculture: its possible application today in the New River floodplain, Belize, C.A.', *Agricultural Ecosystems and Environment* 11: 67–84

Leslie, L. (1970) 'Canals date back to 1810', *The Reporter*, 21 August 1970, Belize City

McMinn, S. (1992) 'The impact of tourism on the environment, Ambergris Caye, Belize', unpublished M. Phil, University of Cambridge

McShane, F. (1991) 'Environmental and social implications of mangrove clearances around Belize City', unpublished M.Sc. thesis, Heriot-Watt University, Edinburgh

Ministry of Economic Development (1991) *Concessions Granted for Period 1985–1991* (Belmopan, Belize)

Morris, D. (1883) *The Colony of British Honduras and its Prospects; with Particular Reference to its Indigenous Plants and Economic Production* (Edward Stanford, London)

Olson, G. W. (1977) 'Significance of physical and chemical characteristics of soils at the San Antonio archeological site', *Journal of Belizean Affairs* 5 (Sept.): 22–36

Pohl, M. D. (ed.) (1990) *Ancient Maya Wetland Agriculture: Excavations on Albion Island, Belize* (Westview Press, Boulder, Colorado)

Pyburn, K. A. (1989) *Prehistoric Maya Community and Settlement at Nohmul, Belize*, International Series 509 (BAR, Oxford)

Rice, D. S and Culbert, T. P. (1990) 'Historical contexts for population reconstruction in the Maya lowlands', in Culbert, T. P and Rice, D. S. (eds) *Precolumbian Population History in the Maya Lowlands* (University of New Mexico Press, Albuquerque)

Santley, R. S., Killon, T. W. and Lycett, M. T. (1986) 'On the Maya collapse', *Journal of Anthropological Research* 42: 123–59

Scarborough, V. L. (1983) 'A Preclassic Maya water system', *American Antiquity* 48: 720–44

Setzekorn, W. D. (1981) *Formerly British Honduras: a Profile of the New Nation of Belize* (Ohio University Press, Athens)

Shafer, H. J and Hester, T. R. (1983) 'Ancient Maya chert workshops in northern Belize, Central America', *American Antiquity* 48: 519–43

Siemens, A. H. (1983) 'Wetland agriculture in prehispanic Mesoamerica', *Geographical Review* 73: 166–181

Stanton Robins and Co. Inc. (1960) *Recommendations for Tourist Development in British Honduras*, (New York)

Stephens, J. L. (1841) *Incidents of Travel in Central America, Chiapas and Yucatan* (London)

Turner, B. L. (1976) 'Population density in the Classic Maya lowlands: new evidence for old approaches', *Geographical Review* 66, 73–82

Turner, B. L. (1983) *Once Beneath the Forest* (Westview Press, Boulder, Colorado)

Turner, B. L and Harrison, P. D. (1978) 'Implications from agriculture for Maya Prehistory, in Harrison, P. D. and Turner, B. L (eds) *Pre-Hispanic Maya Agriculture* (University of New Mexico Press, Alburquerque) pp. 337–73

Turner, B. L and Harrison, P. D. (eds) (1983) *Pulltrouser Swamp: Ancient Maya habitat, Agriculture and Settlement in Northern Belize* (University of Texas Press, Austin)

Vail, G. (1988) *The Archaeology of Coastal Belize*, International Series 463 (BAR, Oxford)

Waight, J. A. (1927) 'Survey of the proposed Sibun Canal', MP 1002–27, Belize Archives Department, Belmopan, April

White, W. B. (1988) *Geomorphology and Hydrology of Karst Terrains (Oxford University Press, New York)*

Wilson, A. M. (1936) 'The logwood trade in the 17th and 18th centuries', in McKay, D. C. (ed.) *Essays in the History of Modern Europe* (New York)

Winzerling, E. O. (1946) *The Beginning of British Honduras, 1506–1765* (North River Press, New York)

Wiseman, F. M. (1983) 'Analysis of the pollen from the fields at Pulltrouser Swamp', in Turner, B. L. and Harrison, P. D. (eds) *Pulltrouser Swamp: Ancient Maya Habitat, Agriculture and Settlement in Northern Belize* (University of Texas Press, Austin) pp. 105–19

16

# Invasion and Transformation of California's Valley Grassland, a Mediterranean Analogue Ecosystem

Mark A. Blumler


ABSTRACT

Consequent upon 1492, transoceanic plant invasions have produced severe impacts on natural ecosystems. Invasions are often unidirectional, with species from one region, such as the Mediterranean, now dominating areas of analogous climate overseas. Explanation focuses on agropastoralism, but natural evolutionary, ecological and geographical aspects should not be ignored. Natural palaeoinvasions also were sometimes markedly unidirectional, with invasion from larger to smaller land masses being the rule. Annual plants, usually regarded as disturbance indicators, may dominate without disturbance in seasonally dry climates, so their success in invasion does not prove human influence. Climate seems to explain the overseas invasion success of Mediterranean species better than anthropogenic factors such as the increase in grazing pressure resulting from the introduction of domestic livestock. An illuminating example is California's valley grassland, which has undergone the most dramatic transformation of any plant community on a continental land mass: typically, alien (primarily Mediterranean) species account for 80–90 per cent of the vegetative cover, even where undisturbed. The pace of environmental change was so rapid that reconstruction of the original vegetation is now extremely problematic, despite several types of biogeographical (pollen, seed remains in adobe bricks, isozymes and ecological experiments) and historical data. The data suggest that native annuals were more important than is generally realized. Some aliens did not require human disturbance to oust natives, due apparently to non-convergent evolution in similar environments, while geographical bottlenecks played a major role in determining which species successfully migrated overseas, and fostered a certain randomness in distributions. For instance, distributions of *Avena* species and genotypes, often regarded as due to natural selection, more likely result from the vagaries of dispersal.

## INTRODUCTION

One of the more intransigent aspects of the study of ecological relationships in historical times is the separation of human impacts from natural environmental or ecological change (Roberts 1989, Suffling 1993). Climate and other features of the physical environment change continuously, while evolution produces never-ending adjustments in the ecological relationships between species. Consequently, there also are likely to be feedbacks of environmental change upon human adaptations, with often fascinating and significant outcomes (e.g. Byrne 1987). All too often, however, environmental historians and historical geographers treat their subject as if humans dominate and transform (usually negatively) a passive and timeless, if occasionally constraining, nature.

To what extent is recent ecological change due to humans, and to what extent is it natural? This theme is discussed in this chapter with reference to an important biogeographical topic – invasion. Humans have intentionally and unintentionally transported an enormous number of species across oceans that formerly served as barriers to the dispersal of terrestrial plants and animals. In particular, the development after 1492 of trade between the Eastern and Western Hemispheres, as well as with Australia and various island groups, has facilitated a massive species exchange, producing, to some degree, a homogenization of biota and reduction in diversity (Elton 1958, Crosby 1986). Although introduced species often die out or remain restricted to highly disturbed habitats, they sometimes escape, proliferate spectacularly and oust natives, in the process altering ecosystem functioning (Vitousek et al. 1987; d'Antonio and Vitousek 1992) and posing major difficulties for the conservation of wilderness areas (Blumler 1992c, Cheater 1992, Edwards 1992). Scholars often assume that human disturbance is responsible none the less – given that our impacts today are so manifold and far-reaching, extending even into the most isolated areas (Fox and Fox 1986), and since at a minimum humans are responsible for the transport of most alien species overseas.

A major issue, then, is the extent to which anthropogenic disturbance is necessary for alien species' subsequent success. An alternative to human impacts is non-convergent evolution (the evolution of dissimilar adaptations in similar environments). If a species evolves a superior adaptation to a given environment on one land mass, it may outcompete native organisms once it is transported to analogous ecosystems overseas. Related to these theoretical questions are more practical concerns such as the prediction of the sorts of organisms most likely to invade, the ecosystems most likely to be vulnerable to invasion, and the effects of invasion upon human adaptation and economy (for example, the havoc that Old World diseases wreaked upon American Indians; Crosby 1972, 1986).

Here, I focus upon one of the most spectacular cases known, the plant invasion of California grasslands. First, I discuss briefly global patterns of invasion and critically evaluate the hypotheses most often proffered in explanation. Then, I examine the California case in detail – placing current invasions in their natural biogeographical context – and argue that both natural and human processes are at work, and deeply intertwined.

PATTERNS OF INVASION

Oceanic islands have proven most vulnerable to invasion: for instance, it is estimated that 50 per cent of Hawaii's flora is endangered (Wagner et al. 1985). On the continents, invasion has most severely impacted those regions experiencing mediterranean-type climates (other than the Mediterranean itself). Species from the Mediterranean/Near East have penetrated into grassland, savanna and other herbaceous associations in regions of analogous climate so successfully that the aliens often dominate over great swaths of the landscape. Annual plants have been especially successful. Regions of cool, humid climate, such as the Argentinean pampas, have also undergone invasion, but to a lesser degree (Blumler 1984, 1992a; see below). California's valley grassland, in which today aliens typically account for 80–90 per cent of the cover (Heady 1977, Bartolome 1979), has suffered perhaps the most drastic transformation of any continental ecosystem. Introduced species often dominate even on sites that are left ungrazed and apparently undisturbed for decades (Biswell 1956, Sauer 1988, Blumler 1992b). The alteration of valley grassland vegetation after Spanish settlement in 1769 was so rapid and so complete that we can only speculate as to the original species composition. In contrast, invasion of the Mediterranean and Near East has been limited, and has taken place especially in irrigated fields and other summer-moist environments (Dafni and Heller 1982, di Castri et al. 1990). Some species often classed as invaders, such as prickly pear cactus (*Opuntia* spp.), are widely cultivated in semi-natural habitats but display little tendency to spread outward from the cultivation sites (e.g. Barbera et al. 1992). In short, invasion of mediterranean-type regions has been markedly unidirectional, from the Mediterranean/Near East to analogous regions on other continents (as well as the Canary Islands).

EXPLANATIONS

Why is this invasion so one-sided? Most scholars have assumed that human disturbance is the key factor. An underlying, though generally unstated, assumption is that the one-sided character of the invasion is puzzling; that is, not to be expected under natural circumstances (cf. the discussion of insect dispersals to North America from Europe; Buckland et al., ch. 12, this volume). Also, annual plants are generally regarded as ruderals (disturbance adapted), and their success as invaders is therefore seen as strong evidence for disturbance. Moreover, European agropastoralism derives from the Near East, where Mediterranean annuals are very important. The many millennia of agropastoralist impacts are presumed to have produced a flora that is coevolved with and adapted to human disturbance (Zohary 1962, 1973; Baker and Stebbins 1965; Naveh and Whittaker 1979; but cf. Danin and Yom-Tov 1990; Blumler 1994; Blumler and Byrne 1991). Invasion took place as this Old World agroecosystem spread to other parts of the globe. Crosby (1986) went so far as to assert that Europeans would not have succeeded in settling overseas if their coevolved weedy plants had not invaded along with them, thereby creating an environment conducive to the flourishing of their (also coevolved) livestock.

Among ecologists, the most popular explanation for the invasion of mediterranean-type and other temperate grasslands and steppes is the grazing pressure hypothesis. The argument runs that many grasslands experienced only light natural grazing pressure, and therefore were vulnerable to invasion by grazing-adapted, European or Mediterranean plants once domestic livestock began to exert more severe pressure (Evans and Young 1972, Mack and Thompson 1982). The great exception, in this view, was the Great Plains – grazed heavily by bison and therefore resistant to invasion by Old World species (Baker 1978, Stebbins 1981, Mack and Thompson 1982). Variants of this hypothesis are generally accepted by Californian ecologists, environmental historians and historical geographers. Because introduced annuals dominate today, and are assumed to be successful because of grazing and other disturbances, it is also argued that perennials must have dominated the original (pre-Spanish) grassland. However, actual evidence in support of this conclusion is lacking, and native grassland annuals are far more diverse than native perennials.

## CRITIQUE OF THE EXPLANATIONS

There are several reasons to suspect that these prevailing views are oversimplified. First, *natural* invasions are often one-sided (in this context, it is well to remember that all species must invade new regions from time to time, or go extinct; Hengeveld 1990). For instance, when the formation of the Panamanian land bridge allowed an exchange of fauna between North and South America (the 'Great American Interchange'), a majority of species migrated from north to south, leading to a massive extinction of native South American mammals (Webb 1978, Simpson 1980). Floristic migration over the land bridge is not so well understood, but it is believed that invasion was primarily in the reverse direction. On the other hand, plants exhibiting 'amphitropic' distributions – that is, occurring in temperate regions of North and South America (primarily in California and vicinity, and in Chile and neighbouring parts of Argentina) – have migrated mainly from north to south (Raven 1963). The huge disjunctions in the ranges of these species suggest that migration of birds along the Pacific flyway was responsible in most cases for their dispersal. Raven concluded that 85 of the 130 amphitropic species had migrated from north to south and 15 in the opposite direction, while 30 species are of dubious origin. Dividing the uncertain species equally among the two groups would give a result of 77 per cent invasion from north to south. This case is particularly interesting as it involved invasion from one mediterranean-climate region to another, and since most invaders were self-pollinating annual species (presumably because of Baker's (1955) rule that establishment after long-distance dispersal is easiest for self-compatible taxa). Yet it was an entirely natural, pre-Columbian invasion.

North America and Eurasia also were connected at times in the geological past by natural routes of migration, primarily at high latitudes. Consequently,

**Table 16.1** Some important 'native' western North American taxa of probable Eurasian origin

| *Taxon* | *Common name* |
|---|---|
| *Artemisia* | Sagebrush |
| *Agrostis* | Bentgrass |
| *Bromus* | Bromegrass |
| *Festuca/Vulpia* | Fescue |
| *Poa* | Bluegrass |
| *Stipa* (in part) | Needlegrass |
| Triticeae (*Elymus, Hordeum,* etc.) | Wheat tribe |
| *Astragalus* | Locoweed |
| *Lotus* | Trefoil |
| *Trifolium* | Clover |
| Vicieae (*Vicia, Lathyrus*) | Vetch, sweet pea |
| Cardueae (*Cirsium*) | Thistle tribe |
| Lactuceae (*Agoseris, Crepis*) | Dandelion tribe |
| Chenopodiaceae (in part) | Goosefoot family |

*Source*: Blumler (1992c).

exchanges of plant species took place. Dispersals of tree species are difficult to trace, as most lineages are ancient; but more recently evolving herbaceous or semi-woody groups, such as those that presumably dominated California and other western steppes and grasslands prior to European settlement, are better understood. Invasion in these groups was almost entirely unidirectional, from Eurasia to North America (table 16.1 lists some of the more important of these 'palaeoinvaders'). One might debate the origins of a few of the taxa, but the basic conclusion of overwhelmingly one-way migration would still stand (Blumler 1992c). It appears that only a few, relatively unimportant groups, such as *Arnica* and *Antennaria*, have invaded in the reverse direction (G. L. Stebbins, personal communication). Interestingly, many of the palaeoinvaders are in the same phyletic lineages as today's most successful invaders (Blumler 1992c). Consequently, in California alien species are currently replacing native species that often are rather closely related.

In each case discussed above, invasion occurred primarily from the larger land mass to the smaller. An 'exception that proves the rule' is the floristic invasion from South to North America over the Panamanian land bridge. While North America is the larger continent, floristic exchange involved tropical taxa, for the most part – and South America has a much greater areal extent of tropical climate. On the other hand, a comparison of total continental land mass size is more appropriate for mammals, because they typically have broad climatic tolerances. The Old World mediterranean-type region, which extends from Iberia and Morocco to the Tien Shan and Pamirs, is far greater in area than any of the climatically analogous regions on other continents. It is also located at the junction of three continents and, hence, has long been a centre of evolutionary interaction and exchange. In contrast, the other mediterranean regions are isolated on continental margins, often surrounded by high mountains and deserts. Darwin (1859) postulated that evolution should proceed most rapidly in the centres of large continents, because more species will interact with each other there than on continental margins. Thus, even in the absence of humans, one would predict that invasion would proceed from the Mediterranean and Near East to the analogue regions rather than vice versa, if suitable natural migration routes were available.

Second, as I have discussed elsewhere (Blumler 1984, 1992b,c, 1993, 1994), the rule that annuals are ruderals breaks down where a wet season alternates with a long, hot, dry season (as in mediterranean regions). Annuals are more diverse there than in other climates (Blumler 1984); they avoid the dry season as dormant seed – the most drought resistant plant organ – and compete vigorously with perennial seedlings in the wet season. Therefore, some annual species may even be 'climax' (Blumler 1992b, 1993). For instance, Litav demonstrated that wild oats (*Avena sterilis*) outcompetes batha shrubs on fertile soils in Israel, and concluded that grazing and other disturbances favour many perennials more than tall annual grasses (Litav 1965, 1967; Litav et al. 1963). Similarly, data from a recent study of Israeli pastures show no significant effect of grazing on perennial cover, but a severe negative impact on tall annual grasses (Noy-Meir et al. 1989; Noy-Meir 1990; Blumler 1992b, 1993). Blumler (1992b, 1993) found that tall annual grasses increased in an Israeli grazing exclosure, until they accounted for over 90 per cent of the cover away from rock outcrops (where a few perennials were sheltered from the vigorous competition of the annuals). In California, Australia and other mediterranean-climate regions too, numerous studies show that wild oats (*Avena* spp.) and other annual grasses are strongly competitive against native perennials (Biswell 1956; Rossiter 1966; McNaughton 1968; Blumler 1984, 1992b). In fact, Californian ecologists and preserve managers are concluding, reluctantly, that only some form of *disturbance*, such as mowing, grazing or fire, can maintain or favour perennials in the face of this competition (Kay et al. 1981, Edwards 1992, Menke 1992, Stebbins 1992). In short, the fact that most invaders of mediterranean regions are annuals is not proof that disturbance was responsible for the invasions.

Finally, severity of invasion correlates best with climate, not natural grazing pressure. Mediterranean herbs tend to grow during the cool (rainy) season, while neighbouring European species also typically have growth optima at cool temperatures because summers are relatively mild in Europe (in comparison, say, with the eastern United States; Blumler 1992a). Therefore, European grasses tend to employ the $C_3$ photosynthetic pathway, while the dominant grasses in the eastern half of the United States are mostly $C_4$. Not surprisingly, then, Mediterranean and some European species have invaded other continents wherever there are cool-season rains. Invasion has been especially severe, however, where there also is severe summer drought. California has the most severe summer drought of any mediterranean region outside the Old World (Blumler 1992b); this is probably the reason why it has also suffered the most severe invasion.

A comparison of climate, grazing and invasion severity illustrates the point (table 16.2). Table 16.2 is oversimplified, in that each region actually has a range of climates: for instance, interior central California has a far more severe

**Table 16.2** The Relationship between climate, natural grazing pressure and severity of invasion by Mediterranean and/or European plants

| *Region* | *Cool-season rains* | *Summer drought* | *Natural large grazers* | *Invasion of wild areas* |
|---|---|---|---|---|
| *USA* | | | | |
| California | Long | Very severe | Elk, pronghorn, deer | Very severe |
| Great Basin | Long | Less severe | Pronghorn, deer | Moderately severe |
| Great Plains | Short | Slight | Bison, elk, etc. | Slight |
| Eastern US | Moderate | Very slight | Elk, deer | Slight to moderate |
| *South America* | | | | |
| Central Chile | Long | Severe | Camellids, deer | Severe |
| Southern Chile | Long | None | Camellids, deer | Moderate? |
| Eastern pampas | Long | None | Camellids, deer | Moderate |
| Western pampas | Very short | Slight | Camellids, deer | Slight |
| *South Africa* | | | | |
| Cape | Long | Moderate | Rhinoceros, etc. | Moderate |
| Eastern veldts | Short | Slight | Numerous | Slight |
| *Australia* | | | | |
| Mediterranean type | Long | Moderate | Kangaroos | Moderate to severe |
| Humid temperate | Long | Slight | Kangaroos | Moderate? |
| Northern | Very short | Slight | Kangaroos | Slight |
| *New Zealand* | Long | None | Moas | Moderate |

*Source*: Blumler (1992a).

summer drought than the northwestern coast (which also has suffered less invasion). Edaphic conditions also are important, in that European and Mediterranean herbs invade most successfully on fertile soils, while they have had little impact on the vegetation of the extremely poor soils that characterize parts of Australia and the Cape Province of South Africa. Such soils are naturally vegetated by shrubs, not herbaceous vegetation (Specht 1969; Blumler 1992b). Moreover, the data on severity of invasion are of variable quality and often difficult to compare. This is particularly true where there are gradations in climate within a country, such as in Chile and Australia, because review papers tend to discuss trends for the country as a whole rather than examining regional variation in invasion in relation to climate. Finally, of course, there is some uncertainty as to the former abundance and species composition of animal grazers. I have included only large herbivores in the interests of brevity, and because the grazing pressure hypothesis presumes that large herbivores exert more severe pressure than smaller animals, and that trampling with hooves is an important aspect of the pressure (literally and figuratively) exerted (Mack and Thompson 1982).

Severity of invasion is given in terms of invasion into natural or semi-natural grassland habitats. Intensive pastoralism, including the systematic development of sown pastures, has radically transformed some regions, such as New Zealand and parts of Australia. Under such management invasion is rather severe, and therefore the accounts of vegetation change (in New Zealand, especially) tend to suggest massive invasion. However, it appears that the vegetation usually reverts to a relatively high native cover, although with some persistent admixture of aliens, when intensive management ceases (Crosby 1986, Blumler 1992a). The same appears to be true of the humid Argentinean pampas, which also are intensively managed (Leon et al. 1984, Blumler 1992a). Such regions are classified as suffering 'moderate' invasion, to distinguish them from regions such as California which remain dominated by aliens indefinitely after the cessation of all human impacts.

Within continents, invasion severity appears to correlate with climate, despite only slight variation in large herbivore species (table 16.2). North America is the most interesting example. The eastern United States was apparently grazed by herbivores similar to those present west of the Rockies, though minus the pronghorn (Hall and Kelson 1959), and yet alien plants are largely restricted to disturbed habitats. The Great Plains, where the bison were most abundant, were less severely invaded – but under a climatic regime less suitable to invasion than the eastern US. In the Plains, there is a much more rapid and dramatic seasonal shift from the cold temperatures of winter to hot temperatures in summer, and a tendency to drought during the short cool periods. Interestingly, the Plains are not noticeably more free from invasion than similar climatic regions of Argentina, where there were no bison (e.g. Lewis et al. 1985, Cheater 1992). The Great Basin is thought to have been grazed primarily by pronghorn, with elk and perhaps deer largely restricted to the mountains (Mack and Thompson 1982). On the other hand, lowland interior California supported herds of 2000 or more elk and a general abundance of game, remarked upon by all early explorers and settlers; yet invasion of California has been considerably more severe than into the Great Basin.

Comparisons between land masses are also instructive. New Zealand had no native mammals, and yet has not been invaded as severely as some regions elsewhere that did. Most striking of all, the *renosterveld* (grassy shrubland) of South Africa's Cape region was invaded after the arrival of Europeans, despite the fact that it had already been grazed for at least a millennium by domestic livestock, and naturally before then by various large game animals such as the black rhinoceros (Kruger 1981, Cody et al. 1983, Morrow et al. 1983, Vlok 1988, Kruger et al. 1989). In short, it is surprisingly difficult to find evidence for *any* association between natural grazing pressure and invasion! Although one hesitates to rule out grazing pressure entirely, and while it is possible that there are other, unknown factors that exert important influences, it seems on present evidence that climate is the most plausible determinant of Mediterranean/European plant species invasion success. If so, one implication is that the invaders may be better adapted to European, and especially to mediterranean-type climates than natives in the analogous climatic regions overseas. Another implication is that overgrazing after European settlement was possibly not nearly so severe nor so destructive as most scholars have presumed (elsewhere, I have argued that agropastoralism in the Old World has had far more complicated but less negative environmental impacts than generally realized: Blumler 1993; see also Blumler 1994). This is further suggested by a closer examination of the historical geography of invasion into California grasslands.

## THE CALIFORNIA INVASIONS

The Spanish settled along the Pacific Coast of California, beginning in 1769 (San Diego), and built their northernmost mission (at Sonoma, north of San Francisco) in 1823. Settlement of the Central Valley and adjacent foothills began about 1840. The prevailing view is that a perennial-dominated grassland was eliminated and replaced by introduced annuals principally during the period 1845–65 due to overgrazing and severe, protracted drought (similar to the drought that has affected the state recently), although additional factors such as ploughing and changes in fire frequency are sometimes mentioned (Heady 1977). This interpretation is due primarily to Burcham (1956), whose study was received uncritically (but see Wester 1975, 1981). Surprisingly, however, Burcham provided no evidence in support of his conclusions. On the contrary, the evidence indicates that perennials were never dominant in many areas, that the replacement of native vegetation was substantially complete much earlier than Burcham believed, and that the severe grazing and drought of the mid-century tended, if anything, to allow natives to come back temporarily (Blumler 1992d).

Table 16.3 lists the sorts of evidence that can be brought to bear upon the reconstruction of the original composition of the grassland, and upon the historical geography of invasion and replacement of native species. Various sorts of fossil evidence (phytoliths, pollen, seed remains in adobe bricks and archaeological sites) provide only limited data on the prior composition of the grassland, and only slightly more information on the dates of appearance and subsequent spread of aliens. The recent discovery of filaree (*Erodium cicutarium*) pollen in varved sediments securely dated to prior to Spanish settlement demonstrates that at least one alien was able to invade without the benefit of Europeans or livestock (Mensing and Byrne 1993). An additional, and under-utilized, source of information is that from isozymes (see below). Ecological experiments have demonstrated that native perennials have great difficulty competing with introduced annuals, but tell us little about the comparative competitive abilities of *native* annuals and perennials, and still less about the natural species composition of the grassland. They do demonstrate, however, that some introduced annuals do not require human impacts to thrive here (Blumler 1992b,c). They also demonstrate that most, if not all, native species require disturbance to maintain themselves in the face of competition from introduced species (Kay et al. 1981, Blumler 1992b, Edwards 1992, Menke 1992, Stebbins 1992). Taxonomic and distributional data give some indication of the natural tolerances of native species, and allow speculations concerning the original vegetation mosaic (e.g. Beetle 1947, Crampton 1974). Changes in abundance and range since the publication of the first state flora (Brewer and Watson 1880) also can be determined, although imprecisely.

**Table 16.3** Data sources for reconstructing the original composition of California grasslands, and for examining the historical geography of invasion

| *Data source* | *Character of evidence* | *References* |
|---|---|---|
| Phytoliths | Limited evidence that perennials were more important formerly | Bartolome et al. (1986) |
| Pollen | Limited information on rates of spread of aliens | Mensing and Byrne (1993) |
| Adobe bricks | Seeds contained within bricks give limited information on dates of appearance of aliens | Hendry and Bellue (1925), Hendry (1931) |
| Ecological experiments | Native perennials sometimes increase in importance in exclosures, but more often they do not; because of dominance by aliens, it is difficult to do field experiments between natives only, to compare their competitive abilities and ecological tolerances | Schultz et al. (1955), Biswell (1956), White (1967), Bartolome and Gemmill (1981)), Blumler (1992b), Edwards (1992) |
| Taxonomic, distributional data | Information on the potential 'players' in the original grassland; information on the shrinkage in range of natives, and the expansion of aliens, since the late nineteenth century | Brewer and Watson (1880), Jepson (1901), Robbins (1940), Beetle (1947), Munz and Keck (1959) |
| Historical accounts | Some regions clearly dominated by native annual plants at an early date, elsewhere perhaps a mix of perennials and annuals with dominance unclear; tantalizing but limited information on identities of the important native species; considerable information on spread of aliens | Burcham (1956), Wester (1975), Timbrook et al. (1982), Blumler (1992d), and numerous primary sources |
| Isozymes | Some data on the number of times individual taxa were introduced, and on the geographical patterns of introduction and subsequent spread | Clegg and Allard (1972), Allard et al. (1978), Garcia et al. (1989), Blumler (1992b) |

Finally, the vast collection of early historical accounts, including diaries, journals, letters, scientific reports and books, contains a great deal of anecdotal information concerning the nature and composition of the grassland, and the changes taking place as a result of introductions and human impacts. The historical record is particularly informative during the mid-nineteenth century, when US citizens began arriving in considerable numbers, and numerous descriptions were published in response to the great interest in California conditions (suitability for agriculture, and so on) amongst prospective settlers. Historical accounts suggest a rather complicated vegetation mosaic, with native annuals dominating in at least some parts of the state (e.g. Wester 1975, 1981, Blumler 1992d). Here, I discuss the evidence with respect to Burcham's (1956) thesis. It should be borne in mind, however, that the bulk of the historical information still awaits examination by someone with botanical training.

**Table 16.4** The sum total of Burcham's (1956) evidence for vegetation change from native perennial bunchgrasses to introduced annuals, 1845–65, due to overgrazing and drought

| *Sources* | *Quotation* |
|---|---|
| Brewer, June 1864, San Joaquin Valley (1949: 510) | '...not a vestige of herbage... wire grass or alkali grass... gnawed... almost into the earth' |
| Bidwell, 1865 (1866: 212) | '...the indigenous grasses... are fast disappearing ... Weeds spring up...' |
| McGarvey 1865 (1866: 227) | '...pasture... exceedingly scarce' |
| Hittell (1863: 105) | '...the wild oats have been eaten... by cattle... [and] killed out' |
| Perkins 1863 (1864: 139) | 'wild oats... stunted...'; '...depasturage... in the great oat ranges...' |
| Bolander (1866: 138) | '[Wild oats is fast disappearing]... on account of the country having been overstocked' |
| Brewer (1883: 963) | '[Wild oats] was most abundant between 1845 and 1855, when hundreds of thousands of acres were clothed with it thick as a meadow' |

The date in parentheses is the date of publication, if different from the date of observation.

There is abundant and unimpeachable evidence that wild oats and filaree were already dominant over large areas by 1840, including regions that Europeans had not yet or had only recently settled (Fremont 1845, Mack 1989, Blumler 1992d). For instance, wild oats were so dominant in Napa and Solano Counties in the 1840s that James Clyman gave this description:

> wild oats – ...This is the greatest oat field (in) perhaps on the globe containing tow or three hundred thousand acres of land and what is most remarkable scarcely a bunch of grass or a weed to be seen... notwithstanding this immence Quantity of native grow[n] oats yet you never see a grain fed to an animal... all is suffered to fall off when ripe to seed the earth for another crop or to feed the millions of water fowl that resort here in the winter or rainy season... at this season it has a beautiful appearanc the earth being thickly clad in deep green foilage as regular as a well set meadow – (Camp 1925–6: 273; orthography as in the original)

Similar descriptions may be found in Wilkes (1845), Bryant (1848), Newberry (1857) and Perkins (1864), among others. Yet Napa and Solano Counties were essentially unsettled until 1840 (Donley et al. 1979). It is difficult to believe that grazing could have had such a drastic impact so quickly. Statewide, other reports suggest still earlier dominance by wild oats (Camp 1923, Quaife 1934, Pattie 1962), but these are sketchier and perhaps less reliable. All reports agree that wild oats *declined* in abundance during the 1845–65 period, as a result of the increase in grazing pressure that occurred at that time. *Avena* never again achieved its former dominance, except in some ungrazed areas. It further declined during the extreme drought of 1862–4 (Perkins 1864, McGarvey 1866, Blumler 1992d).

Table 16.4 summarizes the only descriptions of vegetation and vegetation change that Burcham (1956) presented in support of his thesis that alien annuals replaced native perennials during 1845–65. None of the statements discusses specifically native plants except for one reference to wetland species (Brewer 1949). Bidwell's (1866) mention of 'indigenous' grasses is ambiguous, since both native and uncultivated alien species were included under this rubric in the mid-nineteenth century. The quote from McGarvey (1866) also is ambiguous; but Burcham neglected to mention that McGarvey stated that wild oats were disappearing, while native perennial bunchgrasses were doing very well, as a result of heavy grazing and drought (Blumler 1992d). All other quotes refer to destruction of wild oats! Contrary to Burcham's thesis, in the mid-nineteenth century there was general agreement that California's summer drought favours annuals, that annuals therefore naturally dominate the grasslands, that *Avena* had achieved widespread dominance prior to 1845, and that overgrazing has its major impact on palatable annuals, through destruction of their seeds. On the other hand, native perennials were considered to be tolerant of and perhaps even favored by (winter) drought and overgrazing (McGarvey 1866, Cronise 1868, Blumler 1992d). I have encountered no dissent from any of these views in the early accounts (Blumler 1992d). For instance, John Muir pointed out:

> because so long a period of extreme drought succeeds the rainy season, most of the vegetation is composed of annuals.. – (Muir 1988: 252)

Muir was concerned to preserve the spectacular native wildflower fields of the Central Valley, but he was in complete agreement with his pro-development contemporaries concerning the effect of climate on life form. William Brewer (1883), regarded as perhaps the most astute and judicious of nineteenth-century observers, believed that overgrazing had facilitated invasion of alien species (although he did not single out 1845–65 as a particularly significant period), but he specifically related California's susceptibility to the fact that '... a very considerable portion of the forage is of annual plants...'. That is, he assumed that the grassland had been annual-dominated naturally, and that this allowed rapid species turnover. In sum, Burcham and other ecologists, historians and geographers have imposed their conclusions upon the historical data rather than letting the early accounts speak for themselves. While the perceptions of nineteenth-century observers were not necessarily correct, their reports are so completely at variance with modern interpretations that one must suspect that annuals played a far more important role in the grassland naturally than is generally appreciated, that some aliens were able to penetrate and dominate in the absence of any significant agropastoralist impacts, and that they did so with phenomenal rapidity. On the other hand, of course, many later invaders were undoubtedly successful because of anthropogenic disturbance.

## INVASION AS A GEOGRAPHICAL PROCESS: THE EXAMPLE OF WILD OATS

Analysis of electrophoretic data can allow one to classify a given alien species into distinct multilocus genotypes, and to map their geographical distribution. Spatial analysis then allows one to determine the probable sites of introduction and routes of diffusion of genotypes, if the pattern is not too complex. For instance, this sort of analysis was employed to trace patterns of introduction of water hyacinth (*Eichhornia crassipes*) into Jamaica, and of rice weeds from California to Australia (Barrett and Shore 1989, Barrett and Husband 1990). Electrophoretic data are available for over 200 California populations of the important invader, slender wild oat (*Avena barbata*); in addition, geneticists have compiled information on morphological correlates and ecological tolerances of the isozyme genotypes (e.g. Jain and Marshall 1967; Marshall and Jain 1967, 1968, 1969a,b, 1970; Allard et al. 1972, 1978; Clegg and Allard 1972; Hamrick and Allard 1972, 1975; Hamrick and Holden 1979; Miller 1977; Garcia et al. 1989). However, the geneticists ignored crucial geographical aspects of the invasion process and consequently may have misinterpreted their data.

Thirty-five isozyme genes and morphological characters define two distinct, multilocus genotypes of *A. barbata* (figure 16.1) – termed 'mesic' and 'xeric' because they correlate both regionally and at the microhabitat scale with moisture availability. Also in California are a third ('Geyserville') genotype and some rare variants that may be the result of hybridization of the xeric and mesic genotypes; that is, they do not necessarily represent separate introductions. On the other hand, the species is more genetically diverse in the Old World, where multilocus genotypes are far less distinctive, and less clearly correlated with environment. Allard and his colleagues assumed (without any supporting evidence) that the species successfully dispersed to California many times; in other words, that a large, random sample of Old World genotypes was established in California at an early date, but was subsequently reduced to only two or three genotypes by intense natural selection. They further argued that the isozyme genes must function as coadapted complexes (Allard et al. 1972, Clegg and Allard 1972).

The geographical pattern (figure 16.1) suggests the alternative hypothesis that only a few genotypes were introduced and that present distributions reflect the historical geography of genotype movement more than selection. Specifically, I suspect that a single introduction (the xeric genotype) arrived at an early date, possibly in southern California, and spread rapidly over the entire state; later (possibly much later), the mesic and perhaps the Geyserville genotypes were introduced, probably through San Francisco Bay (since it was the major port of entry throughout much of the nineteenth and early twentieth centuries). Because they arrived comparatively recently, they have not yet completed their spread; moreover, there has been only limited time for gene exchange with the xeric genotype, and therefore for the evolution of new genotypes which might be more fit. Hedrick and Holden (1979) demonstrated that in highly self-pollinated annuals such as *A. barbata* it may take centuries for hybrids to form in large numbers, even if favoured by selection. Therefore,

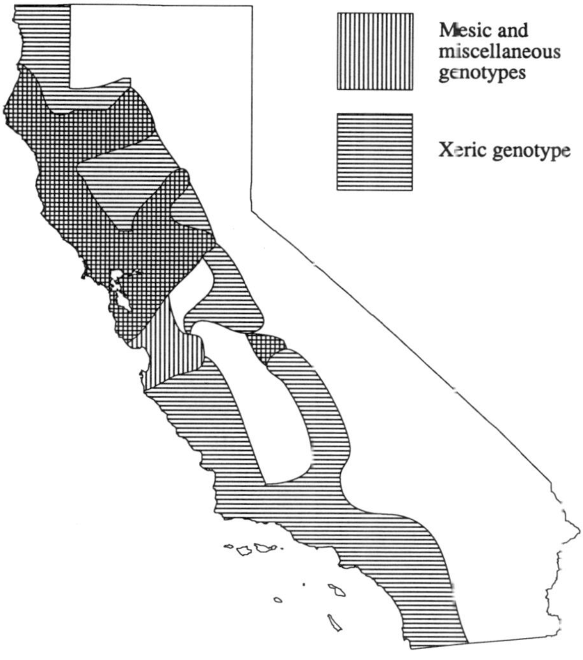


Figure 16.1 The distribution of multilocus genotypes of *Avena barbata* in California (from Miller 1977, Allard et al. 1978).

the predominance of the xeric and mesic genotypes in California is not proof that they are better adapted than other possible gene combinations, nor that any selection on isozyme loci has occurred. On the contrary, the isozyme genes could be merely 'hitchhiking' along with some other selected gene(s) (Hedrick and Holden 1979; Hedrick 1982).

Several lines of evidence suggest that this hypothesis is more plausible than Allard's adaptive explanation. Northwestern California is the most mesic part of the state, yet only the xeric genotype grows there; even mesic habitats in southern California have only the xeric genotype; while xeric areas in central California close to San Francisco do not have the xeric genotype (figure 16.1). That is, the distribution is patterned geographically more than ecologically. Also, much of the region around San Francisco is dominated by possible hybrids (Miller 1977, Allard et al. 1978). This is contrary to Allard's co-adaptation hypothesis, but is what one would predict if the mesic genotype was recently introduced through, and has spread from, San Francisco Bay (for the theoretical basis for this assertion, see Barton 1983, Barton and Hewitt 1985, Golenberg 1989). Furthermore, *A. barbata* is seldom a crop weed, and its seed dispersal by grazing animals is limited (Shmida and Ellner 1984); therefore, it is unlikely to be dispersed overseas as frequently as many other species. All early American taxonomic descriptions of *A. barbata* describe the xeric genotype morphology only (Dewey 1897, Jepson 1901, Abrams 1923). However, Allard's hypothesis *requires* that multiple introductions took place.

In addition, some morphological traits, such as the late flowering of the xeric and early flowering of the mesic genotype, seem maladaptive in their respective habitats (Blumler, unpublished); if so, appropriate hybrid combinations are likely to be more fit than the parental genotypes.

Finally, the phenomenally rapid spread of the species in California would seem to preclude any significant role for selection initially; that is, until populations built up to the point at which intraspecific competition would have become important. On the contrary, it appears that *A. barbata* must have been overwhelmingly competitive against native vegetation. It is difficult to believe that the xeric and mesic genotypes would have been so competitive while all other Old World gene combinations would have been completely unfit. It is particularly difficult to believe that natural selection would have favoured only a single genotype over the entire southern half of the state, where *A. barbata* is abundant in a wide variety of habitats, especially given that the species is genetically diverse throughout its natural Old World range. Occam's razor requires that Allard's selection hypothesis be rejected.

The other dominant wild oat species in California, *A. fatua*, is highly polymorphic. Geneticists have contrasted the relatively simple genetic structure of *A. barbata* with the complicated pattern in *A. fatua*, and offered adaptive explanations (e.g. Jain and Marshall 1967, Marshall and Jain 1968). It is more likely due to geography. *A. fatua* is one of the world's worst crop weeds (Price-Jones 1976), almost certainly introduced many times as a contaminant of grain (Goss 1939). Since it is a sibling species of the cultivated oat (*A. sativa*) (Ladizinsky and Zohary 1971), it also exchanges genes with the latter in cultivated fields in California. Much of the genetic diversity in *A. fatua* probably is just random noise resulting from multiple introductions and introgression, whereas *A. barbata* is flourishing despite its very depauperate genetic representation in California.

Interestingly, *A. fatua* is rare in the Mediterranean, where it is replaced by the closely related *A. sterilis*, which differs in its large, multiseeded (synaptermous) diaspores. The latter are poorly adapted to human dispersal. *A. fatua* diaspores are approximately the same dimensions as wheat, and therefore are frequent contaminants in grain shipments (e.g. Goss 1939), but *A. sterilis* diaspores are so large that they are removed in the sieving of the grain or easily picked out and discarded later (G. C. Hillman, personal communication; see Hillman (1984) for a discussion of traditional grain-cleaning operations and their effects on weed contaminants). *A. sterilis* diaspores are also too bulky to disperse long distances in fur – for discussions of the generally poor dispersal of synaptosperms, see Zohary (1937), Ellner and Shmida (1981) and Blumler 1994. Perhaps due to its poor dispersal capability, *A. sterilis* occurs in California only around Berkeley (Hickman 1993), where it was apparently introduced in about 1900 through the University of California Botanical Garden. The garden was actively engaged in importing Mediterranean species for possible range and pasture improvement at that time (Hilgard 1890, Jones and Love 1945). *A. sterilis* has spread from the campus to a very limited extent, in exclusively mesic microsites (shade, ditches, north slopes, irrigated plantings). In these microsites it frequently forms pure stands, while *A. fatua* typically dominates contiguous but less mesic habitats.

Although classed taxonomically as separate species, *A. sterilis* and *A. fatua* are as interfertile as are the mesic and xeric genotypes of *A. barbata* (Florell 1931, Ladizinsky and Zohary 1971, Vecchio et al. 1982). In this respect the *A. sativa* – *A. sterilis* – *A. fatua* complex is similar to many domesticated cereals and their wild and weedy relatives, which are split arbitrarily into several taxonomic 'species' based on sharp morphological discontinuities but entirely lack genetic barriers to interbreeding (Harlan and de Wet 1971, Harlan 1975). Not surprisingly, *A. sterilis* × *fatua* hybrids are often found where the two species come in contact in California, indicating that gene exchange is taking place (Hickman 1993, Blumler unpublished). Thus, *A. sterilis* and *A. fatua* can be thought of as distinct, multilocus genotypes (since several gene loci determine the differences in diaspore morphology that define the two species taxonomically), with distributions in California that parallel those of the mesic and xeric genotypes of *A. barbata*. *A. sterilis* is found in mesic microsites in a relatively mesic community (Berkeley), while *A. fatua* is much more widespread, including areas that for the most part are more xeric (*A. fatua* is rare in the rainy, northwestern part of the state).

In the Old World, however, the distributions are reversed! *A. sterilis* is widespread in hot, summer-dry mediterranean environments and penetrates into the adjacent deserts, while *A. fatua* is mostly restricted to the more mesic northern Mediterranean and temperate Eurasia (Zohary 1971, Baum et al. 1972, Baum 1977, Guillerm 1991). This pattern reversal is best explained as due to the vagaries of dispersal. The single introduction of *A. sterilis* into California happened to be of a mesic ecotype that consequently has occupied only mesic habitats, but has not yet been able to disperse to the enormous number of sites elsewhere in California, to which presumably it is well-adapted, that are as mesic as those in Berkeley. *A. fatua*, on the other hand, was introduced many times and may have picked up adaptations to relatively hot, mediterranean environments as a result of introgression with oat varieties cultivated by the Spanish and later settlers. The present distribution of the two species in California is almost certainly *not* due to selection on the characters that separate them taxonomically – otherwise, one would expect an analogous distribution in the Old World. Rather, the genes for *A. sterilis* and *A. fatua* type morphology are probably hitchhiking along with other genes that determine the adaptation to mesic and comparatively xeric habitats, respectively; and they will continue to do so until introgression transfers the mesic genes to *A. fatua* type plants and the xeric genes to *A. sterilis* type plants.

Of course, this does not prove that a similar explanation is in order for *A. barbata*, but it does illustrate that genetic hitchhiking is likely to be important when relatively few introductions of a taxon have occurred, and too short a time has elapsed to allow completion of migration, recombination and selection – if, indeed, those processes ever become complete. This analysis also illustrates that dispersal bottlenecks strongly influence the invasion process. Transoceanic migration and establishment is clearly a difficult and chancy proposition, for otherwise the invasion of California grasslands would have reached equilibrium long ago. Yet Mediterranean species and genotypes continue to arrive, spread and become dominant, even in natural habitats (Burcham 1956, Heady 1977, Sauer 1988, Blumler 1992b).

Species and genotypes that happened to arrive and invade successfully only shortly after the first opening up of a region to European settlement may have gained thereby an advantage over later arrivals. I call this the 'first spreads most' hypothesis. *A. fatua* and the xeric genotype of *A. barbata* arrived in California early, and probably were able to establish easily in a wide range of habitats because of a lack of interspecific and intraspecific competition; this facilitated their spread. By now, they are perhaps somewhat locally adapted even to habitats to which they were not particularly well-suited in the Old World. Any later arrival, such as *A. sterilis* or the mesic genotype of *A. barbata*, can succeed only if introduced to a habitat in which it can outcompete the earlier invaders, or in which the earlier aliens are absent. That is, the probability of success of individual introductions should decline with time (but is counterbalanced by an increasing introduction rate), even if later arrivals are more fit when considered over the full range of Californian environments. To put it another way, early introductions might spread as a wavefront, while later invaders might be more likely to confront a patchy environment, and spread in island-hopping fashion. Then, the real-life landscape of invasion becomes a true analogue to Wright's (1932) evolutionary landscape.

## CONCLUSIONS

The view that humans have almost entirely negative impacts on nature, which is widespread among environmental historians, historical geographers, ecologists and environmentalists, paradoxically perpetuates the old Western stereotypes of humans as active, dominating and separate from a nature that is passive and static. A view more in line with twentieth-century empirical data and current ecological theory would emphasize that human-nature relationships are interactive, and embedded within a kaleidoscopic natural world in which nothing is permanent (Blumler 1992a-c, 1993; see also Botkin 1990; Worster 1990; Sprugel 1991). Invasion is a natural process that has caused evolution, extinction and changes in diversity throughout the history of life – although of course the current invasions are strongly mediated by human action. Patterns of invasion are naturally affected by dispersal bottlenecks, differences in evolutionary rates (non-convergent evolution) in geographically separate but climatically equivalent environments and, probably, as the discussion of wild oats suggests, by a significant random component. The spectacular success of (true) Mediterranean species is undoubtedly due in part to coevolution with agropastoralism, but also suggests that they may be better adapted on the whole to winter-wet, summer-dry conditions than the native species on other continents (Blumler 1992b,c). Plants such as wild oats seem to be the products of non-convergent evolution, in that there are no equally vigorous and tall annual grasses in the native floras of the mediterranean analogue regions. Moreover, given that vigorous annual grasses (specifically, wild wheats and barley) may have been the plants first cultivated in the Near East, one can turn the coevolution argument on its head (Blumler and Byrne 1991, Blumler 1994). The superior adaptation of vigorous annuals to winter-wet, summer-dry environments may have allowed those plants to domesticate humans in an agricultural relationship, eventually allowing the plants to be transferred overseas – a case of long-time, long-distance dispersal!

Some biologists (e.g. Clegg and Allard 1972, Jackson 1985) treat invasion as if all species and genotypes disperse almost instantaneously overseas and into all habitats to which they are adapted. This is geographically shortsighted in that dispersal bottlenecks clearly have major effects on the comparative success of aliens, both in transoceanic transport and in finding suitable habitat once introduced abroad, with the result that distribution is to some extent a lottery. Many biologists and other scholars also emphasize the 'weediness' of invaders because of their association with Old World agropastoralism (e.g. Baker and Stebbins 1965). Most aliens must tolerate some sort of agropastoralist disturbance to be successfully transported overseas in the first place, but this does not mean that they are not also capable of outcompeting natives in natural habitats. In fact, there are numerous well-documented examples of aliens outcompeting natives (reviewed in d'Antonio and Vitousek 1992). Furthermore, *natural* disturbance is so pervasive that wild species also must be tolerant of some sort of disturbance (Blumler 1992c, 1993). A popular hypothesis is that introduced invading annuals sport 'general-purpose genotypes' that fit them for plastic responses to a variety of disturbed habitats, but leave them poorly suited to compete in truly wild conditions (Baker 1965); but this has never been substantiated (see Newsome and Noble 1986), and it is more likely that the low genetic diversity of many alien species is due to limited numbers of introductions rather than being a genetic strategy. After all, if the xeric *A. barbata* genotype is such an excellent general-purpose genotype, then why is it not important in the Old World? Disturbance is certainly pervasive in its native range.

The history of American attitudes and adaptations to the peculiar environment and vegetation of California remains to be written. Burcham (1956) and others imposed their ecological theory upon the early accounts, rather than letting the latter speak for themselves. Interestingly, mid-nineteenth century observers understood important aspects of California grassland ecology, and some implications for proper rangeland management, that twentieth-century range managers did not recognize until recently because of the influence of unsound ecological theory (Blumler 1992d). Further study of early historical accounts may reveal additional information of practical value in preserving endangered species and improving rangeland productivity, but historians need to be more ecological, and biologists more geographical, fully to understand the valley grassland transformation.

ACKNOWLEDGEMENTS

The manuscript benefited from the critical comments of R. Whittaker.

REFERENCES

Abrams, L. R. (1923) *All Illustrated Flora of the Pacific states: Washington, Oregon, and California* (Stanford University Press, Stanford)

Allard, R. W., Miller, R. D. and Kahler, A. L. (1978) 'The relationship between degree of environmental heterogeneity and genetic polymorphism', in Freysen, A. J. and Woldendorp, J. W. (eds) *Structure and Functioning of Plant Populations* (North-Holland, New York) pp. 49–73

Allard, R. W., Barbel, G. R., Clegg, M. T. and Kahler, A. L. (1972) 'Evidence for coadaptation in *Avena barbata*', *Proceedings of the National Academy of Sciences, USA* 69: 3043–8

Baker, H. G. (1955) 'Self-compatibility and establishment after 'long-distance' dispersal', *Evolution* 9: 347–8

Baker, H. G. (1965) 'Characteristics and modes of origins of weeds', in Baker, H. G. and Stebbins, G. L. (eds) *The Genetics of Colonizing Species* (Academic Press, New York) pp. 147–72

Baker, H. G. (1978) 'Invasion and replacement in California and neotropical grasslands', in Wilson, J. R. (ed.) *Plant Relations in Pastures* (CSIRO, East Melbourne) pp. 368–84

Baker, H. G. and Stebbins, G. L. (1965) *The Genetics of Colonizing Species* (Academic Press, New York)

Barbera, G., Carimi, F. and Inglese, P. (1992) 'Past and present role of the Indian-fig prickly-pear (*Opuntia ficus-indica* (L.) Miller, Cactaceae) in the agriculture of Sicily', *Economic Botany* 46: 10–20

Barrett, S. C. H. and Husband, B. C. (1990) 'The genetics of plant migration and colonization', in Brown, A. H. D., Clegg, M. T., Kahler, A. L. and Weir, B. S. (eds) *Plant Population Genetics, Breeding, and Genetic Resources* (Sinauer, Sunderland, Massachusetts) pp. 254–77

Barrett, S. C. H. and Shore, J. S. (1989) 'Isozyme variation in colonizing plants', in Soltis, D. E. and Soltis, R. S. (eds) *Isozymes in Plant Biology* (Dioscorides Press, Portland, Oregon) pp. 106–26

Bartolome, J. W. (1979) 'Germination and seedling establishment in California annual grassland', *Journal of Ecology* 67: 273–81

Bartolome, J. W. and Gemmill, B. (1981) 'The ecological status of *Stipa pulchra* (Poaceae) in California', *Madroño* 28: 172–84

Bartolome, J. W., Klukkert, S. E. and Barry, W. J. (1986) 'Opal phytoliths as evidence for displacement of native Californian grassland', *Madroño* 33: 217–22

Barton, N. H. (1983) 'Multilocus clines', *Evolution* 37: 454–71

Barton, N. H. and Hewitt, G. M. (1985) 'Analysis of hybrid zones', *Annual Review of Ecology and Systematics* 16: 113–48

Baum, B. R. (1977) *Oats: Wild and Cultivated, a Monograph of the Genus Avena L. (Poaceae)* (Canada Department of Agriculture, Research Branch, Ottawa)

Baum, B. R., Fleischmann, G., Martens, J. W., Rajhathy, T. and Thomas, H. (1972) 'Notes on the habitat and distribution of *Avena* species in the Mediterranean and Middle East', *Canadian Journal of Botany* 50: 1385–97

Beetle, A. A. (1947) 'Distribution of native grasses of California', *Hilgardia* 17: 309–57

Bidwell, J. (1866) 'Annual address', *Transactions of the California State Agricultural Society* 1864–5: 202–13

Biswell, H. H. (1956) 'Ecology of California grasslands', *Journal of Range Management* 9: 19–24

Blumler, M. A. (1984) 'Climate and the annual habit', unpublished M.A. thesis, Department of Geography, University of California, Berkeley

Blumler, M. A. (1992a) ' "Ecological imperialism": a botanical perspective', paper presented at the Symposium – the Impact of Columbus on the Americas, 27th International Geographical Congress, Washington DC, 10 August 1992

Blumler, M. A. (1992b) 'Seed weight and environment in mediterranean-type grasslands in California and Israel', unpublished Ph.D. dissertation, Department of Geography, University of California, Berkeley

Blumler, M. A. (1992c) 'Some myths about California grasses and grazers', *Fremontia* 20(3): 22–7

Blumler, M. A. (1992d) 'The myth of the bunchgrass prairie', paper presented at California Immigrants: People, Plants, and Animals, 1992 California History Institute Symposium, Stockton, California, 23 April 1992

Blumler, M. A. (1993) 'Successional pattern and landscape sensitivity in the Mediterranean and Near East', in Thomas, D. S. G. and Allison, R. J. (eds) *Landscape Sensitivity* (John Wiley, Chichester) pp. 287–305

Blumler, M. A. (1994) 'Evolutionary trends in the wheat group in relation to environment, Quaternary climatic change and human impacts', in Millington, A. C. and Pye, K. (eds) *Effects of Environmental Change in Drylands* (John Wiley, Chichester) pp. 253–69

Blumler, M. A. and Byrne, R. (1991) 'The ecological genetics of domestication and the origins of agriculture', *Current Anthropology* 32: 23–54

Bolander, H. N. (1866) 'The grasses of the state', *Transactions of the California State Agricultural Society* 1864–5: 131–45

Botkin, D. (1990) *Discordant Harmonies* (Oxford University Press, New York)

Brewer, W. H. (1883) 'Pasture and forage plants', in U. S. Census Office, *Report on the Productions of Agriculture as Returned at the Tenth Census* (U. S. Government Printing Office, Washington, DC) vol. 3: pp. 959–64

Brewer, W. H. (1949) in Farquhar, F. P. (ed.) *Up and Down California in 1860–64* (University of California Press, Berkeley)

Brewer, W. H. and Watson, S. (1880) *Geological Survey of California, Botany*, 2 vols (John Wilson and Sons, University Press, Cambridge, Massachusetts)

Bryant, E. (1848) *What I Saw in California* (Appleton, Philadelphia)

Burcham, L. T. (1956) 'Historical geography of the range livestock industry of California', unpublished Ph.D. dissertation, Department of Geography, University of California, Berkeley

Byrne, R. (1987) 'Climatic change and the origins of agriculture', in Manzanilla, L. (ed.) *Studies in the Neolithic and Urban Revolutions. The V. Gordon Childe Colloquium, Mexico, 1986* (BAR, Oxford) pp. 21–34

Camp, C. L. (1923) 'The chronicles of George C. Yount, California pioneer of 1826', *California Historical Society Quarterly* 2: 3–66

Camp, C. L. (1925–6) 'James Clyman, his diaries and reminiscences', *California Historical Society Quarterly* 4: 105–41, 272–83, 307–60; 5: 44–84, 109–38, 255–82, 378–401

Cheater, M. (1992) 'Alien invasion', *Nature Conservancy* 42(5): 24–9

Clegg, M. T. and Allard, R. W. (1972) 'Patterns of genetic differentiation in the slender wild oat species *Avena barbata*', *Proceedings of the National Academy of Sciences, USA* 69: 1820–4

Cody, M. L., Breytenbach, G. J., Fox, B., Newsome, A. E., Quinn, R. D. and Siegfreid, W. R. (1983) 'Animal communities: diversity, density and dynamics', in Day, J. A. (ed.) *Mineral Nutrients in Mediterranean Ecosystems* (South African National Science Progress Report, Pretoria) pp. 91–110

Crampton, B. (1974) *Grasses in California* (University of California Press, Berkeley)

Cronise, T. F. (1868) *The Natural Wealth of California* (H. H. Bancroft, San Francisco)

Crosby, A. W. (1972) *The Columbian Exchange: Biological and Cultural Consequences of 1492* (Greenwood Press, Westport, Connecticut)

Crosby, A. W. (1986) *Ecological Imperialism: the Biological Expansion of Europe, 900–1900* (Cambridge University Press, Cambridge)

Dafni, A. and Heller, D. (1982) 'Adventive flora of Israel – phytogeographical, ecological and agricultural aspects', *Plant Systematics and Evolution* 140: 1–18

Danin, A. and Yom-Tov, Y. 1990. 'Ant nests as primary habitats of *Silybum marianum* (Compositae)' *Plant Systematics and Evolution* 169: 209–17

d'Antonio, C. M. and Vitousek, P. M. (1992) 'Biological invasions by exotic grasses, the grass/fire cycle, and global change', *Annual Review of Ecology and Systematics* 23: 63–87

Darwin, C. (1859) *On the Origin of Species* (Watts, London)

Dewey, L. H. (1897) 'The genus *Avena* on the Pacific Coast', *Erythea* 5: 29

di Castri, F., Hansen, A. J. and Debussche, M. (1990) *Biological Invasions in Europe and the Mediterranean Basin* (Kluwer Academic, Dordrecht)

Donley, M. W., Allan, S., Caro, P. and Patton, C. P. (1979) *Atlas of California* (Pacific Book Center, Culver City, California)

Edwards, S. (1992) 'Observations on the prehistory and ecology of grazing in California', *Fremontia* 20(1): 3–11

Ellner, S. P. and Shmida, A. (1981) 'Why are adaptations for long-range dispersal rare in desert plants?', *Oecologia* 51: 133–44

Elton, C. S. (1958) *The Ecology of Invasions by Animals and Plants* (Methuen, London)

Evans, R. A. and Young, J. A. (1972) 'Competition within the grass community', in Youngner, V. B. and McKell, C. M. (eds) *The Biology and Utilization of Grasses* (Academic Press, New York) pp. 230–46

Florell, V. H. (1931) 'Inheritance of type of floret separation and other characters in interspecific crosses in oats', *Journal of Agricultural Research* 43: 365–86

Fox, M. D. and Fox, B. J. (1986) 'The susceptibility of natural communities to invasion', in Groves, R. H. and Burdon, J. J. (eds) *Ecology of Biological Invasions: an Australian Perspective* (Australian Academy of Science, Canberra) pp. 57–66

Fremont, J. C. (1845) 'Report of the exploring expedition to the Rocky Mountains in the year 1842 and to Oregon and California in the years 1843–1844', 28th Congress, 2nd Session, Senate Document 174

Garcia, P., Vences, F. J., Perez de la Vega, M. and Allard, R. W. (1989) 'Allelic and genotypic composition of ancestral Spanish and colonial Californian gene pools of *Avena barbata*: evolutionary implications', *Genetics* 122: 687–94

Golenberg, E. M. (1989) 'Migration patterns and the development of multilocus associations in a selfing annual, *Triticum dicoccoides*', *Evolution* 43: 595–606

Goss, W. L. (1939) 'Weed seeds found in California grown barley, wheat, and oats', *California Department of Agriculture Bulletin* 29: 436–55

Guillerm, J. L. (1991) 'Weed invasion in agricultural areas', in Groves, R. H. and di Castri, F (eds) *Biogeography of Mediterranean Invasions* (Cambridge University Press, Cambridge) pp. 379–92

Hall, E. R. and Kelson, K. R. (1959) *The Mammals of North America* (The Ronald Press, New York)

Hamrick, J. L. and Allard, R. W. (1972) 'Microgeographical variation in allozyme frequencies in *Avena barbata*', *Proceedings of the National Academy of Sciences, USA* 69: 2100–4

Hamrick, J. L. and Allard, R. W. (1975) 'Correlations between quantitative characters and enzyme genotypes in *Avena barbata*', *Evolution* 29: 438–42

Hamrick, J. L. and Holden, L. R. (1979) 'Influence of microhabitat heterogeneity on gene frequency distribution and gametic phase disequilibrium in *Avena barbata*', *Evolution* 33: 521–33

Harlan, J. R. (1975) *Crops & Man* (American Society of Agronomy, Madison, Wisconsin)

Harlan, J. R. and De Wet, J. M. J. (1971) 'Towards a rational classification of cultivated plants', *Taxon* 20: 509–17

Heady, H. F. (1977) 'Valley grassland', in Barbour, M. G. and Major, J. (eds) *Terrestrial Vegetation of California* (John Wiley, New York) pp. 491–514

Hedrick, P. W. (1982) 'Genetic hitchhiking: a new factor in evolution?', *BioScience* 32: 845–53

Hedrick, P. W. and Holden, L. R. (1979) 'Hitch-hiking: an alternative to coadaptation for the barley and slender wild oat examples', *Heredity* 43: 79–86

Hendry, G. W. (1931) 'The adobe brick as an historical source', *Agricultural History* 8: 64–71

Hendry, G. W. and Bellue, M. K. (1925) 'The plant content of adobe bricks', *California Historical Society Quarterly* 4: 361–73

Hengeveld, R. (1990) *Dynamic Biogeography* (Cambridge University Press, Cambridge)

Hickman, J. C. (1993) *The Jepson Manual: Higher Plants of California* (University of California Press, Berkeley)

Hilgard, E. W. (1890) *Report on the Agricultural Experiment Stations of the University of California with Descriptions of the Regions Represented [1888/89]* (J. P. Young, Sacramento)

Hillman, G. C. (1984) 'Traditional husbandry and processing of archaic cereals in modern times: part I, the glume wheats', *Bulletin of Sumerian Agriculture* 1: 114–52

Hittell, J. S. (1863) *The Resources of California* (A. Roman, San Francisco)

Jackson, L. E. (1985) 'Ecological origins of California's Mediterranean grasses', *Journal of Biogeography* 12: 349–61

Jain, S. K. and Marshall, D. R. (1967) 'Population studies in predominantly self-pollinating species. X. Variation in natural populations of *Avena fatua* and *A. barbata*', *American Naturalist* 101: 19–33

Jepson, W. L. (1901) *A Flora of Western Middle California* (Encina, Berkeley)

Jones, B. J. and Love, R. M. (1945) 'Improving California ranges', *California Agricultural Extension Service Circular* 129:1–47

Kay, B. L., Love, R. M. and Slayback, R. D. (1981) 'Revegetation with native grasses. I. A disappointing history', *Fremontia* 9(1): 11–15

Kruger, F. (1981) 'Use and management of mediterranean ecosystems in South Africa – current problems', in *Proceedings of the Symposium on Dynamics and Management of Mediterranean-type Ecosystems* (U. S. Department of Agriculture, Forest Service, PSW, Berkeley) pp. 42–8

Kruger, F. J., Breytenbach, G. J., Macdonald, I. A. W. and Richardson, D. M. (1989) 'The characteriztics of invaded mediterranean-climate regions', in Drake, J. A., Mooney, H. A., Di Castri, F., Groves, R. H., Kruger, F. J., Rejmanek, M. and Williamson, M. (eds) *Biological Invasions, a Global Perspective* (John Wiley, New York) pp. 181–213

Ladizinsky, G. and Zohary, D. (1971) 'Notes on species delimitation, species relationships and polyploidy in *Avena*', *Euphytica* 20: 380–95

Leon, R. J. C., Rusch, G. M. and Oesterheld, M. (1984) 'Pastizales pampeanas – impacto agropecuario', *Phytocoenologia* 12: 201–8

Lewis, J. P., Collantes, M. B., Pire, E. F., Carnevale, N. J., Boccanelli, S. I., Stofella, S. L. and Prado, D. E. (1985) 'Floristic groups and plant communities of south-eastern Santa Fe, Argentina', *Vegetatio* 60: 67–90

Litav, M. (1965) 'Effects of soil type and competition on the occurrence of *Avena sterilis* L. in the Judean Hills (Israel)', *Israel Journal of Botany* 14: 74–89

Litav, M. (1967) 'Micro-environmental factors and species interrelationships in three batha associations in the foothill region of the Judean hills', *Israel Journal of Botany* 16: 79–99

Litav, M., Kupernik, G. and Orshan, G. (1963) 'The role of competition as a factor determining the distribution of dwarf shrub communities in the Mediterranean territory of Israel', *Journal of Ecology* 51: 467–80

Mack, R. N. (1989) 'Temperate grasslands vulnerable to plant invasions: characteriztics and consequences', in Drake, J. A., Mooney, H. A., Di Castri, F., Groves, R. H., Kruger, F. J., Rejmanek, M. and Williamson, M. (eds) *Biological Invasions, a Global Perspective* (John Wiley, New York) pp. 155–179

Mack, R. N. and Thompson, J. N. (1982) 'Evolution in steppe with few large, hooved mammals', *American Naturalist* 119: 757–73

Marshall, D. R. and Jain, S. K. (1967) 'Cohabitation and relative abundance of two species of wild oats', *Ecology* 48: 656–59

Marshall, D. R. and Jain, S. K. (1968) 'Phenotypic plasticity of *Avena fatua* and *A. barbata*', *American Naturalist* 102: 457–67

Marshall, D. R. and Jain, S. K. (1969a) 'Genetic polymorphism in natural populations of *Avena fatua* and *A. barbata*', *Nature* 221: 276–8

Marshall, D. R. and Jain, S. K. (1969b) 'Interference in pure and mixed populations of *Avena fatua* and *A. barbata*', *Journal of Ecology* 57: 251–70

Marshall, D. R. and Jain, S. K. (1970) 'Seed predation and dormancy in the population dynamics of *Avena fatua* and *Avena barbata*', *Ecology* 51: 886–91

McGarvey, W. P. (1866) '[Report of the county assessor for 1865], Monterey County', *Transactions of the California State Agricultural Society* 1864–1865: 224–31

McNaughton, S. J. (1968) 'Structure and function of California grasslands', *Ecology* 49: 962–72

Menke, J. (1992) 'Grazing and fire management for native perennial grass restoration on California grasslands', *Fremontia* 20(2): 22–5

Mensing, S. A. and Byrne, R. (1993) 'New evidence for pre-mission transformation of the California grassland', Abstracts of the American Association of Geographers 89th Annual Meeting, Atlanta, 6–10 April 1993

Miller, R. D. (1977) 'Genetic variability in the slender wild oat *Avena barbata* in California', unpublished Ph.D. dissertation, Department of Genetics, University of California, Davis

Morrow P. A., Day, J. A., Fox, M. D., Frost, P. G. H., Jarvis, J. U. M., Milewski, A. V., Norton, P. M. and Fuentes, E. R. (1983) 'Interactions between plants and animals', in Day, J. A. (ed.) *Mineral Nutrients in Mediterranean Ecosystems* (South African National Science Progress Report, Pretoria) pp. 111–24

Muir, J. (1988) *The Mountains of California* (Fulcrum, Golden, Colorado)

Munz, P. A. and Keck, D. D. (1959) *A California Flora* (University of California Press, Berkeley)

Naveh, Z. and Whittaker, R. H. (1979) 'Structural and floristic diversity of shrublands and woodlands in northern Israel and other mediterranean countries', *Vegetatio* 41: 171–90

Newberry, J. S. (1857) 'Report upon the botany of the route', in U. S. War Department, *Reports of Explorations and Surveys, to Ascertain the most Practicable and Economical Route for a Railroad from the Mississippi River to the Pacific Ocean.* (Beverley Tucker, Washington, DC) vol. VI, part III: 9–19

Newsome, A. E. and Noble, I. R. (1986) 'Ecological and physiological characters of invading species', in Groves, R. H. and Burdon, J. J. (eds) *Ecology of Biological Invasions: an Australian Perspective* (Australian Academy of Science, Canberra) pp. 1–20

Noy-Meir, I. (1990) 'The effect of grazing on the abundance of wild wheat, barley and oat', *Biological Conservation* 51: 299–310

Noy-Meir, I, Gutman, M. and Kaplan, Y. (1989) 'Responses of Mediterranean grassland plants to grazing and protection', *Journal of Ecology* 77: 290–310

Pattie, J. O. (1962) *The Personal Narrative of James O. Pattie: the 1831 Edition, Unabridged* (J. B. Lippincott, Philadelphia)

Perkins, J. E. (1864) 'Sheep husbandry in California', *Transactions of the California State Agricultural Society* 1863: 134–45

Price-Jones, D. (ed.) (1976) *Wild Oats in World Agriculture* (Agricultural Research Council, London)

Quaife, M. M. (1934) *Narrative of the Adventures of Zenas Leonard* (The Lakeside Press, Chicago)

Raven, P. H. (1963) 'Amphitropical relationships in the floras of North and South America', *Quarterly Review of Biology* 38: 151–77

Robbins, W. W. (1940) 'Alien plants growing without cultivation in California', *California Agricultural Experiment Station Bulletin* 637: 1–128

Roberts, N. (1989) *The Holocene* (Basil Blackwell, Oxford)

Rossiter, R. C. (1966) 'Ecology of the mediterranean annual-type pasture', *Advances in Agronomy* 18: 1–56

Sauer, J. D. (1988) *Plant Migration: the Dynamics of Geographic Patterning in Seed Plant Species* (University of California Press, Berkeley)

Schultz, A. M., Launchbaugh, J. L. and Biswell, H. H. (1955) 'Relationship between grass density and brush seedling survival', *Ecology* 36: 226–38

Shmida, A. and Ellner, S. (1984) 'Seed dispersal on pastoral grazers in open Mediterranean chaparral, Israel', *Israel Journal of Botany* 32: 147–59

Simpson, G. G. (1980) *Splendid Isolation: the Curious History of South American Mammals* (Yale University Press, New Haven)

Specht, R. L. (1969) 'A comparison of the sclerophyllous vegetation characteristic of mediterranean type climates in France, California, and southern Australia. I. Structure, morphology, and succession', *Australian Journal of Botany* 17: 277–92

Sprugel, D. G. (1991) 'Disturbance, equilibrium, and environmental variability – what is natural vegetation in a changing environment?', *Biological Conservation* 58: 1–18

Stebbins, G. L. (1981) 'Coevolution of grasses and herbivores', *Annals of the Missouri Botanical Garden* 68: 75–86

Stebbins, G. L. (1992) [Letter to the editor], *Fremontia* 20(3): 31–2

Suffling, R. (1993) 'Multiple explanations of historical ecological trends in Northwestern Ontario, Canada', paper presented at the IBG Annual Conference, Egham, January 1993

Timbrook, J., Johnson, J. R. and Earle, D. D. (1982) 'Vegetation burning by the Chumash', *Journal of California and Great Basin Anthropology* 4: 163–86

Vecchio, V., Gasquez, J. and Compoint, J. P. (1982) 'Variabilité morphologique et enzymatique chez une population mixte d'*Avena fatua* L. et *A. sterilis*', *Weed Research* 22: 263–9

Vitousek, P. M., Walker, L. R., Whittaker, L. D., Mueller-Dombois, D. and Matson, P. A. (1987) 'Biological invasion by *Myrica faya* alters ecosystem development in Hawaii', *Science* 238: 802–4

Vlok, J. H. J. (1988) 'Alpha diversity of lowland fynbos herbs at various levels of infestation by alien annuals', *South African Journal of Botany* 54: 623–7

Wagner, W. L., Herbst, D. R. and Yee, R. S. N. (1985) 'Status of the native flowering plants of the Hawaiian Islands', in Stone, C. P. and Scott, J. M. (eds) *Hawaii's Terrestrial Ecosystems: Preservation and Management* (Cooperative National Park Resources Study Unit, University of Hawaii, Honolulu) pp. 23–74

Webb, S. D. (1978) 'A history of savanna vertebrates in the New World. Part II: South America and the Great Interchange', *Annual Review of Ecology and Systematics* 9: 393–426

Wester, L. L. (1975) 'Changing patterns of vegetation on the west side and south end of the San Joaquin Valley during historic time', unpublished Ph.D. dissertation, Department of Geography, University of California, Los Angeles

Wester, L. L. (1981) 'Composition of native grasslands in the San Joaquin Valley, California', *Madroño* 28: 231–41

White, K. L. (1967) 'Native bunchgrass (*Stipa pulchra*) on Hastings Reservation, California', *Ecology* 48: 949–55

Wilkes, C. (1845) *Narrative of the United States Exploring Expedition During the Years 1838, 1839, 1840, 1841, and 1842* (Lea and Blanchard, Philadelphia, 5 vols)

Worster, D. (1990) 'The ecology of order and chaos', *Environmental History Review* 14(1–2): 1–18

Wright, S. (1932) 'The roles of mutation, inbreeding, crossbreeding, and selection in evolution', *Proceedings of the 11th International Congress on Genetics* 1: 356–66

Zohary, D. (1971) 'Origin of south-west Asiatic cereals: wheats, barley, oats and rye', in Davis, P. H., Harper, P. T. and Hedge, I. (eds) *Plant Life of South-West Asia* (Royal Botanical Society, Edinburgh) pp. 235–60

Zohary, M. (1937) 'Die verbreitungsekologischen Verhältnisse der Flora Palaestinas. I. Die anti-telechorischen Erscheinungen', *Beihefte zum Botanischen Zentralblatt* 56A: 1–155

Zohary, M. (1962) *Plant Life of Palestine* (Ronald Press, New York)

Zohary, M. (1973) *Geobotanical Foundations of the Middle East* (Gustav Fischer Verlag, Stuttgart, 2 vols)

# Index

# Related Titles: List of IBG Special Publications

1 Land Use and Resource: Studies in Applied Geography (A Memorial to Dudley Stamp)
2 A Geomorphological Study of Post-Glacial Uplift
*John T. Andrews*
3 Slopes: Form and Process
*D. Brunsden*
4 Polar Geomorphology
*R. J. Price and D. E. Sugden*
5 Social Patterns in Cities
*B. D. Clark and M. B. Gleave*
6 Fluvial Processes in Instrument Watersheds: Studies of Small Watersheds in the British Isles
*K. J. Gregory and D. E. Walling*
7 Progress in Geomorphology: Papers in Honour of David L. Linton
*Edited by E. H. Brown and R. S. Waters*
8 Inter-regional Migration in Tropical Africa
*I. Masser and W. T. S. Gould with the assistance of A. D. Goddard*
9 Agrarian Landscape Terms: A Glossary for Historical Geography
*I. H. Adams*
10 Change in the Countryside: Essays on Rural England 1500–1900
*Edited by H. S. A. Fox and R. A. Butlin*
11 The Shaping of Southern England
*Edited by David K. C. Jones*
12 Social Interaction and Ethnic Segregation
*Edited by Peter Jackson and Susan J. Smith*
13 The Urban Landscape: Historical Development and Management (Papers by M. R. G. Conzen)
*Edited by J. W. R. Whitehand*
14 The Future for the City Centre
*Edited by R. L. Davies and A. G. Champion*
15 Redundant Spaces in Cities and Regions? Studies in Industrial Decline and Social Change
*Edited by J. Anderson, S. Duncan and R. Hudson*

16 Shorelines and Isostasy
*Edited by D. E. Smith and A. G. Dawson*

17 Residential Segregation, the State and Constitutional Conflict in American Urban Areas
*R. J. Johnston*

18 River Channels: Environment and Process
*Edited by Keith Richards*

19 Technical Change and Industrial Policy
*Edited by Keith Chapman and Graham Humphrys*

20 Sea-level Changes
*Edited by Michael J. Tooley and Ian Shennan*

21 The Changing Face of Cities: A Study of Development Cycles and Urban Form
*J. W. R. Whitehand*

22 Population and Disaster
*Edited by John I. Clarke, Peter Curson, S. L. Kayastha and Prithvish Nag*

23 Rural Change in Tropical Africa: From Colonies to Nation-States
*David Siddle and Kenneth Swindell*

24 Teaching Geography in Higher Education
*John R. Gold, Alan Jenkins, Roger Lee, Janice Monk, Judith Riley, Ifan Shepherd and David Unwin*

25 Wetlands: A Threatened Landscape
*Edited by Michael Williams*

26 The Making of the Urban Landscape
*J. W. R. Whitehand*

27 Impacts of Sea-level Rise on European Coastal Lowlands
*Edited by M. J. Tooley and S. Jelgersma*

28 The Challenge for Geography
*Edited by R. J. Johnston*

29 Geography and National Identity
*Edited by David Hooson*

30 Geography and Empire
*Edited by Anne Godlewska and Neil Smith*

31 Diffusing Geography
*Edited by A. D. Cliff, P. R. Gould, A. G. Hoare and N. J. Thrift*

32 Ecological Relations in Historical Times
*Edited by Robin A. Butlin and Neil Roberts*

Also published by Blackwell for the IBG

Atlas of Drought in Britain
*Edited by J. C. Doornkamp and K. J. Gregory*